D1608962

Nonrelativistic Quantum Mechanics

Anton Z. Capri
Physics Department, University of Alberta
Edmonton, Alberta, Canada

The Benjamin/Cummings Publishing Company, Inc.
Menlo Park, California • Reading, Massachusetts
Don Mills, Ontario • Wokingham, U.K.• Amsterdam • Sydney
Singapore • Tokyo • Mexico City • Bogota • Santiago • San Juan

To Kim, Karin, Irene and Skaidrite

for giving me time and love.

Sponsoring Editor: Richard W. Mixter
Production Editor: Karen Gulliver

Library of Congress Cataloging in Publication Data

Capri, Anton Z.
Non relativistic quantum mechanics.

Includes index.
1. Quantum theory. 2. Perturbation (Quantum dynamics)
I. Title.
QC174.12.C35 1985 530.1'2 85-14933
ISBN 0-8053-1505-3

BCDEFGHIJ-MA-89876

The Benjamin/Cummings Publishing Company, Inc.
2727 Sand Hill Road
Menlo Park, California 94025

Editor's Foreward

Everyone concerned with the teaching of physics at the advanced undergraduate or graduate level is aware of the continuing need for a modernization and reorganization of the basic course material. Despite the existence today of many good textbooks in these areas, there is always an appreciable time-lag in the incorporation of new viewpoints and techniques which result from the most recent developments in physics research. Typically these changes in concepts and material take place first in the personal lecture notes of some of those who teach graduate courses. Eventually, printed notes may appear, and some fraction of such notes evolve into textbooks or monographs. But much of this fresh material remains available only to a very limited audience, to the detriment of all. This series aims at filling this gap in the literature of physics by presenting occasional volumes with a contemporary approach to the classical topics of physics at the advanced undergraduate and graduate level. Clarity and soundness of treatment will, we hope, mark these volumes, as well as the freshness of the approach.

Another area in which the series hopes to make a contribution is by presenting useful supplementing material of well-defined scope. This may take the form of a survey of relevant mathematical principles, or a collection of reprints of basic papers in a field. Here the aim is to provide the instructor with added flexibility through the use of supplements at relatively low cost.

The scope of both the lecture notes and supplements is somewhat different from the "Frontiers in Physics" series. In spite of wide variations from institution to institution as to what comprises the basic graduate course program, there is a widely accepted group of "bread and butter" courses that deal with the classic topics in physics. These include: Mathematical methods of physics, electromagnetic theory, advanced dynamics, quantum mechanics, statistical mechanics, and frequently nuclear physics and/or solid-state physics. It is chiefly these areas that will be covered by the present series. The listing is perhaps best described as including all advanced undergraduate and graduate courses which are at a level below seminar courses dealing entirely with current research topics.

Finally, because the series represents a continuing experiment on the part of the editors and the publisher, suggestions from interested readers as to format, contributors, and contributions will be most welcome.

DAVID PINES

Preface

Most textbooks start as a set of lecture notes which expand and undergo numerous revisions over the years. The present version is no exception and is the culmination of many revisions of such lecture notes. I have taught quantum mechanics at several different levels and thus attempted to make this book of sufficiently broad scope for different courses. The book includes material for undergraduate courses as well as a one-semester graduate course. As a guide for possible division of the material in this book I have included a few pages in the following section entitled "How to use this Book".

It is a long way from lecture notes to a text-book and on this journey I have been greatly helped by many people. My first debt is to my students who forced me to clarify and expand my lecture notes. Without their prodding this book would never have been started.

Dr. M. Razavy has also been most generous with his time and ideas. Not only did he test this book in his quantum mechanics classes, but he willingly proof-read all the present version. He also helped with many pertinent references and comments; his criticism of chapter 21 was particularly useful. Without his constant encouragement and support my enthusiasm would have flagged many times and this book might never have been written.

Preface

During the early stages, Dr. W. Brouwer also helped considerably with his detailed criticisms of the first fourteen chapters. I thank him for his generous contribution of time and ideas.

The first version of this book was expertly typed by Mrs. M. Yiu. Had she not been here to read my scrawl, I am sure this book would never been written. I thank her for her patience and tolerance.

The diagrams are due to the heroic efforts of Mrs. J. Hube. Her attention to detail turned some rather skimpy sketches into actual diagrams. I am most grateful for her skills.

Finally, I acknowledge with many thanks the expertly accomplished job of preparing the final manuscript. This was ably handled by Christine Fischer and Laura Heiland. Their friendly attitude did much to make a painful job easier.

I would also like to thank all of my colleagues who, in one way or another, influenced me during our coffee-room discussions. Their contributions, although less tangible, are nevertheless very real.

It is my hope that in writing this book I may have helped a few students to discover the beauty of quantum mechanics.

Anton Z. Capri

Edmonton, Alberta
April, 1985

How to Use This Book

This book is designed for use in any of the following programs:

A) A one-semester introductory course for undergraduates;

B) A two-semester introductory course for undergraduates;

C) A one-semester graduate courses.

To indicate which material is suitable for inclusion in any of the above programs I have added the appropriate letter or letters "A,B,C" after the title in the Table of Contents. Thus, for example, all section headings for Chapter 1, except 1.10 and 1.11, are followed by the letters "AB" to indicate that this material is suitable for both a one-semester introductory course as well as a two-semester course. Sections 1.10 and 1.11 are labelled only B and should be omitted in a one-semester undergraduate course.

The material to be included will also vary from class to class. Thus it is not feasible, nor should it be attempted, to include all the sections marked with a "B" in a two-semester undergraduate course. For example, I have found that I could include either the Rayleight-Ritz variational approach or a somewhat watered-down version of the WKB approximation in a two-semester undergraduate course, but not both.

How to Use This Book

Similarly in a one-semester graduate course a selection from the material marked "C" is necessary. This will, of course, depend on the special interests of the students. The point is that not all the material in any given course division (A,B or C) can be covered in the time normally allotted for such a course. The interests of classes vary from year to year and so I have included extra topics from which a selection can be made.

Contents

1	**The Breakdown of Classical Mechanics**	1
1.1	Blackbody Radiation (AB)	2
1.2	Stability of Atoms - Discrete Spectral Lines (AB)	6
1.3	Photoelectric Effect (AB)	10
1.4	Wave Particle Duality (AB)	12
1.4.1	Reflection	13
1.4.2	Refraction	13
1.5	De Broglie's Hypothesis (AB)	14
1.6	The Compton Effect (AB)	15
1.7	The Davisson-Germer Experiment (AB)	18
1.8	The Franck-Hertz Effect (AB)	21
1.9	Planck's Radiation Law (AB)	23
1.10	Einstein's Model for Specific Heat (B)	24
1.11	Bohr Theory and the Hydrogen Atom (B)	28
2	**Review of Classical Mechanics**	34
2.1	Classical Mechanics of a Particle in One Dimension (AB)	34
2.2	Lagrangian and Hamiltonian Formulation (B)	41
2.3	Contact Transformations and Hamilton-Jacobi Theory (B)	44
2.4	Interpretation of the Action-Angle Variable (B)	52

Contents

2.5 Hydrogen Atom for Bohr-Sommerfeld Quantization (B) 54
2.6 The Schrödinger Equation (B) 59

3 **Elementary Systems** 65
3.1 Plane Wave Solutions (AB) 66
3.2 Conservation Law for Particles (AB) 67
3.3 Young's Double Slit Experiment (A) 69
3.4 The Superposition Principle and Group Velocity (AB) 70
3.5 Formal Considerations (AB) 74
3.6 Ambiguities (AB) 76
3.6.1 Use of Different Coordinate Systems 77
3.6.2 Noncommutativity 78
3.7 Interaction with an Electromagnetic Field (B) 80

4 **One-Dimensional Problems** 85
4.1 Introduction (AB) 85
4.2 Finite Square Well - Bound States (AB) 87
4.3 Parity (AB) 91
4.4 Scattering from a Step-Function Potential (AB) 95
4.5 Infinite Square Well - Particle in a Box (AB) 110
4.6 Time Reversal (B) 111

5 **More One Dimensional Problems** 116
5.1 General Considerations (AB) 116
5.2 Tunneling Through a Square Barrier (AB) 121
5.3 The Simple Harmonic Oscillator (AB) 125
5.4 The Delta Function (AB) 131
5.5 Attractive Delta Function Potential (AB) 134
5.6 Repulsive Delta Function Potential (AB) 136
5.7 Finite Square Well - Scattering and Phase Shifts (B) 138

6 **Mathematical Foundations of Quantum Mechanics** 145
6.1 Geometry of Hilbert Space (AB) 145
6.2 L_2 - A Model Hilbert Space (AB) 150

6.3 Operators on Hilbert Space - Mainly Definitions (BC) 153
6.4 The Cayley Transform and Self-Adjoint Operators (BC) 161
6.5 Some Properties of Self-Adjoint Operators (AB) 166
6.6 Classification of Symmetric Operators (BC) 170
6.7 Spontaneously Broken Symmetry (BC) 179

7 Physical Interpretation of Quantum Mechanics 187
7.1 Assumption (A1) - Physical States (ABC) 187
7.2 Assumption (A2) - Observables (ABC) 188
7.3 Assumption (A3) - Probabilities (ABC) 189
7.4 Assumption (A4) - Reduction of the Wave Packet (ABC) 194
7.5 Example (AB) 195
7.6 Compatibility Theorem and Uncertainty Principle (ABC) 197
7.7 The Heisenberg Microscrope (ABC) 199
7.8 Assumption (A5) - The Schrödinger Equation (ABC) 203
7.9 Time Evolution of Expectation Values and Constants of the Motion (BC) 205
7.10 Time-Energy Uncertainty Relation (BC) 206
7.11 Time Evolution of Probability Amplitudes (BC) 210

8 Distributions, Fourier Transforms, and Rigged Hilbert Spaces 219
8.1 Functionals (BC) 219
8.2 Fourier Transforms (BC) 227
8.3 Rigged Hilbert Space (BC) 229

9 Algebraic Approach to Time-Independent Problems 236
9.1 Simple Harmonic Oscillator (AB) 236
9.1.1 Expectation Values 244
9.2 The Rigid Rotator (ABC) 247
9.3 Rigid Rotator in Three Dimensions - Angular Momentum (BC) 254
9.4 Algebraic Approach to Angular Momentum (BC) 257
9.5 Rotations and Rotational Invariance (BC) 267
9.6 Spin Angular Momentum (BC) 272

Contents

10 Central Force Problems 281

10.1 The Radial Equation (ABC) 282

10.2 Infinite Square Well (ABC) 286

10.3 Simple Harmonic Oscillator: Separation in Cartesian Coordinates (ABC) 289

10.3.1 Degeneracy 291

10.4 Simple Harmonic Oscillator: Separation in Spherical Coordinates (BC) 292

10.5 The Hydrogenic Atom (ABC) 296

10.6 Reduction of the Two-Body Problem (BC) 306

11 Transformation Theory 311

11.1 Rotations in a Vector Space (BC) 312

11.2 Example 1 - Fourier Transform of Hermite Functions (BC) 315

11.3 Dirac Notation (BC) 316

11.4 Example 2 - Angular Momentum (BC) 323

11.5 The Schrödinger Picture (BC) 327

11.6 Heisenberg Picture (BC) 328

11.7 Dirac or Interaction Picture (C) 333

12 Time-Independent Non-Degenerate Perturbation Theory 339

12.1 Rayleigh-Schrödinger Perturbation Theory for Stationary States (BC) 340

12.2 First Order Perturbations (BC) 343

12.3 Example 1 - Anharmonic Oscillator (BC) 345

12.4 Example 2 - Ground State of Helium-like Ions (BC) 346

12.5 Second Order Perturbations (BC) 349

12.6 Example 3 - Displaced Simple Harmonic Oscillator (BC) 350

12.7 Non-degenerate Perturbations to All Orders (C) 353

12.8 Sum Rule for Second Order Perturbation Theory (BC) 357

12.9 Linear Stark Effect (BC) 361

13 **Time-Independent Degenerate Perturbation Theory** 367
13.1 Two Levels: Rayleigh-Schrödinger Method for Degenerate Levels (BC) 368
13.2 General Rayleigh-Schrödinger Method for Degenerate Levels 372
13.3 Example: Spin Hamiltonian (BC) 374
13.3.1 Exact Solution 375
13.3.2 Rayleigh - Schrödinger Solution 377

14 **Further Approximation Methods** 381
14.1 Rayleigh-Ritz Method (BC) 383
14.2 Example 1 - Simple Harmonic Oscillator (BC) 386
14.3 Example 2 - He Ground State (BC) 387
14.4 The W.K.B. Approximation (BC) 389
14.4.1 Turning Points 395
14.5 Example 3 - W.K.B. Applied to a Potential Well (BC) 401
14.6 Special Boundaries (BC) 404
14.7 Example 4 - W.K.B. Approximation for Tunneling (BC) 405

15 **Time-Dependent Perturbation Theory** 411
15.1 Formal Considerations (C) 411
15.2 Direct Computations of Transition Amplitudes (BC) 416
15.3 Periodic Perturbation of Finite Duration (BC) 418
15.4 Photo-Ionization of Hydrogen Atom (BC) 422
15.5 The Adiabatic Approximation (BC) 427
15.6 The Sudden Approximation (BC) 433
15.7 Dipole in a Time-Dependent Magnetic Field (BC) 435
15.7.1 Oscillatory Perturbation 437
15.7.2 Slowly Varying Perturbation 439
15.7.3 Sudden Approximation 441

16 **Applications** 447
16.1 Gauge Transformations (C) 447

Contents

16.2 Motion in a Uniform Magnetic Field - Landau Levels (C) 452

16.3 The Quantum Hall Effect (C) 454

16.4 Motion in a Uniform Magnetic Field
- Heisenberg Equations (C) 457

16.5 Spin and Spin-Orbit Coupling (C) 469

16.6 Alkali Spectra (C) 471

16.7 Addition of Angular Momenta (C) 474

16.8 Example 1 - Two Spin 1/2 States (C) 479

16.9 Example 2 - Spin 1/2 and Orbital Angular Momentum (C) 480

16.10 The Weak-Field Zeeman Effect (C) 481

17 Scattering Theory: Time Dependent Formulation 488

17.1 Classical Scattering Theory (BC) 489

17.2 Asymptotic States in the Schrödinger Picture (C) 491

17.3 The Moller Wave Operators (C) 493

17.4 Green's Functions and Propagators (C) 496

17.5 Integral Equations for the Propagators:
Asymptotic States (C) 500

17.6 Cross-sections (BC) 502

17.7 The Lippmann-Schwinger Equations (C) 503

17.8 The S-Matrix and the Scattering Amplitude (C) 506

18 Scattering Theory: Time Independent Formulation 513

18.1 The Scattering Amplitude (BC) 513

18.2 Green's Functions and The Lippmann-Schwinger Equations (C) 516

18.3 The Born Approximation (BC) 521

18.4 Example - The Yukawa Potential (BC) 524

18.5 The Free Schrödinger Equation in Spherical Coordinates (BC) 526

18.6 Partial Wave Analysis (BC) 534

18.7 Phase Shifts (BC) 537

18.8 The Optical Theorem - Unitarity Bound (BC) 539

19 Further Topics in Potential Scattering 544

19.1 Example - The Square Well (BC) 545

19.2 Partial Wave Analysis of the Lippmann-Schwinger Equation (C) 549
19.3 Effective Range Approximation (C) 551
19.4 The Glauber or Eikonal Approximation (C) 557

20 Systems of Identical Particles 569
20.1 Two Identical Particles (BC) 570
20.2 The Hydrogen Molecule (BC) 572
20.3 N Identical Particles (C) 579
20.4 Non-Interacting Fermions (C) 582
20.5 Non-Interacting Bosons (C) 584
20.6 Occupation Number Space and Second Quantization for Bosons (C) 586
20.7 Occupation Number Space and Second Quantization for Fermions (C) 589
20.8 Field Operators (C) 593
20.9 Representation of Operators (C) 597
20.10 Heisenberg Picture (C) 601

21 Quantum Statistical Mechanics 608
21.1 Introduction (C) 608
21.2 The Density matrix (C) 610
21.2.1 The Microcanonical Ensemble. 610
21.2.2 The Canonical Ensemble 612
21.2.3 Grand Canonical Ensemble 613
21.3 The Ideal Gases (C) 610
21.4 General Properties of the Density Matrix (C) 623
21.5 The Density Matrix and Polarization (C) 626
21.6 Composite Systems (C) 630
21.7 The Quantum Theory of Measurement (C) 634
21.8 Conclusion (C) 639

Index 644

Chapter 1

The Breakdown of Classical Mechanics

During the nineteenth century many of the great advances in physics of the eighteenth century were consolidated and extended. In addition, the theory of electromagnetism was completed by J.C. Maxwell. Except for a few unexplained effects or anomalies there seemed little more in terms of fundamental physics to be done by the beginning of the twentieth century. Yet it is precisely in the year 1900 that quantum theory starts with Planck's formula for blackbody radiation.

Soon there were a host of experimental results, both new ones and earlier ones that again attracted attention. All of these pointed to flaws in the physics of the nineteenth century. In almost all cases these anomalies resulted when Newtonian mechanics and electromagnetism were simultaneously involved. In trying to elucidate these various experimental facts a new theory of physics, quantum theory, was born.

In the next few sections we briefly examine several of these experiments and discuss them with some modern hindsight. First we consider blackbody radiation from the prequantum or classical point of view. We then turn to a consideration of the stability of the classical Rutherford atom. Although Rutherford had experimentally demonstrated the planetlike structure of the electrons in atoms, his model caused a lot of theoretical problems.

An even older effect, dating back to Hertz in 1887, the photoelectric effect, provides another example of the complete breakdown of classical physics. After considering the effect we discuss the elastic scattering of light off electrons, the Compton effect, which also clearly demonstrates the corpuscular nature of light. On the other hand, the Davisson-Germer experiments demonstrate very clearly the wavelike nature of light and are discussed in the following section.

As a final example we consider the Franck-Hertz effect. This is a beautiful experiment demonstrating that only certain definite quanta of energy can be absorbed by atoms in inelastic collision with electrons.

1.1 Blackbody Radiation

Blackbody radiation refers to the equilibrium radiant energy to be found inside a cavity whose walls are completely opaque and held at a fixed temperature T. Such a cavity is called a blackbody cavity. The interest in such radiation is because this radiation is independent of the nature of the walls of the cavity (their material properties, or their geometry); the spectral properties of the radiation depend only on the temperature of the walls. A very simple proof of this fact, utilizing only the second law of thermodynamics, was given by Kirchhoff (1.1).

The fact that the radiation depends only on the temperature of the walls means that it is somehow universal. Most glowing bodies such as a hot piece of iron or our sun are good approximations of a blackbody. By measuring the spectrum of their radiant energy we can determine their temperature.

Since blackbody radiation was so simple, a theory for its spectrum was soon derived from classical mechanics and electromagnetism. The resultant spectral formula, called the <u>Rayleigh-Jeans Law</u> proved to fail completely at high frequencies.

We now derive this formula. Completely opaque walls for a cavity can be described mathematically by assuming that all the radiation is reflected at the walls. Thus we want standing waves inside the cavity.

If the cavity is a cube with sides of length L, then the components of the wavelength of the radiation in each direction must exactly fit into L. Thus, we have for the component of the wavelength of the lowest frequency mode in the x direction.

$$\lambda_{x,1} = 2L.$$

The next mode has:

$$\lambda_{x,2} = \frac{2L}{2}.$$

The 3rd mode has:

$$\lambda_{x,3} = \frac{2L}{3}$$

and so on. The n'th mode has

$$\lambda_{x,n} = \frac{2L}{n_x} \qquad n_x = 1,2,3... \tag{1.1}$$

A similar discussion holds for the y and z directions.

It is useful to convert these formulas into formulas involving wavenumbers because then the integers n_x , n_y , n_z appear in the numerator. The wavenumbers k_x , k_y , k_z are related to the corresponding wavelengths by

$$k_x = \frac{2\pi}{\lambda_x} , \quad k_y = \frac{2\pi}{\lambda_y} , \quad k_z = \frac{2\pi}{\lambda_z} . \tag{1.2}$$

The different vibration modes are therefore characterized by three integers (n_x,n_y,n_z) giving the total wavenumber

$$k = \sqrt{k_x^2 + k_y^2 + k_z^2} \tag{1.3}$$

Using 1.1 and 1.2 we get

$$k^2 = \frac{(2\pi)^2}{4L^2}\left(n_x^2 + n_y^2 + n_z^2\right). \tag{1.4}$$

The wavenumber k is also related to the frequency ν by

$$\nu = \frac{ck}{2\pi} \tag{1.5}$$

where c is the speed of light.

Now each mode of vibration of the electromagnetic field can be considered as a degree of freedom of the field; the different vibration modes are independent of each other. However, according to the equipartition principle of statistical mechanics we have for a temperature T an amount of energy $k_B T$ for each degree of freedom of the field. Here T is the temperature of the cavity wall and k_B is Boltzmann's constant.

From the equipartition principle we can therefore write the formula: The energy dU in a frequency interval between ν and $\nu + d\nu = k_B T$ × (number of modes of oscillation in this interval).

So to obtain the blackbody spectrum requires that we count the number of modes of oscillation corresponding to a frequency interval between ν and $\nu + d\nu$. It is easier to first obtain all modes up to a frequency ν. This is simply the number of points (the volume) inside one quadrant (1/8 since n_x, n_y, n_z are all positive) of a sphere whose equation according to 1.4 is:

$$n_x^2 + n_y^2 + n_z^2 = \frac{4L^2}{c^2}\nu^2 . \tag{1.6}$$

The result is

$$N = \frac{1}{8} \times \frac{4}{3}\pi\left(\frac{2L\nu}{c}\right)^3 \times 2 . \tag{1.7}$$

The last factor of 2 is due to the fact that for light two independent polarizations for each vibration mode are possible. Thus, the number of degrees of freedom is increased by this factor of 2.

The number of modes dN in the frequency interval between ν and $\nu + d\nu$ is now given by:

$$dN = \pi\left(\frac{2L}{c}\right)^3 \nu^2 d\nu \ . \tag{1.8}$$

Hence, using our formula for the energy per unit volume in this frequency interval we get:

$$du = \frac{1}{V} k_B T \ dN = \frac{8\pi k_B T}{c^3} \nu^2 d\nu \tag{1.9}$$

where we have used the fact that the volume $V = L^3$.

So by applying the classical equipartition theorem and classical electromagnetic theory we obtain for the energy density the unambiguous result

$$\rho(\nu) = \frac{du}{d\nu} = \frac{8\pi k_B T}{c^3} \nu^2 \ . \tag{1.10}$$

For low frequencies this result agrees splendidly with the experimental spectrum, but for high frequencies, as shown in figure 1.1, this result fails miserably.

Furthermore, on purely theoretical grounds alone, equation 1.10 must be wrong since it predicts that for a cavity of volume V at a temperature T the total radiant energy in this volume is given by:

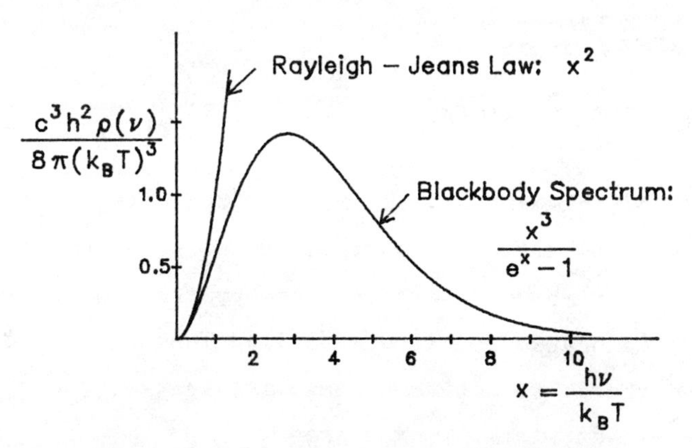

Fig. 1–1 Blackbody Radiation

$$E = V \int_0^\infty \rho(\nu)d\nu = \frac{8\pi k_B TV}{c^3} \int_0^\infty \nu^2 d\nu = \infty. \qquad 1.11$$

Clearly this is an utterly nonsensical result.

Exactly in the year 1900, Planck, who had previously found an analytic formula that fit the experimental results, gave a theoretical derivation of his formula. To obtain his formula Planck applied the same tools as we did above, but in addition he made the very radical assumption that radiation of frequency ν carries energy

$$E = h\nu \qquad 1.12$$

where, as determined by experiment, it was found that $h = 6.63\times10^{-34}$ Joule-seconds and is now known as Planck's constant and plays a fundamental role in all of modern physics.

Planck's energy-frequency relation (equation 1.12) was a completely new and revolutionary equation in physics. Until Planck, the energy of a wave could be any number and was proportional to the square of the amplitude of the wave. Suddenly the energy was quantized in lumps proportional to the frequency. This was a very exciting new development, but it took a while to make an impact on the physics community. In section 1.9 we derive Planck's radiation formula by using his energy-frequency relation.

1.2 Stability of Atoms - Discrete Spectral Lines

In the early part of this century, atoms were beginning to be taken seriously by physicists as actual physical objects and not just models. As a consequence experiments to explore the structure of atoms were begun. The very reasonable model, due to J.J. Thomson, of an atom as a cloud of positive charge with bits of negative charge (the electrons) interspersed was proven wrong by experiment.

In a long series of brilliant scattering experiments Rutherford showed conclusively in 1911 that the atom consisted of a tiny positive core of magnitude about 10^{-12} cm, called the nucleus, with electrons whirling about this nucleus at a much greater distance of about 10^{-8} cm. This planetary model caused a crisis because atoms were known to be stable lasting for millions of years. But, as we now show, such planetary atoms are intrinsically unstable on the basis of classical physics. In fact, given this planetary model we can even obtain a qualitative picture of the classical radiation spectrum. This also turns out to be wrong when compared with experiment.

From classical electromagnetic theory we find that a charge of magnitude e (in Coulombs) undergoing an acceleration radiates energy at the rate

$$S = \frac{2}{3}\frac{e^2a^2}{c^3}\left[\frac{1}{4\pi\varepsilon_o}\right] \text{ watts} \tag{1.13}$$

where a is measured in m/s^2 and c, the speed of light, is measured in m/s. The charge of an electron in Coulombs is

$$e = 1.6021917\times10^{-19} \text{ C.}$$

If we use e.s.u. (electrostatic units) then the factor $\left[\frac{1}{4\pi\varepsilon_o}\right]$ must be dropped. Later we shall use e.s.u. exclusively but in this chapter we give the formulas in MKS rationalized units with a square [ ] bracket for the factor to be dropped to get the corresponding formula in electrostatic units.

Now consider an hydrogen atom consisting of an electron in a spherical orbit about a proton. To a good approximation the center of mass is located at the center of the proton. The acceleration is given by

$$a = \frac{v^2}{R} \tag{1.14}$$

where v is the speed of the electron and R the radius of its orbit.

Equating the centrifugal force and force of electrostatic attraction we get

$$\frac{mv^2}{R} = \frac{e^2}{R^2} \cdot \left[\frac{1}{4\pi\varepsilon_o}\right] . \tag{1.15}$$

Combining 1.14 and 1.15 yields

$$a = \frac{e^2}{mR^2} \cdot \left[\frac{1}{4\pi\varepsilon_o}\right]$$

so that

$$S = \frac{2}{3} \frac{e^6}{m^2 c^3 R^4} \cdot \left[\frac{1}{4\pi\varepsilon_o}\right]^3 . \tag{1.16}$$

From this we can estimate the time t that it would take for an electron to lose all its kinetic energy and spiral into the proton according to

$$t \simeq \text{kinetic energy}/S.$$

Thus,

$$\begin{aligned} t &\simeq \frac{e^2}{2R} \cdot \left[\frac{1}{4\pi\varepsilon_o}\right] \cdot \frac{3}{2} \frac{m^2 c^3 R^4}{e^6} \cdot [4\pi\varepsilon_o]^3 \\ &= \frac{3}{4} \frac{m^2 c^3 R^3}{e^4} \cdot [4\pi\varepsilon_o]^2 \end{aligned} \tag{1.17}$$

The mass of an electron in MKS units is about

$$m_e = 9.1\times10^{-31} \text{ kg} \tag{1.18}$$

and a good approximation for the hydrogen atom radius is about 10^{-10} m.

Substituting all of the above numbers in equation (1.17) yields

$$t \simeq 4\times10^{-10}\ \text{s} . \tag{1.19}$$

Compared even to only 1 yr = 3.1×10^{7} s this prediction is wrong by an incredible factor of 10^{17}. In fact millions of years or 10^{14} s are more reasonable estimates so that the classical prediction is wrong by at least a factor of 10^{24}. Clearly classical physics contradicts the stability of atoms.

There is a second difficulty with the classical result. It has to do with the radiation spectrum. The radiation frequency ν is, in fact, determined by the angular frequency ω of rotation of the electron in its orbit according to

$$\omega = 2\pi\nu . \tag{1.20}$$

Now using that the acceleration of the electron in its orbit is given by

$$a = \omega^2 R \tag{1.21}$$

and using 1.12 we get

$$S = \left[\frac{1}{4\pi\varepsilon_o}\right]\cdot \frac{2}{3} \cdot \frac{e^2R^2}{c^3} (2\pi)^4\nu^4 . \tag{1.22}$$

So we conclude that the spectrum of the radiated energy is continuous. This also contradicts the experimental fact, namely that atomic spectra consist of discrete series of very sharp lines.

Such lines had been studied for half a century and were well classified by the turn of this century. After studying the spectra of many atoms, Rydberg and Ritz (1.2) independently discovered a very important result. They found that the discrete frequencies observed could be expressed more simply. In fact all frequencies could be described by

$$\nu_{nm} = A_n - A_m \qquad n,m = 1,2,3,\ldots \tag{1.23}$$

that is, a difference of two terms. Thus, far fewer terms A_n than frequencies are required. This so-called Rydberg-Ritz Combination Principle provided an important clue in the development of quantum mechanics.

1.3 Photoelectric Effect

A schematic diagram of the experimental arrangement for studying the photoelectric effect is shown in figure 1.2.

We shine light of a fixed frequency, ν on a clear metal surface. The retarding voltage is then increased to a voltage V_c at which the current, as measured by the galvanometer, ceases. We now summarize some of the experimental results that are observed.

1. The frequency of the light ν must be greater than some critical frequency ν_o (even for zero retarding voltage) in order that photoelectrons be emitted.

2. For light intensities as low as 10^{-10} watts/m^2 the delay in time for the photo-current to reach a steady state is less than 10^{-9} s.

3. For a fixed frequency, the photo-current is proportional to the intensity of the light.

4. The energy of the photoelectrons increases linearly with the frequency of the light.

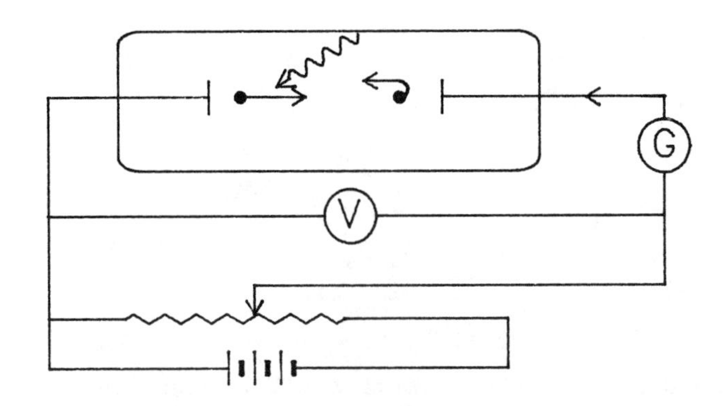

Fig. 1–2 The Photoelectric Effect

We now consider this experiment in purely classical terms and see that the experimental results are in violent disagreement with the conclusion we obtain. Classically, the time-averaged energy density of an electromagnetic wave is given by

$$u = [4\pi\varepsilon_o]\,\frac{1}{8\pi}\,|\vec{E}|^2 \qquad 1.24$$

and is independent of frequency. Here $\vec{E}$ is the amplitude of the electric field in volts/m. Furthermore, classically, all electrons in the surface of the metal absorb energy uniformly. Thus, all the electrons eventually acquire enough energy to be emitted. It is therefore completely impossible to explain results 1 and 4. The experimental result 3 has some hope of being explained this way but it is not clear. The time delay, which is measured as 10^{-9} s or less for an intensity of 10^{-10} watt/m^2 can be estimated however. To do this we need a few numbers. The typical binding energy of an electron in the surface of a metal is 1-2 eV or about 10^{-19} Joules. The typical size of an atom in a metal is 1 Å = 10^{-10} m. Thus in 1 m^2 of surface we have about 10^{20} atoms each of which requires an energy of 10^{-19} Joules. Thus about 10 Joules of energy must be absorbed by 1 m^2 of metal surface before photoelectrons will be emitted. Since we are illuminating the surface with 10^{-10} watts/m^2 the time t required for photo-emission to occur is given by

$$10^{-10}\, t = 10$$

or

$$t = 10^{11}\ \text{s} \simeq 3{,}000\ \text{years}.$$

The discrepancy between theory and experiment is an incredible factor of 10^{20}.

Einstein (1.3) succeeded in explaining this discrepancy by making the same assumption as Planck, namely that light comes in quanta of energy given by

$$E = h\nu\ . \qquad 1.25$$

Thus, a given electron either absorbs all the energy E or none. Hence if ν is large enough so that E exceeds the binding energy ϕ of the electron, then photoelectrons will be emitted almost instantaneously. If ν, however, is too small, then no photoelectrons will be emitted at all. Thus the experimental results 1 and 4 are explained immediately. Result 3 also follows since the number of photons in the beam is what determines the number of photoelectrons emitted, but the number of photons N is given by

$$N = I/h\nu \tag{1.26}$$

where I is the intensity of the beam.

The final result 4) is also easily explained now. We simply equate the energy absorbed to the kinetic energy of the electron plus the energy ϕ required for the electron to break loose from the surface. This gives

$$\frac{1}{2} mv^2 + \phi = h\nu \tag{1.27}$$

which represents the linear relation between frequency and kinetic energy of the photoelectrons observed experimentally.

These results seem to indicate that photons are somehow particle-like, carrying a definite amount of energy, and are localized in space so that one atom can absorb a whole photon. In the subsequent section we discuss several more ideas and experiments that conclusively established the particle-like nature of photons.

1.4 Wave Particle Duality

Even in classical physics, light was not always considered as a wave motion. In fact, Newton formulated a completely corpuscular theory of light. The laws of reflection and refraction (Snell's Law) can then be derived purely on the basis of conservation of momentum and energy. The situation is as shown in figure 1.3. A particle with momentum p_1

is incident from the left in a homogeneous medium described by a constant potential V_1 and is either reflected or transmitted into a homogeneous medium described by a constant potential V_2. The particle thus experiences a force only at the interface between the media. This force is normal to the interface, since in a direction parallel to the interface grad V vanishes. Actually the particle receives an impulse on hitting the interface.

1.4.1 Reflection

Since the force is normal to the interface, the tangential component of momentum is conserved. Furthermore, since energy is conserved, the magnitude of the momentum $p_1 = \sqrt{2m(E-V_1)}$ is conserved. Hence, we get

$$p_1 \sin \theta_1 = p_1 \sin \theta_1' . \tag{1.28}$$

Thus

$$\theta_1 = \theta_1' \tag{1.29}$$

so that the angle of incidence equals the angle of reflection.

1.4.2 Refraction

In this case we still have

$$E = \frac{p_1^2}{2m} + V_1 = \frac{p_2^2}{2m} + V_2 . \tag{1.30}$$

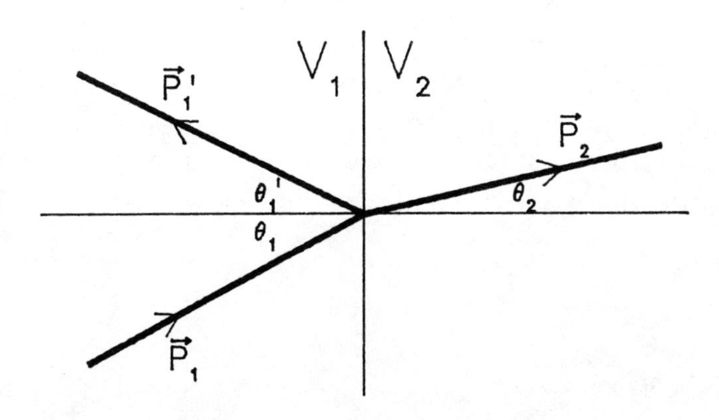

Fig. 1–3 Snell's Law

The tangential component of momentum is still conserved. Thus we have

$$p_1 = \sqrt{2m(E-V_1)}$$

as before, and

$$p_2 = \sqrt{2m(E-V_2)} \ .$$

Equating the tangential components of momentum gives

$$p_1 \sin \theta_1 = p_2 \sin \theta_2 \qquad 1.31$$

so that

$$\frac{\sin \theta_1}{\sin \theta_2} = \frac{p_2}{p_1} = \frac{\sqrt{E - V_2}}{\sqrt{E - V_1}} = \frac{n_2}{n_1} \ . \qquad 1.32$$

Here we have defined the index of refraction to be $\sqrt{E - V}$. This is reasonable since $\sqrt{E - V}$ is simply a number for a given medium and a particle of a given energy.

If one uses wave optics one finds that the index of refraction of a medium is inversely proportional to the wavelength of the light in the medium. Thus

$$\frac{n_2}{n_1} = \frac{\lambda_1}{\lambda_2} \ , \qquad 1.33$$

This suggests that for a particle of momentum p there may be an associated wavelength λ such that

$$\lambda \ \alpha \ \frac{1}{p} \quad .$$

1.5 De Broglie's Hypothesis

De Broglie presented a very simple argument for photons to fix the constant of proportionality in the above relationship. The relationship de Broglie used connects the energy E and momentum p for a plane electromagnetic wave. It states

$$E = cp \quad . \qquad 1.34$$

where c is the speed of light. This equation follows from the fact that photons travel at the speed of light c and have an effective mass $m = E/c^2$. Thus, the momentum p is given by

$$p = mc = (E/c^2)c = E/c$$

as stated in equation 1.34. If we combine this relationship with the Planck energy-frequency relation $E = h\nu$, we get

$$h\nu = cp \quad . \qquad 1.35$$

Thus

$$\lambda = \frac{c}{\nu} = \frac{h}{p} \; . \qquad 1.36$$

This is the famous de Broglie hypothesis, namely that with any particle carrying momentum p one should associate a wave-length

$$\lambda = h/p.$$

1.6 The Compton Effect

The photoelectric effect indicated that photons were somehow granular or particle-like carrying a definite amount of energy given by $E = h\nu$. By scattering X-rays (photons) off free electrons, A.H. Compton (1.4) showed that photons are definitely particle-like, carrying a definite momentum and scattering like point particles. The situation is as depicted in figure 1.4.

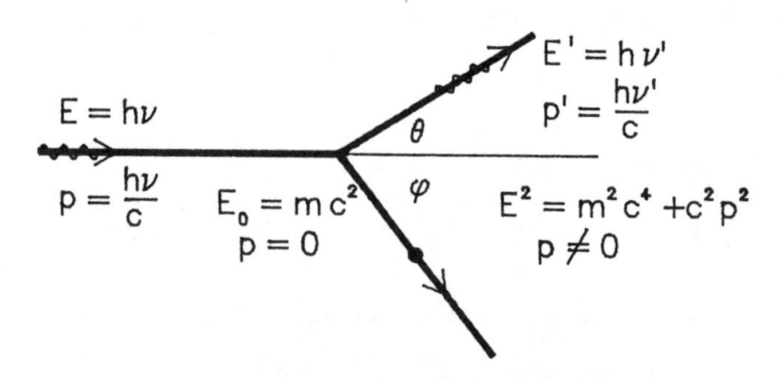

Fig. 1–4 Compton Effect

The incoming photon carries energy $E = h\nu$ and momentum $p = \frac{h\nu}{c}$ in the x-direction, while the electron is initially at rest (energy $E = mc^2$). The scattered photon carries energy $E' = h\nu'$ and total momentum $p' = h\nu'/c$ in the direction given by the angle θ. The electron, which was orginally at rest, recoils with momentum p in the direction given by the angle ϕ. Applying conservation of energy and momentum as for point particles we get

Conservation of Energy:

$$h\nu + mc^2 = h\nu' + E \tag{1.37}$$

Conservation of Momentum:

$$\frac{h\nu}{c} = \frac{h\nu'}{c}\cos\theta + p\cos\phi \tag{1.38}$$

$$0 = \frac{h\nu'}{c}\sin\theta - p\sin\phi\ . \tag{1.39}$$

In his experiments Compton measured the change in wavelength of the scattered X-rays as a function of the scattering angle θ. According to classical electromagnetic theory no change in wavelength should occur. Now in writing down equations 1.37, 1.38 and 1.39 we have treated the X-rays as if they were point particles. Before further discussion we first obtain a formula for the change in wavelength $\Delta\lambda = \lambda' - \lambda$ in terms of θ. By rearranging 1.38 and 1.39, squaring and adding we get:

$$h^2\nu^2 + h^2\nu'^2 - 2h^2\nu\nu' \cos\theta = c^2p^2 . \tag{1.40}$$

Rearranging (1.37) and squaring yields

$$h^2\nu^2 + h^2\nu'^2 - 2h^2\nu\nu' = E^2 + m^2c^4 - 2Emc^2 . \tag{1.41}$$

Subtracting 1.40 from 1.41 and using the energy momentum relation

$$E^2 = c^2p^2 + m^2c^4 \tag{1.42}$$

we get:

$$-2h^2\nu\nu'(1-\cos\theta) = -2mc^2(E-mc^2). \tag{1.43}$$

But from equation 1.37 we have

$$E - mc^2 = h(\nu-\nu').$$

Thus, we finally get:

$$\frac{h}{mc}(1-\cos\theta) = c\left(\frac{\nu-\nu'}{\nu\nu'}\right) = \frac{c}{\nu'} - \frac{c}{\nu} = \lambda' - \lambda .$$

So the increase in wavelength is given by

$$\Delta\lambda = \lambda' - \lambda = \frac{h}{mc}(1-\cos\theta) = \lambda_c(1-\cos\theta). \tag{1.44}$$

Here we have introduced the Compton wavelength of the electron, given by

$$\lambda_c = \frac{h}{mc} . \tag{1.45}$$

Compton's measurements showed that equation (1.44) agreed splendidly with the experimental results. Thus, a photon has particle properties just like a particle has wave properties.

In the next section we discuss an experiment (the Davisson-Germer experiment) that proved conclusively the wave-like nature of particles. So it was found that on the one hand waves sometimes behaved like particles whereas on the other hand particles sometimes behaved like waves.

1.7 The Davisson-Germer Experiment

If we consider the de Broglie relation

$$\lambda = \frac{h}{mv}$$

for an electron with 100 eV energy then we find

$$100 \text{ eV} = 100\times1.6\times10^{-10} \text{ J} = \frac{1}{2} mv^2$$

so

$$mv = 5.40\times10^{-24} \text{ kg m/s}$$

and

$$\lambda = \frac{h}{mv} = 1.23\times10^{-10} \text{ m}$$

or

$$\lambda = 1.23 \text{ Å} .$$

Thus λ is about the same as the wavelength of a hard X-ray.

Now the diffraction of X-rays by crystals had already been observed and explained in 1913 by the father and son team of W.L. and W.H. Bragg (1.5). The planes of a crystal lattice act as a very fine diffraction grating. Secondly, de Broglie during his thesis defence suggested that the matter waves predicted by his formula could be observed in such a manner. The actual experiment, however, was carried out much later in 1925 (1.6) by Davisson and Germer who accidently produced such a diffraction pattern with an electron beam without knowing about the de Broglie hypothesis. (This experiment was also performed by E. Rupp (1.6) using a ruled grating and grazing incidence.)

Davisson and Germer had been scattering electrons off polycrystalline nickel targets when a fortunate accident occurred. The vacuum system broke down and their nickel target oxidized. After repairing the vacuum they tried to expel the oxygen from the nickel by heating. This process changed the polycrystalline nickel to several large crystals. On recommencing the scattering experiment they found that 54 volt electrons incident at 50° to the surface of the nickel, led to extremely strong reflection of these electrons. The result had all the appearance of Bragg reflections. Now from X-ray data the lattice spacing 2a for nickel was known to be 2.15 Å = 2.15×10^{-10} m. The Bragg formula - which we derive immediately - states that for reflection maxima the incident angle should be given by

$$n\lambda = 2a \sin\theta \qquad n = 1,2,3,\ldots \quad . \tag{1.46}$$

Using $n = 1$, $2a = 2.15$ Å, $\theta = 50°$ yields:

$$\lambda = 1.65 \text{ Å}.$$

Computing the wavelength of a 54 volt electron from

$$\lambda = h/p = \frac{h}{[2m\,eV]^{1/2}}$$

we find

$$\lambda = 1.67 \text{ Å} .$$

This is splendid confirmation of the de Broglie hypothesis. We now derive the Bragg formula. The situation is as depicted in figure 1.5. The small spheres represent individual atoms. We begin by computing the path difference $\Delta\ell$ for the two rays reflected as shown. This path difference is given by

$$\Delta\ell = \ell_2 + \ell_3 - \ell_1 \quad . \tag{1.47}$$

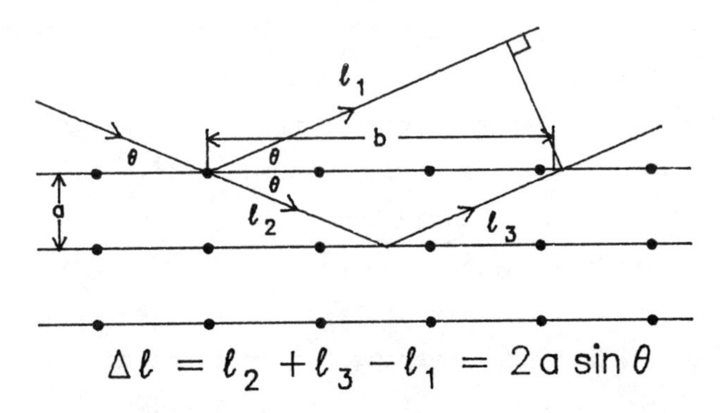

Fig. 1–5 Bragg Reflection

Now

$$\ell_2 = \ell_3 = \frac{a}{\sin\theta} \quad .$$

Also

$$b = \ell_2 \cos\theta + \ell_3 \cos\theta = 2a \cot\theta \ .$$

But

$$\ell_1 = b \cos\theta = 2a \frac{\cos^2\theta}{\sin\theta} \ .$$

$$\therefore \quad \Delta\ell = \frac{2a}{\sin\theta}(1-\cos^2\theta) = 2a \sin\theta. \qquad 1.48$$

For constructive interference (maxima) we need that

$$\Delta\ell = n\lambda \qquad n = 1,2,3,\dots\ . \qquad 1.49$$

Thus, we have arrived at the Bragg formula

$$n\lambda = 2a \sin\theta \ . \qquad 1.50$$

1.8 The Franck-Hertz Effect

This effect was first observed in an experiment in 1914 in which J. Franck and G. Hertz (1.7) looked for the "grainyness" of matter suggested by Planck's radiation law. A later experiment in 1924 by G. Hertz (1.8) verified Bohr's atomic model predictions. We shall describe both experiments at once since the second experiment was simply a more detailed investigation of the first.

A schematic diagram of the experimental set-up is shown in figure 1.6. Electrons are emitted from the hot cathode c and accelerated by the potential difference V between the cathode c and the grid g towards the anode a. The current I reaching the anode a is measured by the ammeter A. The current observed as a function of the accelerating voltage is displayed in figure 1.7.

The first drop in current occurs at 4.9 eV, the second at 9.8 eV and the third at 14.7 eV. Thus it appears that the mercury atoms absorb 1×4.9 or 2×4.9 or 3×4.9 eV of energy from the accelerated electrons. Now the most prominent spectral line emitted by mercury is in the ultraviolet at $\lambda = 2537$ Å.

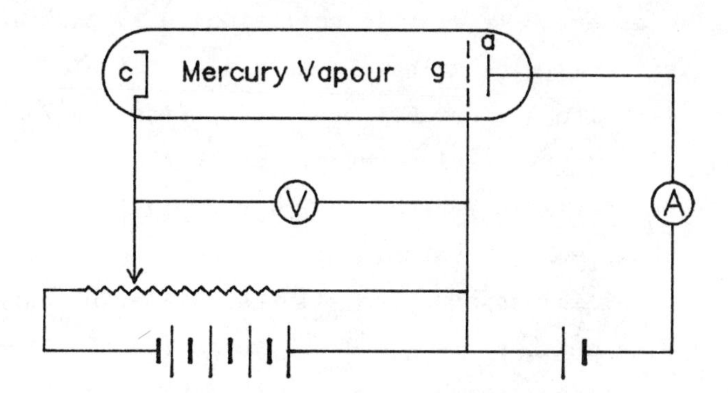

Fig. 1-6 Franck – Hertz Experiment

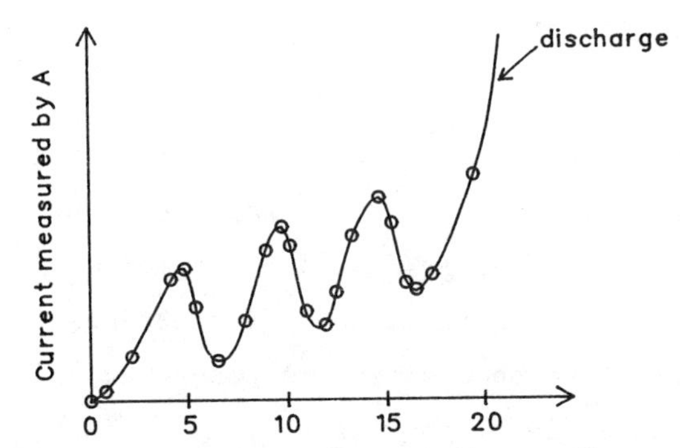

Fig. 1–7 Current – Voltage Curve for Franck – Hertz Experiment

If we use the Planck energy-frequency relation $E = h\nu = \frac{hc}{\lambda}$ we find that this wavelength corresponds to an energy

$$E = \frac{6.63\times10^{-34}\times3.00\times10^{8}}{2.54\times10^{-7}} = 7.83\times10^{-19}\ \text{J}$$

$$= \frac{7.83\times10^{-19}}{1.6\times10^{-19}} = 4.9\ \text{eV}\ .$$

This result can be interpreted as follows. For an energy below 4.9 eV the electrons make essentially elastic collisions with the mercury atoms and the current increases with voltage. At 4.9 eV the electrons can collide ineastically with the mercury atoms to excite the 2537 Å line and thus give up all their kinetic energy. Thus, the current drops. Then as the voltage increases further the current again rises until we reach a voltage of 9.8 = 2×4.9 volts. At this point the electrons can make two inelastic collisions and lose all their kinetic energy. The process again repeats until at 14.7 = 3×4.9 volts the electrons can make three inelastic collisions.

To verify this interpretation G. Hertz (1.8) measured the radiation emitted by the mercury atoms and found that below 4.9 volts the untraviolet radiation corresponding to λ = 2537 Å was not emitted, whereas above 4.9 volts it was emitted strongly. In fact, Hertz was able to correlate the occurrence of many of the spectral lines of mercury with a threshold voltage of the electrons. Thus, the quantized nature of the energy levels of an atom was definitely established.

1.9 Planck's Radiation Law

We now return to a discussion of blackbody radiation. Planck combined classical statistical mechanics with his energy frequency relation. Thus he made the following assumptions:

1. In accordance with classical statistical mechanics, the probablity for an oscillator of energy E to be excited is proportional to the Boltzmann factor e^{-E/k_BT}.

2. The energy of the oscillators is quantized and comes in quanta given by

$$E_n = nh\nu , \quad n = 1,2,3,\ldots \quad .$$

Combining these assumptions we can compute the average energy $\langle E\rangle$ of an oscillator from

$$\langle E\rangle = \sum_i \text{Probability of energy } E_i \times \text{energy } E_i \tag{1.51}$$

so

$$\langle E\rangle = \frac{\sum_n nh\nu\, e^{-nh\nu/k_BT}}{\sum_n e^{-nh\nu/k_BT}} = -\frac{d}{d\left(\frac{1}{k_BT}\right)} \ell n\left(\sum_n e^{-nh\nu/k_BT}\right)$$

$$\langle E\rangle = -\frac{d}{d\left(\frac{1}{k_BT}\right)} \ell n\left(1-e^{-h\nu/k_BT}\right)^{-1} .$$

Thus

$$\langle E\rangle = k_BT \frac{h\nu/k_BT}{e^{h\nu/k_BT} - 1} . \tag{1.52}$$

We next combine this result with the result obtained in section 1.1 and write the equation.

The energy per unit volume du in a frequency interval between ν and $\nu + d\nu = \langle E\rangle \times$ (number of modes of oscillation in this interval).

After dividing by the volume V of the blackbody cavity we find from equation (1.8) that

$$\frac{1}{V}\, dN = \frac{8\pi}{c^3}\, \nu^2 d\nu \ . \qquad 1.53$$

Thus

$$du = \langle E\rangle \cdot \frac{1}{V}\, dN = \frac{8\pi}{c^3} \frac{h\nu^3}{e^{h\nu/k_BT} - 1}\, d\nu \qquad 1.54$$

so

$$\rho(\nu) = \frac{8\pi}{c^3} \frac{h\nu^3}{e^{h\nu/k_BT} - 1} \ . \qquad 1.55$$

This is the famous Planck's blackbody radiation formula shown in fig. 1.1. We notice that for low frequencies namely $h\nu/k_BT \ll 1$, this law just goes over into the Rayleigh-Jeans law since for these frequencies we have $e^{h\nu/k_BT} \simeq 1 + h\nu/k_BT$.

The establishment of this formula and the introduction of the constant h was one of the most revolutionary developments in all of physics.

1.10 Einstein's Model for Specific Heat

Soon after Planck's enunciation of his blackbody law, Einstein in 1907 (1.9) used a very similar argument to explain why the specific heat of solids goes to zero at zero temperature. Einstein's model was only qualitatively correct, a better model was published in 1912 by

P. Debye (1.10). This so-called Debye model remains valid today. We shall briefly discuss this model after considering Einstein's model.

From the classical equipartition theorem one obtains for the internal energy U of one mole of a monatomic gas at temperature T the value

$$U = k_B T \times (\text{number of degrees of freedom})$$

or

$$U = k_B T \cdot 3N_A \qquad 1.56$$

where N_A is Avagadro's number $\simeq 6.02\times10^{26}$/kg mole. But

$$N_A k_B = R = \text{the gas constant} \simeq 2 \text{ k cal/kg mole.}$$

Hence the specific heat at constant volume C_V is given by

$$C_V = \left(\frac{\partial U}{\partial T}\right)_V = 3R \simeq 6 \text{ k cal/kg mole.} \qquad 1.57$$

This result agrees very well at high temperature but fails miserably for low temperatures where C_V goes to zero like T^3.

By assuming that the atoms in a monatomic crystal are harmonically bound (simple oscillators) and applying Planck's assumptions to these oscillators Einstein obtained that the average energy $\langle E\rangle$ of such an oscillator is given by

$$\langle E\rangle = k_B T \cdot \frac{h\nu/k_B T}{e^{h\nu/k_B T} - 1} . \qquad 1.58$$

Now applying the equipartition principle and using the fact that for one mole of a monatomic solid the number of degrees of freedom is given by $3N_A$ he found that the total internal energy U is given by

$$U = 3N_A k_B T \frac{h\nu/k_B T}{e^{h\nu/k_B T} - 1} . \qquad 1.59$$

Clearly at high temperatures $(h\nu/k_BT < 1)$ we again have

$$e^{h\nu/k_BT} \simeq 1 + \frac{h\nu}{k_BT}$$

and thus

$$U \simeq 3N_Ak_BT = 3RT \qquad 1.60$$

in accord with experiment and the classical result. At low temperatures $(h\nu/k_BT > 1)$ we have $e^{h\nu/k_BT} - 1 \simeq e^{h\nu/k_BT}$ and thus

$$U \simeq 3N_Ah\nu\, e^{-h\nu/k_BT} . \qquad 1.61$$

Thus both U and $C_V = \left(\frac{\partial U}{\partial T}\right)_V$ vanish exponentially at low temperature. The agreement with experiment is certainly much better than the classical result.

The reason that U vanishes too rapidly at low T was pointed out by P. Debye 1.10. He assumed that the vibration spectrum of a monatomic solid may be treated, as we have done above, as a homogeneous medium except that the total number of modes is cut off at $3N_A$, to yield the correct number of degrees of freedom of N_A atoms. This yields a cutoff frequency ν_{max}. Now the number of modes per unit volume with frequency less than ν_{max} is given by a formula like 1.6, namely

$$3N_A = \frac{1}{8} \times \frac{4}{3}\pi\left(\frac{2\nu}{v_o}\right)^3 \times 3 . \qquad 1.62$$

Here v_o is an <u>average speed of sound</u> in the solid and the last factor of 3 is due to the fact that there are two transverse modes plus one longitudinal mode of vibration in a solid. Thus the cutoff frequency is given by

$$\nu_{max} = v_o\left(\frac{3N_A}{4\pi}\right)^{1/3} . \qquad 1.63$$

The internal energy per unit volume, dU, for frequencies between ν and $\nu + d\nu$ is now given by the number of modes in this interval times the average energy per mode, namely

$$dU = \frac{12\pi}{v_o^3} \nu^2 d\nu \frac{h\nu}{e^{h\nu/k_B T} - 1} \quad . \tag{1.64}$$

Thus, the total internal energy is given by

$$U = \frac{12\pi}{v_o^3} \int_0^{\nu_{max}} \frac{h\nu^3 \, d\nu}{e^{h\nu/k_B T} - 1} \quad . \tag{1.65}$$

Letting

$$x = \frac{h\nu}{k_B T} \quad , \quad x_{max} = \frac{h\nu_{max}}{k_B T}$$

we find

$$U = \frac{12\pi \, k_B^4 T^4}{h^2 v_o^3} \int_0^{x_{max}} \frac{x^3 \, dx}{e^x - 1} \quad . \tag{1.66}$$

This is Debye's expression. The dimensionless parameter x_{max} can be rewritten as

$$x_{max} = \frac{\Theta_D}{T} \tag{1.67}$$

where the Debye temperature Θ_D is given by

$$\Theta_D = \frac{h v_o}{k_B} \left(\frac{3N_A}{4\pi}\right)^{1/3} \quad . \tag{1.68}$$

The Debye temperature characterizes the specific crystal through the velocity v_o . Typical values of Θ_D range from 150 K to 1000 K.

1.11 Bohr Theory and the Hydrogen Atom

In this section we describe the precursor of quantum mechanics. Bohr extended Planck's hypothesis and made some additional assumptions. We describe these now and apply them to the hydrogen atom in the next section.

1. To get the observed stability of atoms, Bohr assumed that atoms exist only in certain definite states in which they do not radiate. These are the stationary states. The energy is therefore automatically quantized since only these stationary states occur and not all possible states.

2. To get discrete spectra Bohr assumed Planck's law in the form

$$E_j - E_k = h\nu \quad . \tag{1.69}$$

Here E_j, E_k are the energies associated with two stationary states. Thus, he made the implicit further assumption that the energy changes discontinuously from one state to another. This explains discrete spectra both for emission and absorption:

$$\text{Emission occurs if } \; E_j > E_k$$

$$\text{Absorption occurs if } \; E_k > E_j \; .$$

This also gives the Rydberg-Ritz combination principle immediately since from the combination principle

$$\nu = A_j - A_k \quad .$$

Thus the terms A_j can be identified with E_j/h.

3. The correspondence principle was one of Bohr's most useful assumptions. It states: In the limit of large quantum numbers the classical predictions must be recovered at least asymptotically.

A motivation for this assumption can be found in the experimental fact that

$$A_n = \frac{a}{(n+b)^2}$$

for almost all atomic spectra. Thus

$$E_n = \frac{ah}{(n+b)^2} \quad .$$

Then

$$E_n - E_m = ah\left[\frac{1}{(n+b)^2} - \frac{1}{(m+b)^2}\right]$$

$$\xrightarrow[n,m\to\infty]{} ah\left[\frac{m^2-n^2}{m^2n^2}\right] \to 0 \quad .$$

Hence, $\Delta E_n \to 0$ and we get a continuum of energies for large quantum numbers.

We assume the electron in a hydrogen atom is in a circular orbit about a force center given by the Coulomb attraction of the proton. Also we choose the zero of energy to correspond to an unbound electron with zero kinetic energy. Thus in the orbit E is negative.

Classically the frequency ω at which the electron radiates is given by the angular frequency of rotation of the electron in its orbit. Thus

$$\omega = \frac{v}{R} \quad . \qquad 1.70$$

The total energy of the electron in its orbit is

$$E = \frac{1}{2}\, mv^2 - \left[\frac{1}{4\pi\varepsilon_o}\right] \frac{e^2}{R} \, . \qquad 1.71$$

Equating the Coulomb force of attraction and the centrifugal force gives

$$\frac{mv^2}{R} = \left[\frac{1}{4\pi\varepsilon_o}\right]\frac{e^2}{R^2} \qquad 1.72$$

Thus

$$E = -\frac{1}{2}mv^2 = -\frac{e^2}{2R}\cdot\left[\frac{1}{4\pi\varepsilon_o}\right] \qquad 1.73$$

Hence

$$v = \sqrt{\frac{2|E|}{m}} \qquad 1.74$$

Thus,

$$R = \frac{e^2}{2|E|}\cdot\left[\frac{1}{4\pi\varepsilon_o}\right] \qquad 1.75$$

$$\therefore \quad \omega = \frac{v}{R} = \frac{2}{e^2}\left[\frac{2|E|^3}{m}\right]^{1/2}\cdot[4\pi\varepsilon_o] \qquad 1.76$$

so the classical frequency ν_c is given by

$$\nu_c = \frac{\omega}{2\pi} = \frac{1}{\pi e^2}\left[\frac{2|E|^3}{m}\right]^{1/2}\cdot[4\pi\varepsilon_o] \quad . \qquad 1.77$$

By the correspondence principle we have for large n that

$$h\nu = \frac{d|E|}{dn} \,. \qquad 1.78$$

Again using the correspondence principle to equate the two frequencies we get

$$\frac{1}{h}\frac{d|E|}{dn} = \sqrt{\frac{2}{m}}\,\frac{1}{\pi e^2}\,|E|^{3/2}\cdot[4\pi\varepsilon_o] \ .$$

The solution of this differential equation is

$$|E| = \frac{1}{2}\left(\frac{2\pi e^2}{[4\pi\varepsilon_o]hc}\right)^2 mc^2\,\frac{1}{n} \ .$$

Setting

$$\frac{2\pi e^2}{[4\pi\varepsilon_o]hc} = \alpha \tag{1.79}$$

and recalling that E is negative we get

$$E_n = -\frac{1}{2} mc^2\alpha^2 \frac{1}{n^2} . \tag{1.80}$$

Thus we have calculated the terms in the Ritz combination principle for hydrogen. The result agrees with experiment to better than 0.01%.

This completes a brief review of the major developments that led to the modern quantum theory. In the next chapter we briefly review certain aspects of classical mechanics that turned out to be particularly useful in developing an understanding of quantum theory.

Up to this point we have used MKS rationalized electromagenetic units. In atomic physics many calculations are performed in Gaussian units. Thus charge is measured in e.s.u. so that $e \simeq 4.8\times10^{-10}$ e.s.u. and magnetic fields are measured in Gauss. To convert our previous equations to Gaussian units simply drop the square factors $[\frac{1}{4\pi\varepsilon_o}]$ or $[4\pi\varepsilon_o]$ appearing in equations involving electromagnetism.

Notes, References and Bibliography

1.1 A.B. Pippard - The Elements of Classical Thermodynamics - Cambridge University Press (1957) page 78.

1.2 J.R. Rydberg, Report of Intl. Phys. Cong. at Paris, ii, 200 (1900).

W. Ritz, Phys. Z. 9, 521 (1908).

W. Ritz, Astrophys. J. 28, 237 (1908).

1.3 A. Einstein, Ann. Phys. 17, 132 (1905).

1.4 A.H. Compton, Phys. Rev. 22, 409 (1923).

1.5 W.H. Bragg and W.L. Bragg, Proc. Roy. Soc. A88, 428 (1913).
W.L. Bragg, Proc. Roy. Soc. A89, 248 (1913).

1.6 C. Davisson and L.H. Germer, Phys. Rev. 30, 705 (1927).
E. Rupp, Zeits. f. Phys. 52, 8 (1928), also Phys. Zeits. 28, 837 (1928)

1.7 J. Franck and G. Hertz, Verh. Dtsch. Phys. Ges. 16, 512 (1914).

1.8 G. Hertz, Z. Phys. 22, 18 (1924).

1.9 A. Einstein, Ann. d. Physik 22, 180 (1907); 34, 170 (1911).

1.10 P. Debye, Ann. d. Physik 29, 789 (1912).

For further reading on this subject we strongly recommend the following books:

A.P. French and E.F. Taylor - An Introduction to Quantum Physics - W.W. Norton and Co. Inc. (1978).

A.T. Weidner and R.L. Sells - Elementary Modern Physics - Allyn and Bacon, 2nd edition (1969).

M. Born - Atomic Physics - Blackie and Son Ltd., London, 5th edition (1952).

1.11 C. Kittel - Introduction to Solid State Physics - John Wiley and Sons, 2nd edition (1963) Chapter 6.

Chapter 1 Problems

1.1 Calculate the principle quantum number for the earth in its orbit about the sun. What is the energy difference between two neighbouring energy levels?

Hint: For large n, $E_n \approx E_{classical}$.

1.2 What is the wavelength associated with gas molecules at a temperature T? Estimate this wavelength for a typical gas at room temperature and compare it to visible light.

1.3 For a monchromatic beam of electromagnetic radiation (λ = 5000 Å) of an intensity of 1 watt/m^2, calculate the number of photons passing 1 cm^2 of area normal to the beam in one second.

1.4 Show that if one assumes that the circumference of a stationary state orbit of an electron in a hydrogen atom is an integral multiple of the de Broglie wavelength, one also obtains the correct energy levels (eqn. 1.80).

1.5 List several experiments or observational results that may be used to obtain a lower bound on the lifetime of atomic hydrogen in free space.

1.6 Estimate the effect on the specific heat of reducing a crystal to a fine powder of dimensions of about 10^{-6} cm. Hint: Study the Debye model of specific heat (1.11) and realize that the size of the crystal now also imposes an upper limit on the wavelength of the sound waves in the crystal.

1.7 The shortest possible wavelength of sound in sodium chloride is twice the lattice spacing, about 5.8×10^{-8} cm. The sound velocity is approximately 1.5×10^{5} cm/sec.

a) Compute a rough value for the highest sound frequency in the solid.

b) Compute the energy of the corresponding phonons, or quanta of vibrational energy.

c) Roughly what temperature is required to excite these oscillations appreciably?

Chapter 2

Review of Classical Mechanics

Classical mechanics was reformulated in an elegant manner during the nineteenth century. In this chapter we review some of these formulations. To lead into the discussion we consider in section 1 the mechanics of a point particle confined to one dimension and acted on by a conserative force. We use this to give a heuristic introduction to Hamilton's principle and the Euler-Lagrange equations as well as Hamilton's equations.

In section 2, we review the Lagrangian and Hamiltonian formulation of many-particle systems. This section assumes a somewhat deeper knowledge of analytical mechanics. The dicussion of section 2 is further elaborated to Hamilton–Jacobi theory in sections 3 and 4.

In section 5, we then apply some of these classical techniques to find constants of the motion and perform Bohr-Sommerfeld quantization for the hydrogen atom.

Finally we give a heuristic derivation, actually more of a plausibility argument, for the Schrödinger equation in section 6.

2.1 Classical Mechanics of a Particle in One Dimension.

We begin by considering a particle of mass m moving in one dimension and acted on by a conservative force $F(x)$. The equation of motion is given by

$$m \frac{d^2x}{dt^2} = F(x). \qquad 2.1$$

This equation contains all the information about the dynamics (forces) of this system. In addition (since equation 2.1 is second order in time) we need two pieces of initial data, say the position x_o and velocity v_o of the particle at some instant t_o, in order to determine the motion of the system (particle) for all subsequent times. We now restate this simple fact in somewhat more fancy language which will later allow us to compare classical and quantum mechanics more easily.

To begin with define $x(t)$ and $v(t) = \frac{dx}{dt}$ to be dynamical variables. Because equation 2.1 allows us to express $a(t) = \frac{d^2x}{dt^2}$ in terms of $x(t)$, it follows that all dynamical variables such as acceleration, energy, momentum, etc. may be expressed in terms of $x(t)$ and $v(t)$. That we need both $x(t)$ and $v(t)$ follows from the fact that 2.1 is second order in time. We therefore say that $x(t)$ and $v(t)$ form a complete set of dynamical variables.

Specifying a complete set of dynamical variables (i.e. $x(t)$ and $v(t)$) at a given time t_o, specifies the state of the system completely at the instant t_o. This means that if we know $x(t_o)$ and $v(t_o)$ then all physical quantities at $t = t_o$ are, in principle, determined.

To determine the state of the system at any later time $t > t_o$ we need to solve equation 2.1 for $t > t_o$. We want to bring out more clearly this fact that x and v specify the state completely. To this end we want to find a pair of first order differential equations for a complete set of dynamical variables. It turns out that rather than x and v it is more convenient to use x and p. Here p is the linear momentum.*

$$p = mv \quad . \tag{2.2}$$

We now rewrite equation 2.1 using the fact that the force F is conservative. This means that there is a potential $V(x)$ such that

$$F = -\frac{\partial V}{\partial x} \quad . \tag{2.3}$$

Then 2.1 reads

$$m \frac{d^2x}{dt^2} = -\frac{\partial V}{\partial x} \; . \tag{2.4}$$

This equation may be immediately integrated if we multiply by $\frac{dx}{dt}$. Then we get:

$$\frac{1}{2} m \frac{d}{dt} \left(\frac{dx}{dt}\right)^2 = -\frac{\partial V}{\partial x} \cdot \frac{dx}{dt} \tag{2.5}$$

so that integrating between t_o and t we find

$$\frac{1}{2} m v^2(t) + V(x) = \frac{1}{2} m v^2(t_o) + V(x_o) = E. \tag{2.6}$$

Here E is a constant called the total energy. If we solve equation 2.2 for v in terms of p and substitute the result in equation 2.6 we get:

$$\frac{p^2}{2m} + V(x) = E. \tag{2.7}$$

The function of p and x on the left hand side of this equation is

---

* Equation 2.2 defines the momentum correctly for this example. In general a more complicated definition is required. This is done in detail in section 2.2, equation 2.36. See also equation 2.30.

called the hamiltonian $H(x,p)$ of this system.

$$H(x,p) = \frac{p^2}{2m} + V(x). \tag{2.8}$$

If we know the hamiltonian, it is a simple matter to find a pair of equations equivalent to the definition 2.2 and equation of motion 2.4. In fact by straightforward differentiation we see that:

$$\frac{\partial H}{\partial p} = \frac{p}{m} = \frac{dx}{dt}$$

and

$$\frac{\partial H}{\partial x} = \frac{\partial V}{\partial x} = -F = -\frac{dp}{dt}. \tag{2.9}$$

The pair of equations

$$\frac{dx}{dt} = \frac{\partial H}{\partial p} \tag{2.10}$$

$$\frac{dp}{dt} = -\frac{\partial H}{\partial x} \tag{2.11}$$

are called Hamilton's equations of motion. At this stage they are nothing more than a fancy way of rewriting 2.1 and 2.2

In 1834 and 1835, Sir William Rowan Hamilton (2.1) published two papers which gave deep insight into classical physics. We give a heuristic discussion of this Hamilton's principle before we state it. Consider a simple harmonic oscillator (consisting of a point mass m and spring k), which starts from rest with the spring compressed. Thus all the energy is intially potential and none of the energy is kinetic.

$$T(o) = 0 \;, \quad V(o) = E. \tag{2.12}$$

After 1/4 of a cycle τ the spring is completely unstressed and the potential energy is zero. On the other hand the particle is moving with maximum velocity so that the kinetic energy is a maximum, in fact the total energy:

$$T\left(\frac{\tau}{4}\right) = E \ , \quad V\left(\frac{\tau}{4}\right) = 0 \ . \tag{2.13}$$

After 1/2 cycle the situation is again reversed. The particle has overshot the equilibrium and the spring is stretched with the particle momentarily at rest so that

$$T\left(\frac{\tau}{2}\right) = 0 \ , \quad V\left(\frac{\tau}{2}\right) = E \ . \tag{2.14}$$

After 3/4 cycle we have:

$$T\left(\frac{3\tau}{4}\right) = E \ , \quad V\left(\frac{3\tau}{4}\right) = 0, \tag{2.15}$$

and after a full cycle we are back to our initial configuration

$$T(\tau) = 0, \quad V(\tau) = E. \tag{2.16}$$

If we consider this motion we see that as it progresses we alternate between potential energy V and kinetic energy T. Thus the motion is such that the function

$$L = \int_{t_1}^{t_2} (T-V)dt \tag{2.17}$$

of the particle path is minimized. We call L the action and

$$\mathcal{L} = T - V \tag{2.18}$$

the Lagrangian of the system. Hamilton's principle states:

A dynamical system evolves along that path which minimizes or maximizes the action. To see how this works let us consider our one dimensional particle.

$$L = \int_{t_1}^{t_2} [\frac{1}{2} m \dot{x}^2 - V(x)]dt \qquad 2.19$$

where we are writing $\dot{x}$ for $\frac{dx}{dt}$. In extremizing (maximizing or minimizing) L we must choose $\dot{x}(t)$, $x(t)$ such as to pass through $\dot{x}(t_1)$, $x(t_1)$ and $\dot{x}(t_2)$, $x(t_2)$. The state at the end points t_1 and t_2 is fixed.

We now assume that $x(t)$ is that motion which makes L an extremum. To find the equation for $x(t)$ we introduce

$$\bar{x}(t) = x(t) + \varepsilon(t) \qquad 2.20$$

where $\varepsilon(t)$ is a small but arbitrary deviation from $x(t)$. One frequently writes

$$\bar{x}(t) - x(t) = \varepsilon(t) \equiv \delta x(t) \ . \qquad 2.21$$

The fact that the motion is fixed at the end points means

$$\varepsilon(t_1) = \varepsilon(t_2) = 0. \qquad 2.22$$

Substituting 2.20 in the expressions for T and V and keeping only terms of lowest (first) order in $\varepsilon(t)$ we get:

$$T = \frac{m}{2} [\dot{x} + \dot{\varepsilon}]^2 = \frac{1}{2} m\dot{x}^2 + m\dot{x}\dot{\varepsilon} \qquad 2.23$$

$$V = V(x+\varepsilon) = V(x) + \varepsilon \frac{\partial V}{\partial x} \ . \qquad 2.24$$

Thus,

$$\delta L = \int_{t_1}^{t_2} [T(\bar{x}(t))-V(\bar{x}(t))]dt - \int_{t_1}^{t_2} [T(x(t))-V(x(t))]dt$$

$$= \int_{t_1}^{t_2} [m\dot{x}\dot{\varepsilon} - \varepsilon \frac{\partial V}{\partial x}] \ dt. \qquad 2.25$$

For an extremum we must have $\delta L = 0$. So we get:

$$\int_{t_1}^{t_2} \left[m\dot{x}\,\frac{d\varepsilon}{dt} - \varepsilon\frac{\partial V}{\partial x}\right]dt = 0. \qquad 2.26$$

Integrating the first term by parts and using 2.22 we find

$$\int_{t_1}^{t_2} -\varepsilon(t)\left[m\ddot{x} + \frac{\partial V}{\partial x}\right]dt = 0. \qquad 2.27$$

But $\varepsilon(t)$ is arbitrary so whenever $m\ddot{x} + \frac{\partial V}{\partial x}$ is positive we can make $\varepsilon(t)$ negative and vice versa. Thus we can always make the integrand positive. Therefore the only way to ensure that for arbitrary $\varepsilon(t)$ the integrand in 2.22 vanishes is to have

$$m\ddot{x} = -\frac{\partial V}{\partial x}. \qquad 2.28$$

This, however, is just Newton's law of motion with which we started.

If we had carried this computation through for an arbitrary Lagrangian L, equation 2.28 would read

$$\frac{d}{dt}\frac{\partial L}{\partial \dot{x}} - \frac{\partial L}{\partial x} = 0. \qquad 2.29$$

Either of equations 2.28 and 2.29 is known as the Euler-Lagrange equation for this system. Thus, for a general system, to find the equations of motion find the kinetic energy T and potential V and form $L = T - V$. The Euler-Lagrange equation is then the equation of motion.

To obtain Hamilton's equations 2.10, 2.11 we transform variables from $x, \dot{x}$ to x, p. The equation 2.2 defining the momentum p may be written more generally as

$$p = \frac{\partial L}{\partial \dot{x}}. \qquad 2.30$$

We then define the hamiltonian function $H(x,p)$ by

$$H(x,p) = \frac{\partial L}{\partial \dot{x}} \dot{x} - L \tag{2.31}$$

where $\dot{x}$ is replaced by p as obtained from 2.30. In our example 2.30 yields

$$p = \frac{\partial L}{\partial \dot{x}} = m\dot{x} \quad \text{or} \quad \dot{x} = \frac{p}{m}$$

Then

$$H(x,p) = \frac{\partial L}{\partial \dot{x}} \dot{x} - L$$

$$= m\dot{x}\dot{x} - \left[\frac{1}{2} m\dot{x}^2 - V(x)\right] ,$$

or finally

$$H(x,p) = \frac{p^2}{2m} + V(x) . \tag{2.32}$$

This result agrees, of course, with our previous equation 2.8.

In the next section we generalize these results to a system of many particles.

2.2 Lagrangian and Hamiltonian Formulation

In classical mechanics, as we stated before, the state of a system at a given time t is specified by giving the values of a complete set of dynamical varibles at time t. A complete set is one from which all other dynamical variables at time t may be calculated. For example we may specify all generalized coordinate $q_1 \ldots q_N$ and the corresponding velocities $\dot{q}_1 \ldots \dot{q}_N$.

In the Lagrange formulation there is one scalar function

$$\mathcal{L}(q_1,\ldots, q_N , \dot{q}_1 \ldots, \dot{q}_N)$$

from which the equations of motion for all coordinates are determined by the so-called Euler-Lagrange equations

$$\frac{d}{dt}\left(\frac{\partial \mathcal{L}}{\partial \dot{q}_r}\right) - \frac{\partial \mathcal{L}}{\partial q_r} = 0. \qquad 2.33$$

These equations result from finding an extremum of the action integral

$$L = \int_{t_1}^{t_2} \mathcal{L}\, dt. \qquad 2.34$$

Thus, 2.33 follows from

$$\delta \int_{t_1}^{t_2} \mathcal{L}\, dt = 0. \qquad 2.35$$

Where the variations at the end points vanish. By introducing a Legendre transformation one arrives at Hamilton's principle.

Therefore, define the generalized momenta p_r conjugate to q_r according to

$$p_r \equiv \frac{\partial \mathcal{L}}{\partial \dot{q}_r} \qquad 2.36$$

and the hamiltonian function H by the Legendre transformation

$$H = H(q_1 \ldots q_N, p_1 \ldots p_N) = \sum_{r=1}^{N} \frac{\partial \mathcal{L}}{\partial \dot{q}_r} \dot{q}_r - \mathcal{L} \;. \qquad 2.37$$

In H the $\dot{q}_r$ have been eliminated and replaced by the p_r using 2.36.

The resulting equations of motion are

$$\dot{q}_r = \frac{\partial H}{\partial p_r} \qquad \dot{p}_r = - \frac{\partial H}{\partial p_r} \ . \tag{2.38}$$

Notice that classically one may get many different hamiltonians to describe the same system by performing so-called contact transformations about which we will say more shortly.

First, some examples:

Consider

$$\text{a)} \quad L_a = \frac{1}{2} m\dot{q}^2 \ . \tag{2.34a}$$

The equation of motion is

$$\ddot{q} = 0 \ . \tag{2.35a}$$

$$\text{b)} \quad L_b = A \ e^{\beta \dot{Q}} \ . \tag{2.34b}$$

The equation of motion is

$$\frac{d}{dt} (\beta A e^{\beta \dot{Q}}) = 0$$

which yields

$$\ddot{Q} = 0 \ . \tag{2.35b}$$

Thus both give the same equation of motion.

hamiltonians:

$$\text{a)} \quad p = \frac{\partial L_a}{\partial \dot{q}} = m\dot{q} \tag{2.36a}$$

$$H_a = \frac{1}{2} m\dot{q}^2 = \frac{p^2}{2m} \ . \tag{2.37a}$$

$$\text{b)} \quad P = \frac{\partial L_b}{\partial \dot{Q}} = \beta A e^{\beta \dot{Q}} \tag{2.36b}$$

$$\dot{Q} = \frac{1}{\beta} \ln \frac{P}{\beta A} \tag{2.37b}$$

$$H_b = A(\beta \dot{Q} - 1) e^{\beta \dot{Q}} \tag{2.38}$$

$$H_b = (\ln \frac{P}{\beta A} - 1)\frac{P}{\beta} \tag{2.38b}$$

The Hamilton equations of motion are

$$\text{a)} \quad \dot{p} = -\frac{\partial H_a}{\partial q} = 0 \tag{2.39a}$$

$$\dot{q} = \frac{\partial H_a}{\partial p} = \frac{p}{m} \tag{2.40a}$$

$$\text{b)} \quad \dot{P} = -\frac{\partial H_b}{\partial \dot{Q}} = 0 \tag{2.39b}$$

$$\dot{Q} = \frac{\partial H_b}{\partial P} = \frac{1}{\beta} \ln \frac{P}{\beta A} . \tag{2.40b}$$

For classical mechanics it clearly does not matter much which hamiltonian we use, but for quantum mechanics it does.Therefore, we agree henceforth to choose H such that <u>H coincides with the total energy of the system</u>.

2.3 Contact Transformations and Hamilton-Jacobi Theory

In classical mechanics, given the set of variables $\{q_i, p_i\}$ we can introduce new variables

$$\begin{aligned} Q_i &= Q_i(q_r, p_r) \\ P_i &= P_i(q_r, p_r) . \end{aligned} \tag{2.41}$$

Of particlar interest are those transformations that preserve the form of Hamilton's equations. Thus, there must exist a function $K(Q_i, P_i)$ such that the equations of motion read

$$\dot{Q}_i = \frac{\partial K}{\partial P_i} \qquad \dot{P}_i = -\frac{\partial K}{\partial Q_i} \quad . \tag{2.42}$$

Such transformations 2.41 are called <u>contact</u> or <u>canonical</u>. Since Q_i, P_i satisfy 2.42 they must also satisfy a variational principle (2.19 and following). But,

$$L = \sum_i p_i \dot{q}_i - H \tag{2.43}$$

and

$$\delta \int_{t_1}^{t_2} [\sum_i p_i \dot{q}_i - H(p_i, q_i)]dt = 0 \tag{2.44}$$

as well as

$$\delta \int_{t_1}^{t_2} [\sum P_i \dot{Q}_i - K(Q_i, P_i)]dt = 0 \quad . \tag{2.45}$$

Thus, the two integrands can differ at most by a total differential dF/dt since then

$$\int_{t_1}^{t_2} \frac{dF}{dt}\, dt = F(2)-F(1)$$

and the variation

$$\int_{t_1}^{t_2} \frac{dF}{dt}\, dt = 0 \quad .$$

F is called the <u>generating function</u> of the transformation. Naively, one expects F to be a function of the $4N + 1$ variables q_i, p_i, Q_j, P_j, t. However, due to the connecting equations 2.41 only $2N + 1$ are independent. For example

$$F = F_1(q_i, Q_j, t). \qquad 2.46$$

Since the integrands differ only by dF/dt we have

$$\sum p_i \dot{q}_i - H = \sum P_i \dot{Q}_i - K + \frac{dF_1}{dt}. \qquad 2.47$$

But

$$\frac{dF_1}{dt} = \sum \frac{\partial F_1}{\partial q_i} \dot{q}_i + \sum \frac{\partial F_1}{\partial Q_i} \dot{Q}_i + \frac{\partial F_1}{\partial t}. \qquad 2.48$$

Since q_i and Q_i are independent variables we get from 2.47 and 2.48 that

$$p_i = \frac{\partial F_1}{\partial q_i} \qquad 2.49$$

$$P_i = -\frac{\partial F_1}{\partial Q_i} \qquad 2.50$$

$$K = H + \frac{\partial F_1}{\partial t} \qquad 2.51$$

The equations of motion 2.42 in the new variables are particularly easy to solve if

$$\frac{\partial K}{\partial P_i} = \dot{Q}_i = 0 \qquad 2.52$$

and

$$-\frac{\partial K}{\partial Q_i} = \dot{P}_i = 0. \qquad 2.53$$

This is most easily achieved by choosing $K = 0$. Thus F is determined by

$$\frac{\partial F}{\partial t} + H = 0 \ .$$

In this case it is more convenient to choose F as a function of the q_i and P_i. This generating function F_2 is related to F_1 by a Legendre transformation

$$F_2(q_i,P_i,t) = F_1(q_i,Q_i,t) + \sum P_iQ_i \ . \qquad 2.54$$

It is then an easy matter to check that instead of 2.49, 2.50 and 2.51 we get

$$p_i = \frac{\partial F_2}{\partial q_i} \qquad 2.49a$$

$$Q_i = \frac{\partial F_2}{\partial P_i} \qquad 2.50a$$

$$K = H + \frac{\partial F_2}{\partial t} \qquad 2.51a$$

Thus with $H = H(q_i,P_i)$ the equation for $K = 0$ reads

$$H(q_1,\ldots,q_N; \frac{\partial F}{\partial q_1},\ldots,\frac{\partial F}{\partial q_N}; t) + \frac{\partial F}{\partial t} = 0 \qquad 2.55$$

where we have dropped the subscript on F and used 2.49a to replace the p_i by $\frac{\partial F}{\partial q_i}$. This is the celebrated Hamilton-Jacobi equation.

Example:

$$H = \frac{p^2}{2m} \qquad 2.56$$

Then

$$\frac{\partial F}{\partial t} + \frac{1}{2m}\left(\frac{\partial F}{\partial q}\right)^2 = 0. \qquad 2.57$$

A solution of 2.57 is

$$F = \alpha q - \beta t$$

with

$$\beta = \frac{\alpha^2}{2m}\,. \qquad 2.58$$

Before proceeding we do some more formal manipulations that will give us a basis later for a heuristic "derivation" of Schrödinger's equation

Consider the Hamilton-Jacobi equation 2.55 and call S a solution of it. Equation 2.55 is a first order partial differential equation in $N + 1$ variables. However if S is a solution so is $S + \alpha$ for any constant α. Thus S contains only N constants as far as tranformations are concerned. Furthermore,

$$\dot{P}_i = 0. \qquad 2.53$$

Thus the P_i are constants. Hence

$$S = S(q_1 \ldots q_N\,,\ \alpha_1 \ldots \alpha_N\,,\ t)$$

is a solution where

$$P_i = \alpha_i$$

$$p_i = \frac{\partial S(q_i, \alpha_i, t)}{\partial q_i} \qquad 2.59$$

and

$$Q_i = \beta_i = \frac{\partial S(q_i,\alpha_i,t)}{\partial \alpha_i} . \tag{2.60}$$

We can now invert these equations 2.60 and get

$$q_i = q_i(\alpha_r,\beta_r,t) .$$

In our example

$$S = -\frac{\alpha^2 t}{2m} + \alpha q$$

$$Q = \beta = \frac{\partial S}{\partial \alpha} = -\frac{\alpha t}{m} + q$$

so

$$q = \beta + \frac{\alpha t}{m} .$$

Thus, solving the Hamilton-Jacobi equation

$$\frac{\partial S}{\partial t} + H(q_1 \ldots q_N , \frac{\partial S}{\partial q_1} \ldots \frac{\partial S}{\partial q_N} , t) = 0 \tag{2.61}$$

gives a solution of the dynamical problem. Note that we can always write

$$S(q_i,\alpha_i,t) = W(q_i,\alpha_i) - \alpha_1 t .$$

Then 2.61 becomes

$$H(q_1 \ldots q_N, \frac{\partial W}{\partial q_1} \ldots \frac{\partial W}{\partial q_N}) = \alpha_1 \tag{2.62}$$

This equation is t independent. We can separate the equation even further by writing

$$W = \sum_i W_i(q_i, \alpha_1 \ldots \alpha_N) \ . \tag{2.63}$$

Then we get N equations

$$H_i\left(q_i, \frac{\partial W_i}{\partial q_i}, \alpha_1 \ldots \alpha_N\right) = \alpha_1 \ . \tag{2.64}$$

These are first order ordinary differential equations. The momenta p_i are still given by the equations of canonical transformations

$$p_i = \frac{\partial W_i(q_i, \alpha_1 \ldots \alpha_N)}{\partial q_i} \ .$$

If the motion is periodic, then the action integral

$$J_i = \oint p_i \, dq_i = \oint \frac{\partial W_i}{\partial q_i} dq_i$$

is a function only of $\alpha_1 \ldots \alpha_N$. Thus J_i is a constant. The rules for quantization given by Bohr were extended by Sommerfeld to cover such periodic motions according to

$$\oint p_i \, dq_i = nh = 2\pi\hbar n \ . \tag{2.65}$$

These are the so-called Bohr-Sommerfeld quantization rules. In order to understand where they come from we shall first discuss the meaning of the J_i. As example we use our old friend the simple harmonic oscillator.

$$H = \frac{p^2}{2m} + \frac{k}{2} q^2 \ . \tag{2.66}$$

Setting $p = \partial S/\partial q$ we get the Hamilton-Jacobi equation

$$\frac{1}{2m}\left(\frac{\partial S}{\partial q}\right)^2 + \frac{k}{2}q^2 + \frac{\partial S}{\partial t} = 0. \qquad 2.67$$

Now set

$$S(q,\alpha,t) = W(q,\alpha) - \alpha t. \qquad 2.68$$

Then

$$\frac{1}{2m}\left(\frac{\partial W}{\partial q}\right)^2 + \frac{1}{2}kq^2 = \alpha \qquad 2.69$$

so

$$W = \sqrt{mk}\int\sqrt{\frac{2\alpha}{k} - q^2}\,dq \qquad 2.70$$

and

$$S = \sqrt{mk}\int\sqrt{\frac{2\alpha}{k} - q^2}\,dq - \alpha t. \qquad 2.71$$

Also q is given by

$$\beta = \frac{\partial S}{\partial \alpha} = \sqrt{\frac{m}{k}}\int\frac{dq}{\sqrt{\frac{2\alpha}{k} - q^2}} - t$$

or

$$\beta = -t - \sqrt{\frac{m}{k}}\cos^{-1}\left[\sqrt{\frac{k}{2\alpha}}\,q\right] .$$

Thus

$$q = \sqrt{\frac{2\alpha}{k}}\cos\omega(t+\beta) \qquad 2.72$$

where

$$\omega = \sqrt{\frac{k}{m}} \ .$$

To get α and β we have to impose initial conditions. For example if at

$$t = 0 \qquad p = 0 \quad \text{and} \quad q = q_0$$

then

$$\beta = 0 \quad \text{and} \quad \alpha = \frac{k}{2} q_0^2 \ ,$$

Thus,

α = total initial energy.

Then finally

$$q = q_0 \cos \omega t \ , \tag{2.73}$$

a result we could have obtained by much more elementary means.

2.4 Interpretation of the Action-Angle Variable:

In this case

$$J = \oint pdq = \oint \frac{\partial W(q_1 \alpha)}{\partial q} dq$$

$$= \sqrt{mk} \oint \sqrt{\frac{2\alpha}{k} - q^2} \ dq \ .$$

Let

$$q = \sqrt{\frac{2\alpha}{k}} \sin \theta \ .$$

Then

$$J = 2\alpha\sqrt{\frac{m}{k}}\int_0^{2\pi} \cos^2\theta d\theta = 2\pi\alpha\sqrt{\frac{m}{k}} = \frac{2\pi\alpha}{\omega} \qquad 2.74$$

$$\therefore \quad \alpha = H = \frac{J\omega}{2\pi} = J\nu$$

so that

$$\frac{\partial H}{\partial J} = \nu. \qquad 2.75$$

This last result is of more general validity than the derivation indicates.

Now from the correspondence principle we have for $\nu_{classical}$ a function of E as the quantum number $m \to \infty$

$$\frac{E_{m+n} - E_m}{n} \to h\,\nu_{classical} \quad .$$

Thus,

$$\frac{dE}{dn} \to h\,\nu_{classical}$$

$$= 2\pi\hbar\nu_{classical} \ ,$$

and we get as a quantization rule

$$\int_{E_{min}}^{E_{max}} \frac{dE}{\nu_{classical}(E)} = 2\pi n\hbar \quad .$$

But

$$\frac{\partial H}{\partial J_i} = \frac{\partial E}{\partial J_i} = \nu_i \qquad 2.76$$

so

$$\frac{dE}{\nu_i} = dJ_i$$

2.5 Hydrogen Atom for Bohr-Sommerfeld Quantization

Thus

$$\oint dJ_i = \oint p_i \, dq_i = 2\pi n\hbar \; . \tag{2.77}$$

This is one way to arrive at Bohr-Sommerfeld quantization.

As an example of the use of this technique we work out the hydrogen atom including elliptical orbits.

$$\frac{\vec{p}^{\,2}}{2m} - \frac{e^2}{r} = \frac{1}{2m}\left(p_r^2 + \frac{p_\theta^2}{r^2} + \frac{p_\phi^2}{r^2 \sin^2\theta}\right) - \frac{e^2}{r} \tag{2.78}$$

where now e is measured in e.s.u. The quantization is given by

$$\oint p_i \, dq_i = nh. \tag{2.79}$$

Also

$$\begin{aligned} p_r &= m\dot{r} \\ p_\theta &= mr^2\dot{\theta} \\ p_\phi &= mr^2 \sin^2\theta\dot{\phi} \; . \end{aligned} \tag{2.80}$$

The Hamilton-Jacobi equation for W is

$$\frac{1}{2m}\left[\left(\frac{\partial W}{\partial r}\right)^2 + \frac{1}{r^2}\left(\frac{\partial W}{\partial \theta}\right)^2 + \frac{1}{r^2 \sin^2\theta}\left(\frac{\partial W}{\partial \phi}\right)^2\right] - \frac{e^2}{r} = \alpha_1 = E. \tag{2.81}$$

Separating the variables

$$W = W_r(r) + W_\theta(\theta) + W_\phi(\phi) \tag{2.82}$$

we get

$$\frac{\partial W_\phi}{\partial \phi} = \alpha_\phi = \text{constant} \qquad 2.83$$

and

$$\left(\frac{\partial W_\theta}{\partial \theta}\right)^2 + \frac{\alpha_\phi^2}{\sin^2\theta} = \alpha_\theta^2 \qquad 2.84$$

and

$$\left(\frac{\partial W_r}{\partial r}\right)^2 + \frac{\alpha_\theta^2}{r^2} = 2m\left(E + \frac{e^2}{r}\right). \qquad 2.85$$

These last three equations are three conservation statements; 2.83 gives

$$p_\phi = \alpha_\phi \quad \text{(Conservation of } p_\phi\text{)}$$

2.84 gives

$$p_\theta^2 + \frac{p_\phi^2}{\sin^2\theta} = \alpha_\theta^2 \quad \text{(Conservation of total angular momentum)}$$

This last follows from

$$H = \frac{1}{2m}\left(p_r^2 + \frac{L^2}{r^2}\right) - e^2/r \qquad 2.86$$

so

$$L^2 = p_\theta^2 + \frac{p_\phi^2}{\sin^2\theta} \qquad 2.87$$

as stated.

We now apply the quantization.

$$J_\phi = \oint p_\phi \, d\phi = \oint \alpha_\phi \, d\phi = 2\pi\alpha_\phi = 2\pi m'\hbar$$

so

$$\alpha_\phi = m'\hbar \ . \qquad 2.88$$

Then

$$J_\theta = \oint \sqrt{\alpha_\theta^2 - \frac{\alpha_\phi^2}{\sin^2\theta}} \, d\theta \ . \qquad 2.89$$

To evaluate this we use a trick.

Since the equations for generalized coordinates do not involve t explictly we have for the kinetic energy T

$$2T = \sum p_i \dot{q}_i \qquad 2.90$$

so that in polar coordinates

$$2T = p_r \dot{r} + L\dot{\psi} = p_r \dot{r} + p_\theta \dot{\theta} + p_\phi \dot{\phi} \ . \qquad 2.91$$

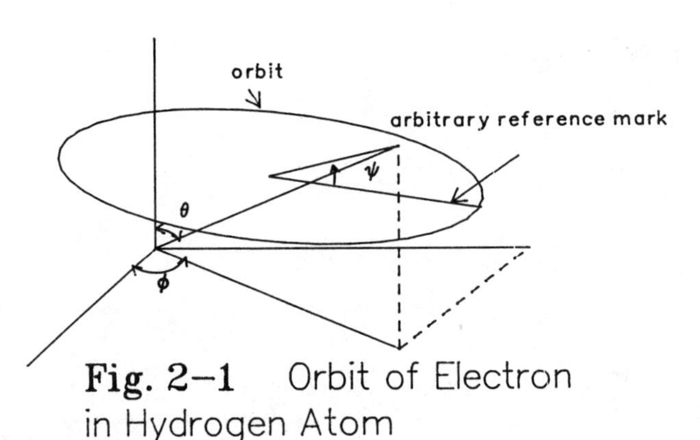

Fig. 2–1 Orbit of Electron in Hydrogen Atom

Here ψ = angle of azimuth of the particle in its orbit. (See figure 2.1). Thus

$$p_\theta \, d\theta = L d\psi - p_\phi \, d\phi \tag{2.92}$$

From 2.92 we get:

$$J_\theta = \oint L d\psi - \oint p_\phi \, d\phi \, . \tag{2.93}$$

Now as θ goes through a complete cycle ψ and ϕ go through 2π. Also,

$$L = \alpha_\theta = \text{constant}.$$

$$p_\phi = \alpha_\phi \, .$$

$$\therefore \quad J_\theta = 2\pi(\alpha_\theta - \alpha_\phi) = 2\pi(\ell - m')\hbar \, . \tag{2.94}$$

Hence,

$$J_r = \oint \left[2mE + \frac{2me^2}{r} + \frac{(J_\theta + J_\phi)^2}{4\pi^2 r^2} \right]^{1/2} dr$$

$$= 2\pi(n-\ell)\hbar \, . \tag{2.95}$$

We evaluate this integral by contour integration. Now

$$J_\theta + J_\phi = 2\pi\ell\hbar \, .$$

Then

$$J_r = \oint \left[2mE + \frac{2me^2}{r} - \frac{\ell^2\hbar^2}{r^2} \right]^{1/2} dr \, . \tag{2.96}$$

The motion is bounded only if $E < 0$. The maximum and minimum values

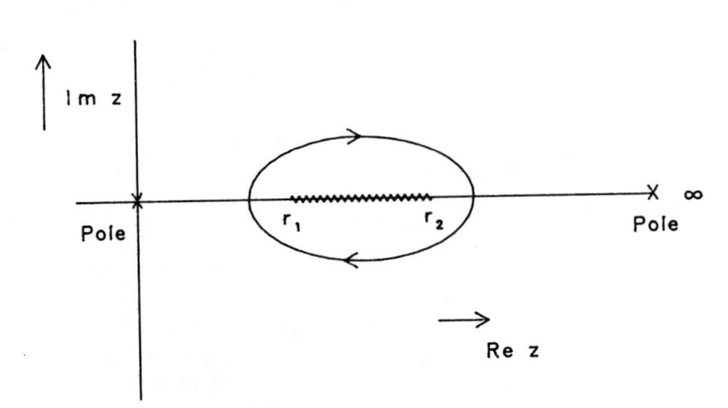

Fig. 2–2 Contour Used in Evaluating J_r

of r are then obtained from

$$-2m|E| + \frac{2me^2}{r} - \frac{\ell^2\hbar^2}{r^2} = 0. \qquad 2.97$$

This gives:

$$r_{1,2} = r_{\substack{max\\min}} = \frac{2me^2 \pm [4m^2e^4 - 8m|E|\ell^2\hbar^2]^{1/2}}{4m|E|} . \qquad 2.98$$

Considered as a function of z

$$\left[-2m|E| + \frac{2me^2}{z} - \frac{\ell^2\hbar^2}{z^2}\right]^{1/2}$$

has branch points at r_1 and r_2 and poles at 0 and ∞. (See figure 2.2).

The integral for J_r is given by the contour shown. However, by considering this contour as enclosing its external rather than its internal part we get the same result if we integrate in the reverse direction (counter clockwise). Then

$$J_r = 2\pi i \times (\text{residue at } 0 \text{ and } \infty)$$

$$= 2\pi(me^2/\sqrt{2m|E|} - \ell\hbar) \qquad 2.99$$

$$= 2\pi(n-\ell)\hbar \quad . \qquad 2.100$$

Thus,

$$|E| = -E = \frac{1}{2}\frac{me^4}{n^2\hbar^2} \qquad 2.101$$

or

$$E = -\frac{1}{2}mc^2\frac{e^4}{\hbar^2c^2}\frac{1}{n^2} = -\frac{1}{2}mc^2\alpha^2\frac{1}{n^2} \quad . \qquad 2.102$$

So

$$R_y = \frac{1}{2}mc^2\alpha^2/\hbar$$

just as before.

On the other hand, the procedure is very clumsy and applies only to periodic motions. We now look for some way to generalize the procedure. To do this we study the connection between geometrical and physical optics to obtain a generalization of classical mechanics.

2.6 The Schrödinger Equation

Consider a one-dimensional system with hamiltonian H which is

1) The total energy;
2) A constant of the motion.

As previously stated we only consider such hamiltonians.

Now the principal function S and the characteristic function W

are related by

$$S(q,P,t) = W(q,P) - Et \qquad 2.103$$

according to 2.61 and 2.62. Furthermore by 2.59

$$p = \frac{\partial S}{\partial q} \ .$$

Thus,

$$p = \frac{\partial S}{\partial q} = \frac{\partial W}{\partial q} \ . \qquad 2.104$$

Furthermore with

$$H = T + V$$

$$= \frac{1}{2m} p^2 + V(q) = E \ , \qquad 2.105$$

the Hamilton-Jacobi equation reads

$$\frac{1}{2m} \left(\frac{\partial W}{\partial q}\right)^2 + V(q) = E \ . \qquad 2.106$$

These equations tell us how to determine the actual motion of a particle if we are given W.

From 2.103 we see that

at $t = 0$ $\quad S \equiv W$

at $t = 1$ the surface $S = 0$ coincides with $W = E$

at $t = 2$ the surface $S = E$ coincides with $W = 2E$

and so on.

Thus as time progresses, a fixed surface S = constant moves over the surfaces W = constant. Therefore, we can say that S describes a wave motion in coordinate space.

What is the speed of propagation of these waves? In going from

(x,t) to $(x+\delta x, t+\delta t)$ the change in S is

$$\delta S = \frac{\partial S}{\partial x} \cdot \delta x + \frac{\partial S}{\partial t} \delta t. \qquad 2.107$$

If this is just due to propagation with velocity u during time t then we have

$$\delta S = 0$$

and

$$\delta x = u \delta t. \qquad 2.108$$

Thus,

$$\frac{\partial S}{\partial t} + u \frac{\partial S}{\partial x} = 0 \qquad 2.109$$

or

$$\frac{\partial S}{\partial t} + u . p = 0. \qquad 2.110$$

Hence,

$$-\frac{\partial S}{\partial t} = pu = mvu \qquad 2.111$$

where v is the particle velocity. Now,

$$\frac{\partial S}{\partial t} = -E$$

and

$$H = \frac{1}{2} mv^2 + V = E$$

so that

$$v = \left[\frac{2(E-V)}{m}\right]^{1/2} \qquad 2.112$$

and we get

$$u = \frac{E}{\sqrt{2m(E-V)}} . \qquad 2.113$$

Also the Hamilton-Jacobi equation can be rewritten in terms of u to read

$$\left(\frac{\partial S}{\partial x}\right)^2 = \frac{1}{u^2}\left(\frac{\partial S}{\partial t}\right)^2 . \qquad 2.114$$

This already has some of the appearance of a wave equation.

It is therefore very plausible for us to think that classical mechanics is some sort of approximation to a wave theory. Hence we write down the simplest possible wave equation

$$\frac{\partial^2 \Psi}{\partial x^2} = \frac{1}{u^2}\frac{\partial^2 \Psi}{\partial t^2} \qquad 2.115$$

where Ψ is to describe this quantum mechanical wave whose interpretation we must look for later. Now since the velocity u depends only on x we can separate out the time by writing

$$\Psi(x,t) = \psi(x)e^{-iEt/\hbar} \qquad 2.116$$

The minus sign in the exponential is conventional. The choice of $E/\hbar$ as the coefficient of t is motivated by Planck's law since $E/\hbar$ is an angular frequency. Substituting 2.116 into 2.115 and using 2.115 we find

$$\frac{\partial^2 \psi}{\partial x^2} = -\frac{2m(E-V)}{\hbar^2}\psi \qquad 2.117$$

or

$$-\frac{\hbar^2}{2m}\frac{\partial^2 \psi}{\partial x^2} + V\psi = E\psi . \qquad 2.118$$

This is the <u>time-independent Schrödinger equation</u>.

To recover the time dependence multiply $\psi(x)$ by $e^{-iEt/\hbar}$. Thus the time-dependent $\Psi(x,t)$ satisfies

$$-\frac{\hbar^2}{2m}\frac{\partial^2\Psi}{\partial x^2} + V\Psi = i\hbar\frac{\partial\Psi}{\partial t}. \qquad 2.119$$

This is the <u>time-dependent Schrödinger equation.</u>

The steps above do not in any way constitute a derivation of these two equations, they are simply plausibility arguments. These equations describe a new set of physical laws and their validity derives from experimental tests of the effects they predict. Our task for most of the rest of this book will be to study the physical interpretation and meaning of these equations.

Notes, References and Bibliography

2.1 W.R. Hamilton - <u>Collected Papers</u> Vol. II Cambridge University Press (1980), pp. 103-211.

Two of the standard reference on classical mechanics are:

H. Goldstein - <u>Classical Mechanics</u>, Addison-Wesley Publishing Co., Inc., Reading, Mass., U.S.A. (1959).

L.D. Landau and E.M. Lifschitz - <u>Mechanics</u>, Pergamon Press - U.S. distributors: Addison-Wesley Publishing Co., Inc., Reading, Mass. (1960).

Chapter 2 Problems

2.1 Find the Lagrangian for an harmonic oscillator. Use the definition of conjugate momentum to find p and H.

2.2 Repeat problem 1 for the simple pendulum. Interpret the momentum P conjugate to the angle variable θ.

2.3 Use Bohr-Sommerfeld quantization to calculate the energy levels of a one-dimensional simple harmonic oscillator.

2.4 Use Bohr-Sommerfeld quantization to calculate the energy levels of a particle confined to a box of length L. For simplicity assume this is a "one-dimensional box".

2.5 Suppose a gyroscope has a magnetic moment $\vec{\mu}$ proportional to its angular momentum $\vec{L}$ according to

$$\vec{\mu} = M\vec{L} \quad .$$

The potential energy due to placing the gyroscope in a magnetic field $\vec{B}$ is $V = -\vec{\mu}.\vec{B}$. Assume $\vec{B}$ is constant and derive the equation of motion for $\vec{L}$. Show that the gyroscope precesses with the angular Larmor frequency $\omega_L = MB$.

2.6 The system of quantization proposed by Bohr in 1913 is not applicable to all systems. To what general kinds of physical systems is Bohr's procedure applicable? For what kinds of systems is it not applicable.

2.7 Consider the Schrödinger equation 2.119 and put $\Psi = A\, e^{iS/\hbar}$, A = constant. Show that in the limit as $\hbar \to 0$ equation 2.119 reduces to the Hamilton-Jacobi equation 2.61.

2.8 In problem 2.7 set $S = W - Et$ and let $W = W_o + \hbar W_1 + \hbar^2 W_2 + \ldots$ for the case of a one-dimensional Schrödinger equation. Find the equations for W_o and W_1 and solve them. This is the so-called Wentzel-Kramers-Brillouin or WKB approximation.

2.9 Find the transformation equations 2.41 relating the variables P, Q to the variables p,q in the examples 2.34a and 2.34b.

Chapter 3

Elementary Systems

In the last chapter (section 2.6) we gave a heuristic derivation of the Schrödinger equation. It is important to note that no valid derivation of this equation from classical mechanics is possible since it represents a new law of physics extending beyond classical mechanics. From this point of view it is clear that section 2.6 was only in the nature of a plausibility argument and nothing was missed if you skipped it.

Just as in classical mechanics one does not derive Newton's laws but simply starts with them, so here we shall simply start with the Schrödinger equation. The purpose of this chapter is to start familiarizing us with this equation. We begin by considering the simplest case of a free particle and consider the solutions of the corresponding Schrödinger equation. We next obtain a conservation law on the basis of the Schrödinger equation. This law provides a guide for obtaining Born's probability interpretation for the solutions to the Schrödinger equation. This interpretation is then examined by considering Young's double slit experiment.

We then proceed to look at some properties of a wave packet solution to see how it describes a particle. There we also introduce the concept of group and phase velocity.

To see how to make the transition from classical to quantum mechanics we go on to consider some purely formal analysis as well as mathematical relations. These lead us to some formal rules for quantizing. Sometimes, however, some of these rules can lead to ambiguities and we discuss some of the most prominent ambiguities.

Finally we consider the hamiltonian function for the very important electromagnetic interaction. The results of this last section will not be used until chapter 15 and may thus be skipped on a first reading.

3.1 Plane Wave Solutions

As a first look at the Schrödinger equation consider the case for a particle in free space, where no forces are acting on it or the potential $V = 0$. Then

$$i\hbar \frac{\partial \Psi}{\partial t} = -\frac{\hbar^2}{2m} \nabla^2 \Psi \; . \tag{3.1}$$

A solution is

$$\Psi_{\vec{p}}(\vec{r},t) = A \exp \frac{i}{\hbar} (\vec{p}\cdot\vec{r} - Et) \tag{3.2}$$

where

$$E = \vec{p}^{\,2}/2m \; . \tag{3.3}$$

This solution describes a wave of frequency

$$\omega = E/\hbar \qquad \text{(Planck's relation)} \tag{3.4}$$

and a wavelength

$$\lambda = \frac{2\pi\hbar}{p} \qquad \text{(de Broglie's relation)} \; . \tag{3.5}$$

Thus, two of our previous quantum mechanical results are automatically given by the Schrödinger equation.

The solution 3.2 is however physically not very realistic since according to it a physical particle has associated with it a wave uniformly spread out through all space. Fortunately this is not a defect of the theory but rather of our treatment.

Since 3.1 is linear, the most general solution of it is an arbitrary linear superposition of solutions of the form 3.2. Thus, we can get solutions of the form

$$\Psi(\vec{r},t) = \int A(\vec{p}) \exp \frac{i}{\hbar} (\vec{p}.\vec{r} - \frac{\vec{p}^2}{2m} t) d^3p \ . \qquad 3.6$$

As we shall see this can describe a localized wave and is called a <u>wave packet</u>.

We shall now try to give a physical interpretation of the wave function Ψ.

3.2 Conservation Law for Particles

So far, we have stressed the similarity between light and particles. But even quantum mechanically there are some very important differences. Light may be absorbed or emitted. Thus, the number of photons changes with time and no conservation law for the number of photons holds. On the other hand, as long as we are in a non-relativistic regime, particles cannot be destroyed or created. Of course particles can form bound states as when an electron and a proton combine to form a hydrogen atom. However, we then still have two particles. Thus, in general, except for such relativistic effects as pair creation or annihilation, the number of particles is conserved. So in a non- relativistic theory of quantum mechanics we should have conservation of particles.

If we consider the Schrödinger equation, 2.97, it is easy to derive a conserved density and current. We have

$$i\hbar \frac{\partial \Psi}{\partial t} = -\frac{\hbar^2}{2m}\nabla^2\Psi + V\Psi \ . \tag{2.97a}$$

If we complex conjugate this expression we get

$$-i\hbar \frac{\partial \Psi^*}{\partial t} = -\frac{\hbar^2}{2m}\nabla^2\Psi^* + V\Psi^* \ . \tag{2.97b}$$

Multiplying 2.97a by $-\frac{i}{\hbar}\Psi^*$ and 2.97b by $\frac{i}{\hbar}\Psi$ and adding we get

$$\frac{\partial \Psi^*}{\partial t}\Psi + \Psi^*\frac{\partial \Psi}{\partial t} = \frac{i\hbar}{2m}(\Psi^*\nabla^2\Psi - \Psi\nabla^2\Psi^*) \ . \tag{3.7}$$

This is of the form

$$\frac{\partial \rho}{\partial t} + \vec{\nabla}\cdot\vec{j} = 0 \tag{3.8}$$

where

$$\rho = \Psi^*\Psi \tag{3.9}$$

and

$$\vec{j} = -\frac{i\hbar}{2m}(\Psi^*\vec{\nabla}\Psi - \Psi\vec{\nabla}\Psi^*) \ . \tag{3.10}$$

Thus, we have an equation of continuity. Calling

$$R = \int_V \rho \, d^3x \tag{3.11}$$

we can derive

$$\frac{dR}{dt} = -\frac{i\hbar}{2m}\int_A (\Psi^*\vec{\nabla}\Psi - \Psi\vec{\nabla}\Psi^*)d\vec{A} = \int_A \vec{j}\cdot d\vec{A} \tag{3.12}$$

This suggests that the rate of change of R in the volume V in 3.11 is given by the flux of the current $\vec{j}$ through the surface A surrounding V. It is therefore reasonable to associate the conserved quantity R with the number of particles. For only one particle $R = 1$ if V is all of space and hence ρ is a density for the particle. However, classically a particle is at some point or it is not; it is not all smeared out, so ρ cannot be a matter density. The most reasonable interpretation is that ρ is a probability density*. This gives us our interpretation of ρ and hence of Ψ.

Thus, Ψ is a probability amplitude. One further comment: In the derivation of 3.7 we assumed that the potential V is real. This is true of course for classical potentials. It is also crucial in our derivation and will be examined later (see problem 3.2).

To justify our interpretation further we consider some experimental results. Very early in the history of quantum mechanics experiments were performed to test experimentally the wave nature of matter. The classic experiments which we have already discussed, are those of Davisson and Germer. Thomson and Rupp ref 3.1, independently used powdered crystals and obtained the analogue of Debye- Scherrer patterns. For a modern version see ref 3.2. We now consider a slightly idealized experiment which was not performed until much later.

3.3 Young's Double Slit Experiment

The experimental set up is well known and is shown below (figure 3.1). The intensity distribution of a wave is given by $|\Psi|^2$, the square of its amplitude. This is also found to be the case for a high intensity beam of particles. On the other hand, when we decrease the beam intensity to the point where we can see individual particles arriving at

---

* Actually, for a system of n particles, ρ is a function in a space of 3n dimensions and can not be interpreted as a physical matter density.

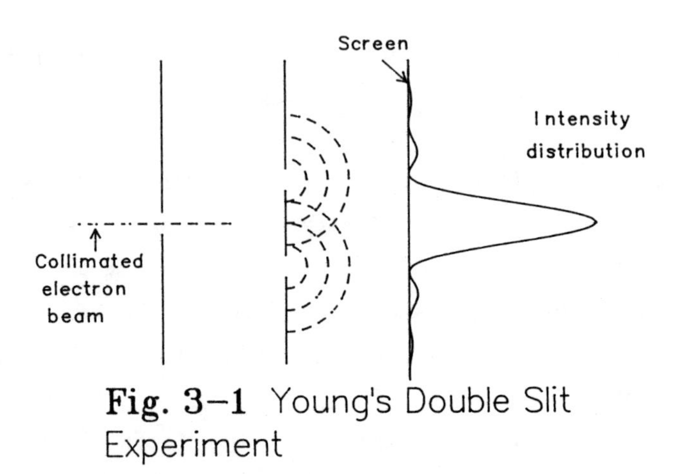

Fig. 3–1 Young's Double Slit Experiment

the screen, we find that they arrive at specific points and not spread out all over the screen. The point at which any given particle arrives is completely unpredictable. However, $|\Psi|^2$ still gives information, in fact, the probability density. Thus, we find experimentally that $|\Psi(x)|^2 dx$ is the probability that the particle arrives at a point lying between x and $x+dx$. Thus, our interpretation of Ψ as a probability amplitude is again justified.*

Incidentally it is also possible to do this experiment with light by decreasing the intensity to the point where individual photons are observed.

3.4 The Superposition Principle and Group Velocity

A fundamental property of the Schrödinger equation is that it is linear in Ψ. This has the following important consequence. If Ψ_1 and Ψ_2 are two different solutions, then $\lambda_1\Psi_1 + \lambda_2\Psi_2$ is also a solution if λ_1 and λ_2 are constants. This is known as the superposition principle.

As an example of the use of this principle we now construct and

---

* The interpretation of Ψ as a probability amplitude is due to Max Born (ref 3.3).

examine a wave packet. We already wrote down the equation for a wave packet (eqn. 3.6). For simplicity we work in one space dimension.

A plane monochromatic wave is given by

$$u_k(t,x) = e^{i(kx-\omega t)} \tag{3.13}$$

where

$$\hbar\omega = \frac{(\hbar k)^2}{2m} \quad \text{for a free particle.} \tag{3.14}$$

Also

$$\lambda = 2\pi/k. \tag{3.15}$$

In general $\omega = \omega(k)$. Forming a superposition of such plane waves we build up a wave packet

$$\Psi(x,t) = \int_{-\infty}^{\infty} f(k)e^{i[kx-\omega(k)t]}\, dk . \tag{3.16}$$

We want to find out where this wave is concentrated in x-space and with what velocity the peak of the wave travels. This is known as the group velocity.

To do this we assume $f(k)$ is a smooth well-behaved function concentrated in a region Δk about $k = k_o$. We further assume that in this region $\omega(k)$ may be expanded in a power series about k_o.

$$\omega(k) = \omega_o + (k-k_o)\frac{d\omega}{dk}\Big|_{k_o} + - - - - - - . \tag{3.17}$$

Substituting this expression in 3.16 we get

$$\Psi(x,t) \simeq e^{i[k_o x-\omega_o t]} \int_{k_o-\Delta k}^{k_o+\Delta k} dk\, f(k)\exp i\Big[(k-k_o)\Big(x- \frac{d\omega}{dk}\Big|_o t\Big)\Big] \tag{3.18}$$

This expression is of the form

$$\Psi(x,t) \simeq e^{i(k_o x-\omega_o t)} F\left(x- \frac{d\omega}{dk}\Big|_o t\right) \tag{3.19}$$

This is an envelope function F, the Fourier transform of f, multiplying a plane monochromatic wave. The phase velocity is given by the monochromatic wave and is

$$v_p = \frac{\omega_o}{k_o} \tag{3.20}$$

The envelope function is unchanged if we replace x and t by

$$x + \delta x \text{ and } t + \delta t \text{ such that}$$

$$\delta x = \frac{d\omega}{dk}\Big|_o \delta t .$$

Thus the group velocity is

$$v_g = \frac{d\omega}{dk}\Big|_{k=k_o} . \tag{3.21}$$

To make these considerations valid, the phase

$$\phi = kx-\omega(k)t$$

of the integrand in 3.16 must not vary too rapidly in the region about k_o, otherwise the positive and negative contributions cancel. In fact we require roughly one oscillation to occur in this region. This gives the condition

$$\Delta k\left| \frac{d\phi}{dk} \right|_o \simeq 1 .$$

But

$$\frac{d\phi}{dk} = x- \frac{d\omega}{dk} t \simeq x- \frac{d\omega}{dk}\Big|_o t \tag{3.22}$$

The center of the packet is given by $x_o = \frac{d\omega}{dk}\Big|_o t$.

$$\therefore \frac{d\phi}{dk} = x_o - \frac{d\omega}{dk}\Big|_o t + (x - x_o) = \Delta x$$

Thus the packet is concentrated in a region Δx such that

$$2\Delta k \Delta x \approx 1 . \qquad 3.23$$

Since $p = \hbar k$ this can be written $\Delta x \Delta p \approx \hbar$. This is an example of Heisenberg's uncertainty relation. We shall have more to say about this relation later on.

The group velocity v_g as previously calculated corresponds to the velocity of the classical particle. We can see this as follows. Classically $v = \frac{dE}{dp}$. Now we have $E = \hbar\omega$, $p = \hbar k$ and $\frac{dE}{dp} = \frac{d\omega}{dk} = v_g$ as stated.

There is one more property of a plane wave that is very enlightening and fortifies our faith in the interpretation of Ψ given so far. This is the value of the current or flux for such a wave. If we take a wave

$$\Psi = Ae^{i(kx-\omega t)} \qquad 3.24$$

and use 3.10 to compute the current we get that

$$j = -\frac{i\hbar}{2m}|A|^2 \left[e^{-i(kx-\omega t)} \frac{d}{dx} e^{i(kx-\omega t)} - e^{i(kx-\omega t)} \frac{d}{dx} e^{-i(kx-\omega t)}\right] = -\frac{\hbar k}{m}|A|^2$$

Since the momentum $p = \hbar k$ and $p/m = v$, the velocity, we have

$$j = v|A|^2 . \qquad 3.25$$

Thus the current consists of the velocity v times the intensity $|A|^2$ of the wave.

3.5 Formal Considerations

We next develop some more mathematics that will be useful later. We introduce the concept of operator. As motivation let us reexamine the Schrödinger equation

$$i\hbar \frac{\partial \Psi}{\partial t} = -\frac{\hbar^2}{2m} \nabla^2 \Psi + V\Psi .$$

This equation was derived for a system with a classical hamiltonian

$$H = \frac{p^2}{2m} + V. \tag{3.26}$$

Now if we replace $\vec{p}$ by $\frac{\hbar}{i}\nabla$, then we get an operator H_{op} such that

$$H_{op}\Psi = -\frac{\hbar^2}{2m}\nabla^2\Psi + V\Psi . \tag{3.27}$$

And the time independent Schrödinger equation reads

$$E\psi = H_{op}\psi . \tag{3.28}$$

So our equation looks much more suggestive this way. In fact this procedure of replacing classical variables by operators has more than formal significance.

Now suppose ψ is a function describing a given state, then $d\psi/dx$ is another function and is obtained by operating on ψ with d/dx. Clearly d/dx can operate only on functions where the first derivative exists. These functions lie in the <u>domain</u> of d/dx. More generally if one has a procedure that assigns to every function of a set $\{\psi_1, \ldots, \psi_n\}$ a unique function of some other set $\{\phi_1, \ldots, \phi_\ell\}$ one says that this mapping is given by an operator and we write

$$A\psi_i = \phi_i . \tag{3.29}$$

The operators of interest in quantum mechanics are in a sense the simplest; they are linear. An operator A is linear if

$$A(\lambda_1\psi_1+\lambda_2\psi_2) = \lambda_1 A\psi_1 + \lambda_2 A\psi_2 \qquad 3.30$$

where λ_1 and λ_2 are arbitrary constants and ψ_1 and ψ_2 are two arbitrary states.

Examples:

The operator d/dx is linear since

$$\frac{d}{dx}(\lambda_1\psi_1+\lambda_2\psi_2) = \lambda_1 \frac{d\psi_1}{dx} + \lambda_2 \frac{d\psi_2}{dx} .$$

On the other hand the operator $\sqrt{}$ is not linear since

$$\sqrt{\lambda_1\psi_1 + \lambda_2\psi_2} \neq \lambda_1\sqrt{\psi_1} + \lambda_2\sqrt{\psi_2} .$$

Another important example of a linear operator is multiplication by a function say $f(x)$ since

$$f(x)(\lambda_1\psi_1(x) + \lambda_2\psi_2(x)) = \lambda_1 f(x)\psi_1(x) + \lambda_2 f(x)\psi_2(x).$$

We shall only consider linear operators from now on unless we expressly state otherwise.

It is also possible to have an algebra of operators. The following operations are defined.

1) Scalar multiplication:

λA is defined by

$$(\lambda A)\psi = \lambda(A\psi) . \qquad 3.31$$

2) Addition:

$$(A+B)\psi = A\psi + B\psi . \qquad 3.32$$

3) Multiplication:

$$(AB)\psi = A(B\psi) \quad . \tag{3.33}$$

These last two satisfy distributive laws. On the other hand, the product is not commutative and in general $AB \neq BA$. For example, if $A = x, \quad B = d/dx$

$$(AB)\psi = x \frac{d\psi}{dx}$$

$$(BA)\psi = \frac{d}{dx}(x\psi) = \psi + x \frac{d\psi}{dx} \; .$$

To express the lack of commutativity we introduce the <u>commutator</u>

$$[A,B] = AB - BA. \tag{3.34}$$

Thus in the example above

$$[\frac{d}{dx}, x] = 1$$

since

$$\frac{d}{dx}(x\psi) - x \frac{d\psi}{dx} = 1.\psi \; .$$

The commutator plays an important role in quantum mechanics. Other familiar examples of operators are grad and curl.

3.6 Ambiguities

As we saw, the time independent Schrödinger equation could be written in the form

$$H_{op}\psi = E\psi \; .$$

The rule for forming H_{op} from the classical hamiltonian H was simply to replace $\vec{p}$ by $\frac{\hbar}{i}\nabla$. This rule, however, is not completely unambiguous. There are at least two sources of ambiguity which we now discuss. The first of these has to do with the use of different coordinate systems and does not present any serious problems in principle. The second is due to the fact that operators do not usually commute. Here the ambiguities must be resolved partly by mathematics and partly by physics.

3.6.1 Use of Different Coordinate Systems (ref. 3.4)

Consider the free particle hamiltonian

$$H = \frac{1}{2m}\vec{p}^{\,2} . \tag{3.35}$$

In Cartesian coordinates this reads

$$H = \frac{1}{2m}(p_x^2 + p_y^2 + p_z^2) \tag{3.36}$$

giving a quantum mechanical operator hamiltonian

$$H_{op} = -\frac{\hbar^2}{2m}\left(\frac{\partial^2}{\partial x^2} + \frac{\partial^2}{\partial y^2} + \frac{\partial^2}{\partial z^2}\right) . \tag{3.37}$$

In cylindrical coordinates 3.35 reads

$$H = \frac{1}{2m}\left(p_r^2 + \frac{1}{r^2}p_\phi^2 + p_z^2\right) . \tag{3.38}$$

If one naively replaces p_r by $\frac{\hbar}{i}\frac{\partial}{\partial r}$ and p_ϕ by $\frac{\hbar}{i}\frac{\partial}{\partial \phi}$ this gives

$$H'_{op} = -\frac{\hbar^2}{2m}\left(\frac{\partial^2}{\partial r^2} + \frac{1}{r^2}\frac{\partial^2}{\partial \phi^2} + \frac{\partial^2}{\partial z^2}\right) , \tag{3.39}$$

whereas if one transforms 3.37 one gets

$$H_{op} = -\frac{\hbar^2}{2m}\left(\frac{\partial^2}{\partial r^2} + \frac{1}{r}\frac{\partial}{\partial r} + \frac{1}{r^2}\frac{\partial^2}{\partial \phi^2} + \frac{\partial^2}{\partial z^2}\right) . \qquad 3.40$$

In fact the last one is correct. The difference is due to the fact that we must distinguish between covariant and contravariant vectors. However, it is possible to get away without knowing this distinction by adhering to the following simple rule. <u>Always write all momenta in Cartesian coordinates and then make the replacements</u>

$$p_x \to \frac{\hbar}{i}\frac{\partial}{\partial x}$$

$$p_y \to \frac{\hbar}{i}\frac{\partial}{\partial y} \qquad 3.41$$

$$p_z \to \frac{\hbar}{i}\frac{\partial}{\partial z} .$$

After that, transform to the desired coordinate system. The problem will now usually become the standard one of writing the Laplacian ∇^2 in the appropriate coordinate system, and this should not give any difficulties.

3.6.2 Non Commutativity (ref. 3.5)

In classical mechanics, since all dynamical variables commute, it does not matter in which order we write them as factors. This is not the case in quantum mechanics. For example,

$$\frac{1}{2m} p^2 = \frac{1}{2m} xp \frac{1}{x^2} px \qquad 3.42$$

classically. But this is definitely not the case in quantum mechanics. In fact, using the relation

$$[g(x),p] = i\hbar \frac{dg(x)}{dx} \qquad 3.43$$

(see problem 3.1) we obtain

$$xp \frac{1}{x^2} px = xp \frac{1}{x^2} xp + xp \frac{1}{x^2} [p,x] \tag{3.44}$$

$$= xp \frac{1}{x} p - i\hbar xp \frac{1}{x^2} \tag{3.45}$$

$$= x \frac{1}{x} p^2 + x[p, \frac{1}{x}]p - i\hbar x(- \frac{2i\hbar}{x^3}) \tag{3.46}$$

$$= p^2 + \frac{i\hbar}{x} p - \frac{i\hbar}{x} p - i\hbar x(- \frac{2i\hbar}{x^3})$$

or

$$xp \frac{1}{x^2} px = p^2 - \frac{2\hbar^2}{x^2} \tag{3.47}$$

$$\neq p^2 .$$

This example was, of course, artificially constructed, but unfortunately there is no rule for obtaining the correct order. This has to come from experiment, although in some cases mathematics can help because all physical observables must have corresponding hermitian operators. Sometimes this helps to eliminate some of the many possible orderings. Fortunately, in almost all cases of physical interest, the order is known. Furthermore, the classical hamiltonian usually has an especially simple form

$$H = T + V \tag{3.48}$$

where

$$T = \text{kinetic energy} = \sum_i \frac{p_i^2}{2m_i}$$

and

$$V = \text{potential energy} = \frac{1}{2} \sum_{i \neq j} V_{ij}(x_i - x_j) .$$

In this case the dynamical terms can be translated directly and no difficulty occurs. There are a few very important exceptions. One is the so-called spin-orbit interaction which we will encounter much later and another is the interaction of a charged particle with an electromagnetic field. In both these cases we get terms of the form $\vec{p}.\vec{A}(x)$. We discuss the electromagnetic interaction next. This discussion is classical but allows a translation to quantum mechanics.

3.7 Interaction with an Electromagnetic Field

Classically the force in this case is the Lorentz Force

$$\vec{F} = e(\vec{E}+ \frac{1}{c} \vec{v}\times\vec{B}) \ . \qquad 3.49$$

The corresponding Lagrangian is

$$L = T - e(\phi - \frac{1}{c} \vec{v}.\vec{A}) \qquad 3.50$$

where

$$\vec{E} = -\text{grad } \phi \qquad 3.51$$

$$\vec{B} = \text{curl } \vec{A} \qquad 3.52$$

and

$$T = \frac{1}{2} m\vec{v}^2 \ .$$

We are considering time-independent electromagnetic fields. If they are time dependent, equations 3.51 and 3.52 have to be suitably modified. The momentum $\vec{p}$ conjugate to $\vec{x}$ is given by

$$p_i = \frac{\partial L}{\partial \dot{x}_i} = \frac{\partial L}{\partial v_i} = mv_i + \frac{e}{c} A_i$$

or

$$\vec{p} = m\vec{v} + \frac{e}{c} \vec{A} \ . \qquad 3.53$$

Hence we get the hamiltonian

$$H = \sum p_i \dot{x}_i - L = \vec{p}.\vec{v} - L$$

$$= mv^2 + \frac{e}{c}\vec{v}.\vec{A} - \frac{1}{2}mv^2 + e\phi - \frac{e}{c}\vec{v}.\vec{A}$$

or

$$H = \frac{1}{2}mv^2 + e\phi .$$

We now have to use 3.53 to replace $\vec{v}$ by $\vec{p}$. Then

$$H = \frac{1}{2m}(\vec{p} - \frac{e}{c}\vec{A})^2 + e\phi \qquad 3.54$$

or expanding

$$H = \frac{p^2}{2m} - \frac{e}{2mc}(\vec{p}.\vec{A} + \vec{A}.\vec{p}) + \frac{e^2}{2mc^2}\vec{A}^2 + e\phi . \qquad 3.55$$

In this form the hamiltonian translates directly into H_{op} by replacing $\vec{p}$ by $(\hbar/i)\nabla$.

We are now ready to start looking at solutions of the Schrödinger equation.

Notes, References and Bibliography

3.1 G.P. Thomson - Proc. Roy. Soc. A117, 600 (1928).
E. Rupp. Ann. d. Physik 85, 981 (1928).

3.2 C. Jönssen, Z. Physik 161, 454 (1961). A translation of this paper is given by D. Brandt and S. Hirschi, Am. J. Phys. 42 5 (1974).

3.3 M. Born, Z. Physik 37, 863 (1926)
M. Born, Nature 119, 354 (1972)

3.4 G.R. Gruber - On Quantum Mechancial Operators in Generalized Coordinates - American Journal of Physics 40, (1972) 1537-1538.

Also

G.R. Gruber - Quantization in Generalized Coordinates - Foundations of Physics 1, 227-234 (1971).

- Quantization in Generalized Coordinates II - International Journal of Theoretical Physics 6, 31-35 (1972).

For a treatement of the construction of quantum mechanical operators and the possible ambiguities see:

Gary R. Gruber - Amer. J. Phys. 40, 1537 (1972).

also

J.R. Shewell - Amer. J. Phys. 27, 16 (1959).

Goldstein - Classical Mechanics - Addison Wesley Publishing Co., Reading, Mass., U.S.A. (1959) Sections 1-5.

Chapter 3 Problems

3.1 a) Verify the identity

$[AB,C] = A[B,C] + [A,C]B.$

b) Using the result above and $[x,p] = i\hbar$ prove that

$$[x^2,p] = 2i\hbar x$$

and

$$[x^n,p] = ni\hbar x^{n-1} .$$

Hence prove that for any function $g(x)$ analytic at the origin

$$[g(x),p] = i\hbar \frac{dg(x)}{dx} .$$

3.2 Assume that the potential V is complex of the form

$V = U + iW$

Show that W corresponds to a sink or source of probability.

Hint: Show that

$$\frac{\partial \rho}{\partial t} + \vec{\nabla}.\vec{j} = \frac{2}{\hbar} W\rho$$

This proves that unless $W = 0$ probability is not conserved

3.3 a) In deep water the phase velocity of water waves of wavelength

λ is

$$v = \sqrt{\frac{g\lambda}{2\pi}} .$$

What is the group velocity?

3.3 b) The phase velocity of a typical electromagnetic wave in a wave guide has the form

$$v = \frac{c}{\sqrt{1-\left(\frac{\omega_o}{\omega}\right)^2}}$$

where ω_o is a certain characteristic frequency. What is the group velocity of such waves?

3.4 Which of the following operators are linear?

a) $K\psi(x) = \int K(x,y)\psi(y),dy.$

b) K^3 where K is defined above.

c) AB if A and B are linear.

d) B^{-1} if B is linear and B^{-1} is defined by

$$B^{-1}B = BB^{-1} = 1.$$

e) $\exp A \equiv \sum_{n=0}^{\infty} \frac{A^n}{n!}$ if A is linear.

f) $A\psi = \exp(\lambda\psi).$

3.5 a) Compute the probability current density $j(t,x)$ as well as the probability density $\rho(t,x)$ for the wave function.

$$\Psi(t,x) = \int_{-\infty}^{\infty} dk\, A(k) e^{-i\left(\frac{\hbar k^2}{2m} t - kx\right)}$$

where

$$A(k) = e^{-L^2 k^2/2} .$$

b) Without using the explicit form of $A(k)$, verify the equation of conservation of probability, namely

$$\frac{\partial \rho}{\partial t} + \frac{\partial j}{\partial x} = 0.$$

Try to interpret the results you get.

3.6 Show that if we write

$$\vec{L} = \vec{r}\times\vec{p} = \frac{\hbar}{i}\,\vec{r}\times\nabla$$

then

$$L_x L_y - L_y L_x = i\hbar\, L_z$$

$$L_y L_z - L_z L_y = i\hbar\, L_x$$

$$L_z L_x - L_x L_z = i\hbar\, L_y \,.$$

These relations will be used later.

3.7 If a beam of free particles moving along the x-axis with velocity v is such that there is one particle in a volume V,

a) What is the corresponding normalized, time- dependent wave function for such a particle?

b) What is the number of particles crossing a unit area, normal to the x-axis, per unit time?

Chapter 4

One Dimensional Problems

The purpose of this chapter is to further build up our familiarity of the Schrödinger equation by solving several problems. We begin by classifying two types of problems. Then we consider a potential well and look for bound states. This leads us naturally to consider the effect of mirror symmetry or parity. The next problem we treat, in considerable detail, is scattering from a potential with a step. This allows us to finally deduce the boundary conditions for an impenetrable wall. The problem of a potential well corresponding to impenetrable walls, the so-called particle in a box, is treated next. Finally we briefly consider the concept of time reversal and how to include it in a quantum mechanical treatment.

4.1 Introduction

Starting with the Schrödinger equation

$$H\Psi = i\hbar \frac{\partial \Psi}{\partial t}$$

we now look for stationary state solutions. These are of the form

$$\Psi = \phi_E \, e^{-\frac{iE}{\hbar}t} . \tag{4.1}$$

Henceforth we suppress the E in ϕ_E , and simply write ϕ . Then ϕ satisfies the time independent Schrödinger equation

$$H\phi = E\phi . \tag{4.2}$$

For one dimensional systems we consider hamiltonians of the form

$$\begin{aligned} H &= T + V \\ &= p^2/2m + V(x) . \end{aligned} \tag{4.3}$$

Thus the Schrödinger equation 4.2 becomes

$$\left(-\frac{\hbar^2}{2m}\frac{d^2}{dx^2} + V(x)\right)\phi = E\phi . \tag{4.4}$$

In order to develop our intuition we study this equation for some simple solvable cases. The problems to be considered can be divided into two cases.

1) Scattering:
In this case $E \geq 0$. The potential may be attractive $V < 0$ or repulsive $V > 0$.

2) Bound States:
In this case $E < 0$. The potential must be attractive. We have used our freedom in defining energy to within an arbitrary constant to choose the zero of energy to correspond to the cases above.

Before proceeding, we rewrite 4.4 in dimensionless form. Call

$$U(x) = \frac{2m}{\hbar^2} V(x) \tag{4.5}$$

and

$$\varepsilon = \frac{2mE}{\hbar^2} . \qquad 4.6$$

Then we get

$$\psi'' + (\varepsilon - U)\psi = 0. \qquad 4.7$$

This is the equation we shall study for a while. To specify our solutions further we state that we are looking for solutions on $-\infty < x < \infty$ such that they are

1. everywhere finite;
2. continuous and differentiable.

Now as we saw before, Problem 3.2, V(x) has to be real in order to ensure conservation of probability. Thus if our boundary conditions are real, we are looking for real solutions of 4.7. This will always be the case for bound state problems. For scattering problems, the boundary conditions will be complex since they will specify some incoming wave of the form 3.16.

4.2 Finite Square Well - Bound States

From a comparison of geometrical and wave optics we expect wave phenomena to be most prominent when the index of refraction changes rapidly. Thus quantum effects will be most prominent if V(x) and hence U(x) varies rapidly in one wavelength. This will always be the case if U(x) has a step. Thus we consider the following simple potential

$$U(x) = \begin{matrix} 0 & |x| > a \\ -U & |x| < a \end{matrix}$$

where $U > 0$. For bound state problems we call

$$\varepsilon = -\kappa_o^2$$

$$\varepsilon + U = \kappa^2$$

since $\varepsilon < 0$. Thus the Schrödinger equation 4.7 becomes

$$\psi'' - \kappa_0^2\psi = 0 \qquad |x| > a$$

$$\psi'' + \kappa^2\psi = 0 \qquad |x| < a \;. \tag{4.8}$$

Therefore,

$$\psi = A_1 e^{\kappa_0} + A_2 e^{-\kappa_0 x} \qquad |x| > a$$

and

$$\psi = B_1 \sin \kappa x + B_2 \cos \kappa x \qquad |x| < a \;. \tag{4.9}$$

Furthermore we want $\int_{-\infty}^{\infty} |\psi(x)|^2 dx$ to represent the probability of finding the particle somewhere since $|\psi|^2$ is the probability density. This requires that the constants A_1 , A_2 and B_1 , B_2 be so chosen that

$$\int_{-\infty}^{\infty} |\psi(x)|^2 dx = 1. \tag{4.10}$$

In order that 4.10 be possible it imposes the restriction that $\psi(x) \underset{x\to\pm\infty}{\longrightarrow} 0$.
Thus, we get that

$$A_1 = 0 \quad \text{for} \quad x > a$$

$$A_2 = 0 \quad \text{for} \quad x < -a \;.$$

From the differential equation 4.8 it follows by integrating about the points $x = \pm a$ that both ψ and $d\psi/dx$ are continuous at these points. Imposing these conditions we obtain

$$A_2 e^{-\kappa_o a} = B_1 \sin \kappa a + B_2 \cos \kappa a$$

$$-\kappa_o A_2 e^{-\kappa_o a} = \kappa(B_1 \cos \kappa a - B_2 \sin \kappa a)$$

$$A_1 e^{-\kappa_o a} = -B_1 \sin \kappa a + B_2 \cos \kappa a$$

$$\kappa_o A_1 e^{-\kappa_o a} = \kappa(B_1 \cos \kappa a + B_2 \sin \kappa a).$$

These can be combined to give

$$2B_1 \sin \kappa a = (A_2 - A_1)e^{-\kappa_o a} \qquad \text{a)}$$

$$2B_2 \cos \kappa a = (A_1 + A_1)e^{-\kappa_o a} \qquad \text{b)}$$

$$2\kappa B_1 \cos \kappa a = \kappa_o(A_1 - A_2)e^{-\kappa_o a} \qquad \text{c)} \qquad 4.11$$

$$2\kappa B_2 \sin \kappa a = \kappa_o(A_1 + A_2)e^{-\kappa_o a} \qquad \text{d)}$$

If $A_1 - A_2 \neq 0$ and $B_1 \neq 0$ then combining a) and c) yields

$$\kappa \cot \kappa a = -\kappa_o \ . \qquad 4.12$$

On the other hand if $A_1 + A_2 \neq 0$ and $B_2 \neq 0$ then combining b) and d) yields

$$\kappa \tan \kappa a = \kappa_o \ . \qquad 4.13$$

The equations 4.12 and 4.13 are not simultaneously possible since they combine to give

$$\tan^2 \kappa a = -1$$

which requires κ to be imaginary and would therefore give a positive energy for a bound state, contrary to our original assumption. Thus there are two classes of solutions possible.

Class 1:

$$A_1 - A_2 = 0 \quad \text{and} \quad B_1 = 0 \quad \text{so that}$$

$$\kappa \tan \kappa a = \kappa_o .$$

Class 2:

$$A_1 + A_2 = 0 \text{ and } B_2 = 0 \text{ so that}$$

$$\kappa \cot \kappa a = -\kappa_o .$$

In either cases it requires numerical or graphical techniques to determine the energy eigenvalues. For example, setting

$$\kappa a = x \quad \text{and} \quad \kappa_o a = y$$

we get

$$\begin{aligned} x^2 + y^2 &= (\kappa^2 + \kappa_o^2)a^2 \\ &= (\varepsilon + U - \varepsilon)a^2 \end{aligned}$$

or

$$x^2 + y^2 = \frac{2ma^2}{\hbar^2} V^2 = R^2 . \tag{4.14}$$

In addition the two eigenvalue equations are

Case 1:

$$x \tan x = y$$

and

Case 2:

$$x \cot x = -y.$$

It is now an easy matter to solve these equations graphically.

If $|x| \ll 1$ we get the approximate solution for case 1.

$$x^2 \simeq y \qquad x^2 + y^2 = R^2$$

so

$$y^2 + y - R^2 = 0$$

and

$$y = -\frac{1}{2} \pm \frac{1}{2}\sqrt{1+4R^2} \quad .$$

Only the + sign applies. Furthermore $R^2 < 1$ so we get

$$y \simeq \frac{1}{2}\,[1 + 2R^2 - 1] \simeq R^2 \; .$$

But

$$y = \kappa_o a = a\sqrt{-\varepsilon}$$

$$\therefore \qquad -\varepsilon = \frac{R^4}{a^2}$$

$$\varepsilon = -\frac{R^4}{a^2}$$

and

$$\varepsilon = -\left(\frac{2maV}{\hbar^2}\right)^2$$

This is an approximation for the lowest eigenvalue. For values of the numerical solution see the book of L. Schiff (ref. 4.1).

4.3 Parity

In the previous example we found two classes of solutions corresponding to

Class 1. $\psi(x) = \psi(-x)$

and

Class 2. $\psi(x) = -\psi(-x)$.

This is a consequence of an interesting property of the hamiltonian.

Consider a general hamiltonian

$$H = \frac{p^2}{2m} + V(x) \tag{4.15}$$

where

$$V(x) = V(-x) \ . \tag{4.16}$$

The Schrödinger equation reads

$$H\psi = E\psi \tag{4.17}$$

and due to the symmetry of $V(x)$ we find that whenever $\psi(x)$ is a solution then so is

$$\psi_e(x) = \frac{1}{\sqrt{2}} [\psi(x) + \psi(-x)] \tag{4.18}$$

and

$$\psi_o(x) = \frac{1}{\sqrt{2}} [\psi(x) - \psi(-x)] \tag{4.19}$$

Of course if ψ is an even function then ψ_o vanishes and if ψ is an odd function then ψ_e vanishes.

These considerations lead us to introduce an operator P called the parity operator such that

$$(P\psi)(x) = \psi(-x). \tag{4.20}$$

That is, P operating on a function $\psi(x)$ gives the same function with a negative argument $\psi(-x)$. In effect if a state is decribed by $\psi(x)$ then $(P\psi)(x)$ describes the mirror image of this state. We shall henceforth simplify the notation and write $(P\psi)(x)$ simply as $P\psi(x)$.

Thus we have as an eigenvalue equation for P

$$P\psi(x) = \lambda\psi(x). \tag{4.21}$$

Then,

$$P^2\psi(x) = \lambda^2\psi(x) . \tag{4.22}$$

But,

$$P^2\psi(x) = P\psi(-x) = \psi(x) . \tag{4.23}$$

So $\lambda^2 = 1$ and the eigenvalues of P are ± 1. Clearly the functions ψ_e, ψ_o are eigenfunctions of P since

$$P\psi_e = 1 \cdot \psi_e \tag{4.24}$$

$$P\psi_o = -1 \cdot \psi_o . \tag{4.25}$$

If we now consider the Schrödinger equation and apply P to it we find since

$$\frac{p^2}{2m} = -\frac{\hbar^2}{2m}\frac{d^2}{dx^2}$$

is even under reflection that

$$PH\psi = HP\psi = EP\psi \tag{4.26}$$

where we have used the fact that

$$PV(x) = V(-x)P = V(x)P. \tag{4.27}$$

Thus we have that if ψ is an eigenfunction of H then so is $P\psi$. Thus it is possible for ψ to be an eigenfunction of H and P simultaneously. In fact, quite generally, as we shall see later, if two observables commute (in this case HP = PH) one can solve the eigenvalue problem for them simultaneously.

Returning to our example of the previous section with the potential

$$U(x) = \begin{cases} 0 & |x| > a \\ -U & |x| < a \end{cases}$$

we see that we could have immediately looked for either purely even or odd solutions. Thus instead of 4.9 we would have had the following two classes of solutions:

Class 1.

$$\begin{aligned} \psi_e &= Ae^{\kappa_o x} & x < -a \\ \psi_e &= Ae^{-\kappa_o x} & x > a \\ \psi_e &= B\cos\kappa x & |x| < a\,. \end{aligned} \tag{4.28}$$

Class 2.

$$\begin{aligned} \psi_o &= Ae^{\kappa_o x} & x < -a \\ \psi_o &= -Ae^{-\kappa_o x} & x > a \\ \psi_o &= B\sin\kappa x & |x| < a\,. \end{aligned} \tag{4.29}$$

These are of course the two classes of solutions we found previously using much more laborious techniques.

In general, whenever the hamiltonian has some special symmetry,

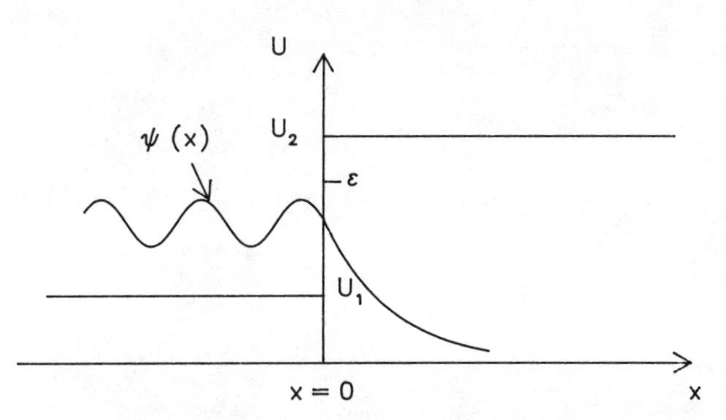

Fig. 4–1 Scattering from a Potential Step $V_1 < E < V_2$

this will reflect itself in the wave-functions. It is of great assistance to recognize and utilize this symmetry from the start as in the case of parity above.

4.4 Scattering from a Step-Function Potential

We still keep the step function. This time, however, we consider a scattering problem. Also we choose

$$U(x) = U_1 \qquad x < 0$$

$$U(x) = U_2 \qquad x > 0 \,. \tag{4.30}$$

Case 1)

$$U_1 < \varepsilon < U_2$$

The physical problem is as follows. We have a particle corresponding to energy ε come from $-\infty$, hit the potential U_2 and reflect back. For $x > 0$ we expect damped exponential behaviour (fig. 4.1). The Schrödinger equation is

$$\frac{d^2\psi}{dx^2} + k_1^2\psi = 0 \qquad x < 0$$

$$\frac{d^2\psi}{dx^2} - k_2^2\psi = 0 \qquad x > 0 \tag{4.31}$$

Here,

$$k_1^2 = \varepsilon - U_1 > 0$$

$$k_2^2 = U_2 - \varepsilon > 0 \ .$$

We can therefore write

$$\psi = A_1 \sin(k_1 x + \phi) \qquad x < 0$$

$$= A_2 e^{-k_2 x} \qquad x > 0 \tag{4.32}$$

We have already imposed the condition that $\psi(x)$ must vanish for $x \to \infty$. Furthermore we have chosen a particular form for the solution for $x < 0$; we could have written

$$\psi = B_1 e^{ikx} + B_2 e^{-ikx}$$

with arbitrary constants B_1 and B_2. Instead, we have the arbitrary constants A_1 and ϕ.

Boundary Conditions

In most scattering problems in one dimension it is more convenient to match logarithmic derivatives $\frac{1}{\psi}\frac{d\psi}{dx}$ rather than $\frac{d\psi}{dx}$ and ψ separately. This is because the normalization constants then cancel. In our case matching $\frac{1}{\psi}\frac{d\psi}{dx}$ at $x = 0$ yields

$$k_1 \cot \phi = k_2 \ . \tag{4.33}$$

This is an equation for ϕ. Furthermore, ϕ is determined up to π since changing ϕ to $\phi + \pi$ is the same as changing A_1 to $-A_1$. So

$$\phi = \arctan k_1/k_2 \qquad (-\tfrac{\pi}{2} \leq \phi \leq \tfrac{\pi}{2}). \tag{4.34}$$

Now matching ψ at $x = 0$ gives:

$$\frac{A_2}{A_1} = \sin\phi = k_1(k_1^2 + k_2^2)^{-1/2} . \tag{4.35}$$

Notice that we still have one free parameter say A_1. This corresponds to the fact that we have to specify the flux of incoming particles, i.e. we have to normalize ψ to yield the correct incoming flux. Before discussing the physics of this case we solve another case. We shall then discuss the physics for both these cases.

Case 2)

$$U_1 < U_2 < \varepsilon \tag{4.36}$$

In this case we can have particles incident both from the left and right unlike case 1) where particles were incident only from the left. Our theory, however, is linear and therefore we can consider the two cases (particles from the left and particles from the right) separately and superimpose the results. Physically this corresponds to the fact that particles coming from the left and from the right scatter independently from the potential. Thus we are neglecting any possible interactions between the particles.

Particles from Left: Suppose we normalize the flux coming from the left so that the incident wave is e^{+ikx}. Then the solution of the Schrödinger equation under condition 4.36 is

$$\begin{aligned} \psi &= e^{ik_1x} + Re^{-ik_1x} \qquad & x < 0 \\ &= Se^{ik_2x} . & x > 0 \end{aligned} \tag{4.37}$$

In this case

$$k_1^2 = \varepsilon - U_1$$

$$k_2^2 = \varepsilon - U_2 \ .$$

Equating logarithmic derivatives at $x = 0$ gives:

$$\frac{ik_1 - ik_1R}{1 + R} = ik_2 \ .$$

Hence

$$R = \frac{k_1 - k_2}{k_1 + k_2} \tag{4.38}$$

or

$$R = \frac{\sqrt{\varepsilon - U_1} - \sqrt{\varepsilon - U_2}}{\sqrt{\varepsilon - U_1} + \sqrt{\varepsilon + U_2}} \ . \tag{4.39}$$

Furthermore, equating the wave functions at $x = 0$ yields

$$S = 1 + R = \frac{2k_1}{k_1 + k_2} \ . \tag{4.40}$$

R gives the amplitude of the reflected wave, S gives the amplitude of the transmitted wave. We have to multiply $|S|^2$ by appropriate velocity factors to get the actual transmitted flux. This we shall do later. Note that a ψ of the form ψ' where

$$\psi' = e^{-ik_1x} + Re^{ik_1x} \qquad x < 0$$

$$= Se^{-ik_2x} \qquad x > 0 \tag{4.41}$$

is also a solution. However, 4.41 does not satisfy our physical requirements.

Particles from Right: In this case we look for a solution of the form

$$\begin{aligned} \psi &= e^{-ik_2x} + Re^{ik_2x} & x > 0 \\ &= Se^{-ik_1x} \, . & x < 0 \end{aligned} \qquad 4.42$$

This solution can of course also be written as a superposition of 4.37 and 4.41. The interpretation of 4.42 is the same as that of 4.37.

We now turn to an examination of the physics involved in these solutions. To do this we rewrite the solutions slightly and compare them with the classical problem.

Case 1.

$$U_1 < \varepsilon < U_2$$

Classically this corresponds to a particle of energy

$$E = \frac{\hbar^2}{2m} \varepsilon$$

incident from the left with velocity $v_1 = \hbar k_1/m$. At $x = 0$ it bounces elastically off the potential U_2 (i.e. it receives an impulse) and starts to travel back towards $x = -\infty$ with the original velocity v_1 .

We now examine the quantum mechanical case. The solution 4.32 can also be written

$$\begin{aligned} \psi_\varepsilon(x) &= e^{ik_1x} - e^{-i(k_1x+2\phi)} & x < 0 \\ &= \frac{2A_2 i}{A_1} e^{-i\phi} e^{-k_2x} \, . & x > 0 \end{aligned} \qquad 4.43$$

We get this from 4.32 by multiplying by $\frac{2i}{A_1} e^{-i\phi}$. This merely corresponds to normalizing the incoming flux to a certain value, i.e. the incoming wave is $e^{ik_1 x}$.

Now as we know, a particle is somehow localized and therefore corresponds to a wave packet and not a plane wave. Thus to get a better comparison with the classical case we form the wave packet

$$\Psi(x,t) = \int_{-\infty}^{\infty} f(k'-k_1)\phi_{\varepsilon'}(x)e^{-iE't/\hbar}\, dk' \ . \qquad 4.44$$

Here

$$\varepsilon' = 2mE'/\hbar^2 = k'^2 + U_1 \quad \text{for} \quad x < 0 \ .$$

Also we assume that $f(k'-k_1)$ is peaked around $k' = k_1$ (the classical momentum was $\hbar k_1$). Now for $x < 0$, $\Psi(x,t)$ is a superposition of <u>two</u> wave packets, an incident packet

$$\Psi_I(x,t) = \int_{-\infty}^{\infty} f(k'-k_1)\ e^{ik'x-iE't/\hbar}\, dk' \qquad 4.45$$

and a reflected packet

$$\Psi_R(x,t) = -\int_{-\infty}^{\infty} f(k'-k_1)\ e^{-i(k'x+2\phi+E't/\hbar)}dk'. \qquad 4.46$$

For the incident packet Ψ_I , the "center of mass" moves with the velocity

$$v_1 = \frac{\hbar k_1}{m}$$

to the right, reaching the point $x = 0$ at $t = 0$. The reflected packet Ψ_R , is centered at

$$x = -v_1 t - 2\,\frac{d\phi}{dk_1} \qquad 4.47$$

and travels to the left with velocity v_1. It arrives at $x = 0$ at the time

$$\tau = -\frac{2}{v_1}\frac{d\phi}{dk_1} = -2\hbar\frac{d\phi}{dE_1} . \qquad 4.48$$

Now $\dfrac{d\phi}{dk_1} = \dfrac{k_2}{k_1^2 + k_2^2}$ since $\phi = \tan^{-1}\dfrac{k_1}{k_2}$. So

$$|\tau| = \frac{2m}{\hbar k_1}\frac{k_2}{k_1^2 + k_2^2} = \frac{2m}{\hbar\sqrt{U_1-\varepsilon}}\frac{\sqrt{U_2-\varepsilon}}{U_1+U_2-2\varepsilon} .$$

For an electron with $V_1 = 10$ eV, $V_2 = 20$ eV, $E = 5$ eV, $|\tau| \simeq 10^{-16}$sec. Thus the motion of the "center" of the packet is almost the same as the motion of the corresponding classical particle except that the reflected wave packet suffers a delay τ when it is reflected.

The considerations just described are sensible as long as the shape of the packet does not change too much during the motion so that the "center" of the packet obtained as above makes sense. We want the center to be very close to the center of mass of the packet and we denote its location by $\bar{x}$. Then $\bar{x}$ will be approximately the center of mass as long as the center of mass is far from the origin $x = 0$ compared to the width Δx of the packet. Thus we need

$$\bar{x} > \Delta x .$$

We want the same condition to hold for the reflected wave. Thus, we require that the width Δk of $f(k)$ be small compared to the region over which ϕ varies as a function of k.

Thus, we require

$$\Delta k \frac{d\phi}{dk_1} < 1. \qquad 4.49$$

But we have

$$\Delta x \sim \frac{1}{\Delta k} \tag{4.50}$$

so we get

$$\Delta x > \frac{d\phi}{dk_1} = \frac{v_1 \tau}{2} \ . \tag{4.51}$$

Thus

$$\Delta x / v_1 > \tau/2 \ . \tag{4.52}$$

In other words, the width Δx of the packet is so large that the time $\Delta x/v_1$ spent passing a giving point is large compared to the delay time τ.

This gives us two limits

$$\bar{x} > \Delta x > v_1 \tau/2 \ .$$

There is another even bigger difference between the classical and quantum mechanical decription. Classically the particle can never enter the region $x > 0$ where $U_2 > \varepsilon$. In this case we would have

$$E = T + V_2 < V_2$$

where T is the kinetic energy. Thus,

$$T = \frac{1}{2} mv^2 < 0$$

and this cannot occur. On the other hand, quantum mechanically we see that $\Psi \neq 0$ for $x > 0$. In fact Ψ is only damped exponentially as follows from 4.28 and 4.29.

Case 2.

Both classically and quantum mechanically there are two possible motions in this case. The particles can be incident from the left or the right. We shall only discuss the former case.

Classically the particle is incident from the left with a velocity

$$v_1 = \frac{\hbar k_1}{m} , \tag{4.53}$$

experiences a sudden change in velocity at $x = 0$ (impulse) and continues to the right at a velocity

$$v_2 = \frac{\hbar k_2}{m} . \tag{4.54}$$

Quantum mechanically this motion is described as follows. We have

$$\begin{aligned} \phi_\varepsilon(x) &= e^{ik_1x} + Re^{-ik_1x} \qquad & x < 0 \\ &= Se^{+ik_2x} & x > 0 \end{aligned}$$

where

$$R = \frac{k_1 - k_2}{k_1 + k_2} .$$

We form a wave packet corresponding to the classical situation.

$$\Psi(x,t) = \int_{-\infty}^{\infty} f(k'-k_1)\phi_{\varepsilon'}(x)e^{-iE't/\hbar}\, dk' . \tag{4.55}$$

This is very similar to the case 1) that we have already considered.

Again for $t < 0$ $\Psi(x,t) \approx 0$ for $x > 0$ and the packet moves to the right with speed v_1 according to $\bar{x} = v_1 t$, reaching the origin at $t = 0$. This is the term corresponding to e^{ik_1x}. The term with e^{-ik_1x} does not contribute appreciably to the integral 4.55 for $t < 0$.

For $t > 0$ this term starts to contribute, so the packet splits into two packets:

a transmitted packet

$$\Psi_T(x,t) = \int_{-\infty}^{\infty} f(k'-k_1)S(k')e^{ik'x}e^{-iE't/\hbar}dk' \qquad 4.56$$

moving according to

$$\bar{x} = v_2 t$$

(the same as the classical particle) and a reflected packet

$$\Psi_R(x,t) = \int_{-\infty}^{\infty} f(k'-k_1)R(k')e^{-ik'x}e^{-iEt/\hbar}dk' \qquad 4.57$$

moving according to

$$\bar{x} = -v_1 t \ .$$

Thus, unlike the classical case, there is now a non-zero probability that the particle is reflected at the discontinuity in the potential. This is similar to light hitting a window pane, some is transmitted and some is reflected.

In order to get a better understanding of the two coefficients R and S we return to our concept of probability current. Recall that the probability density is given by

$$\rho = \Psi^*\Psi = |\Psi|^2 \qquad 4.58$$

and the probability density current is given by

$$\vec{j} = -\frac{i\hbar}{2m}(\Psi^*\nabla\Psi - \Psi\nabla\Psi^*) \ . \qquad 4.59$$

Thus, the probability of finding the particle in a volume V is given by

$$P = \int_V \rho \, dv \quad . \tag{4.60}$$

We also have the Schrödinger equation

$$i\hbar \frac{\partial \Psi}{\partial t} = - \frac{\hbar^2 \nabla^2}{2m} \Psi + V\Psi$$

and its complex conjugate

$$-i\hbar \frac{\partial \Psi^*}{\partial t} = - \frac{\hbar^2 \nabla^2}{2m} \Psi^* + V\Psi^* \quad .$$

From these we get

$$\Psi \frac{\partial \Psi^*}{\partial t} + \Psi^* \frac{\partial \Psi}{\partial t} = \frac{i\hbar}{2m} (\Psi^* \nabla^2 \Psi - \Psi \nabla^2 \Psi^*) \quad . \tag{4.61}$$

But

$$\frac{dP}{dt} = \int_V \frac{\partial \rho}{\partial t} \, dv$$

if we keep the volume fixed. Using 4.58 this becomes

$$\frac{dP}{dt} = \int_V \left(\frac{\partial \Psi^*}{\partial t} \Psi + \Psi^* \frac{\partial \Psi}{\partial t}\right) dv \quad . \tag{4.62}$$

Using 4.61 we get

$$\frac{dP}{dt} = \frac{i\hbar}{2m} \int_A (\Psi^* \nabla^2 \Psi - \Psi \nabla^2 \Psi^*) . d\vec{A}$$

and using Green's theorem this becomes

$$\frac{dP}{dt} = \frac{i\hbar}{2m} \int_A (\Psi^* \vec{\nabla} \Psi - \Psi \vec{\nabla} \Psi^*) . d\vec{A} \tag{4.63}$$

or

$$\frac{dP}{dt} = - \int_A \vec{j} \cdot d\vec{A} \tag{4.64}$$

where we have used 4.59. This is the integral form of the conservation law,

$$\frac{\partial \rho}{\partial t} + \vec{\nabla} \cdot \vec{j} = 0 \ .$$

The meaning of 4.64 is that the rate of change of probability in the volume V is given by the net flux of probability current through the surface A surrounding V.

We are finally in a position to interpret the coefficients R and S. We have

$$\Psi = \begin{cases} \phi_< = e^{ik_1x} + Re^{-ik_1x} & x < 0 \\ \phi_> = Se^{ik_2x} & x > 0 \end{cases}$$

where

$$R = \frac{k_1 - k_2}{k_1 + k_2}$$

$$S = \frac{2k_1}{k_1 + k_2} \ .$$

Throughout this problem we have velocities v_i given by

$$v_i = \frac{\hbar k_i}{m} \ .$$

If we compute the probability density flux j for $x < 0$ we get

$$j(x) = -\frac{i\hbar}{2m}\left(\psi_<^* \frac{\partial\psi_<}{\partial x} - \psi_< \frac{\partial\psi_<^*}{\partial x}\right)$$

$$= -\frac{i\hbar}{2m}\Big[\left(e^{-ik_1x}+R^*e^{ik_1x}\right)ik_1x\left(e^{ik_1x}-Re^{-ik_1x}\right)$$

$$-\left(e^{ik_1x}+Re^{-ik_1x}\right)(-ik_1)\left(e^{-ik_1x}-R^*e^{-ik_1x}\right)\Big]$$

$$= \frac{\hbar k_1}{2m}\Big[1 - Re^{-i2k_1x} + R^*e^{i2k_1x} - |R|^2$$

$$+ 1 + Re^{-i2k_1x} - R^*e^{i2k_1x} - |R|^2\Big]$$

or

$$j(x) = \frac{\hbar k_1}{m}\left[1 - |R|^2\right] = v_1 - v_1|R|^2 \ . \qquad 4.65$$

Thus, for $x < 0$ we have a flux v_1 to the right and a flux $v_1|R|^2$ to the left. Thus $|R|^2$ is the fraction of the flux reflected and hence $|R|^2$ is the reflection probability $\mathcal{R}$.

For $x > 0$ we get

$$j(x) = -\frac{i\hbar}{2m}\left[S^*e^{-ik_2x}(ik_2)Se^{ik_2x} - Se^{ik_2x}(-ik_2)S^*e^{-ik_2x}\right]$$

$$j(x) = \frac{\hbar k_2}{m}|S|^2 = v_2|S|^2 \qquad 4.66$$

or

$$j(x) = \frac{\hbar k_1}{m}\cdot\frac{k_2}{k_1}|S|^2 = v_1\frac{k_2}{k_1}|S|^2 \ . \qquad 4.67$$

∴ The fraction of the incident flux transmitted is

$$\frac{v_1\frac{k_2}{k_1}|S|^2}{v_1} = \frac{k_2}{k_1}|S|^2 \ .$$

Thus the transmission probability T is

$$T = \frac{k_2}{k_1} |S|^2 \ . \tag{4.68}$$

As a check we have that a particle is either transmitted past the potential step or reflected. Thus

$$T + R = 1 \ . \tag{4.69}$$

But this gives

$$\frac{k_2}{k_1} \cdot \left(\frac{2k_1}{k_1+k_2}\right)^2 + \left(\frac{k_1-k_2}{k_1+k_2}\right)^2 = \frac{4k_1k_2 + (k_1-k_2)^2}{(k_1+k_2)^2} = 1$$

as required.

We now consider case 1) of example 2. In this case we choose $U_1 = 0$ and we shall finally let $U_2 \to \infty$. What we are interested in is to derive the appropriate boundary condition at the edge of an infinite potential. The Schrödinger equation is

$$\frac{d^2\psi}{dx^2} + k_1^2\psi = 0 \qquad x < 0$$

$$\frac{d^2\psi}{dx^2} - k_2^2\psi = 0 \qquad x > 0$$

with

$$k_1^2 = \varepsilon \qquad k_2^2 = U_2 - \varepsilon \ .$$

The solutions are as before

$$\psi = A_1 \sin (k_1 x + \phi) \qquad x < 0$$

$$= A_2 e^{-\kappa_2 x} \qquad x > 0$$

where the boundary condition gave

$$\phi = \arctan k_1/k_2$$

and

$$A_2 = A_1 \sin \phi .$$

If we now let $U_2 \rightarrow \infty$ we get

$$\phi = 0$$

and

$$A_2 = 0 .$$

Thus for $U_2 = \infty$ we have

$$\psi(x) = A_1 \sin k_1 x \qquad x < 0$$

$$= 0 \qquad x > 0 . \qquad 4.70$$

Notice that this says that at an infinite potential barrier the wave function vanishes. Although we only derived this for this special one-dimensional problem, it is a result that is generally valid. With this result we are now in a position to solve another extremely simple problem.

4.5 Infinite Square Well - Particle in a Box

We consider a particle in an infinitely deep square well

$$U(x) = \begin{cases} 0 & |x| < a \\ \infty & |x| > a \,. \end{cases} \tag{4.71}$$

Thus we have to solve

$$\frac{d^2\psi}{dx^2} + \varepsilon\psi = 0 \tag{4.72}$$

subject to

$$\psi = 0 \quad \text{at} \quad x = \pm a \,. \tag{4.73}$$

Since $U(-x) = U(x)$, the hamiltonian corresponding to this potential commutes with the parity operator and we can look for purely even or purely odd solutions.

The even solutions are

$$\psi_{e,k}(x) = A_k \cos kx \qquad -a \leq x \leq a \tag{4.74}$$

where

$$k^2 = \varepsilon = \frac{2mE}{\hbar^2} \,. \tag{4.75}$$

Applying the boundary conditions

$$\psi_{e,k}(\pm a) = 0$$

we obtain the eigenvalues

$$k_n a = \frac{2n+1}{2}\pi \qquad n = 0,1,2,\ldots \tag{4.76}$$

Thus the energy eigenvalues are

$$E_n = \frac{\hbar^2 k_n^2}{2m} = \frac{\hbar^2 \pi^2}{8ma^2}(2n+1)^2 \qquad n = 0,1,2,\ldots \tag{4.77}$$

The odd solutions are

$$\psi_{o,k}(x) = B_k \sin kx \tag{4.78}$$

with eigenvalues

$$k_n a = n\pi \qquad n = 1,2,3,\ldots \tag{4.79}$$

In this case the energy eigenvalues are

$$E_n = \frac{\hbar^2 \pi^2}{8ma^2}(2n)^2 \qquad n = 1,2,\ldots \quad . \tag{4.80}$$

The infinite square well is not as unphysical or artificial as it might appear at first. In fact, it is a good mathematical representation of a particle confined to a (one-dimensional) box. In such a box, a measurement of the energy of the particle would yield one of the energies we computed. However, for macroscopic boxes, the energy levels are so closely spaced that they are experimentally indistinguishable from a continuum of energy levels.

4.6 Time Reversal

In Section 3, we found that the use of mirror symmetry, or parity, greatly facilitated the computation. There is another discrete symmetry which is frequently useful, it is called time reversal. Imagine a system such as perfectly elastic billiard balls on a frictionless surface. In viewing a movie of the collisions of these billiard balls we are unable to tell whether the movie is running forwards or backwards; the motions obtained by reversing the direction of the film or time, are equally physically possible. We state this by saying that the system is invariant under time reversal.

To examine this a little more closely we first consider the following classical transformation T which maps t into -t. So

$$Tt = -t \; . \qquad 4.81$$

From this it then follows that

$$T\vec{x}(t) = \vec{x}(-t) \qquad 4.82$$

$$T\vec{p}(t) = -\vec{p}(-t) \qquad 4.83$$

and

$$TV(\vec{x}) = V(\vec{x}) \; . \qquad 4.84$$

Thus if we have a system with a Hamiltonian

$$H = \frac{p^2}{2m} + V(x) \; , \qquad 4.85$$

then changing t to -t according to 4.81 yields, as equations 4.82 to 4.84 show, the same Hamiltonian. This means that if a possible motion of the system (solution of the equations of motion) is given by $(x(t),p(t))$ then another possible motion is given by $(x(-t),-p(-t))$. This is one way to sometimes obtain new solutions from old ones.

We now look at this from the point of view of quantum mechanics. The Hamiltonian 4.85 which is invariant under the transformations 4.81, is our starting point. The time-dependent Schrödinger equation reads:

$$i\hbar \frac{\partial \Psi(t,x)}{\partial t} = H\Psi(x,t) \; . \qquad 4.86$$

If we complex conjugate this equation and then do the transformation

$$t \to t' = -t \tag{4.87}$$

we find

$$i\hbar \frac{\partial \Psi^*(-t',\phi)}{\partial t'} = H\Psi^*(-t',x) \tag{4.88}$$

where we have used the fact that H is invariant under $t \to -t$. Dropping the primes we see that $\Psi^*(-t,x)$ satisfies the same Schrödinger equation as $\Psi(t,x)$. We therefore define the time reversal operator T by

$$(T\Psi)(t,x) = \Psi^*(-t,x), \tag{4.89}$$

and $T\Psi$ is a solution of 4.86 if Ψ is a solution. This operator is <u>anti-linear</u> due to the complex conjugation. Thus if λ is a complex number we find

$$T(\lambda\Psi) = \lambda^* T\Psi \ . \tag{4.90}$$

The operator T acting on other operators such as x and p has the following property

$$Tx = xT \tag{4.91}$$

$$Tp = -pT \tag{4.92}$$

and for the Hamiltonian 4.84, as we saw

$$TH = HT \ . \tag{4.93}$$

The particular way of implementing time-reversal discussed above is called Wigner (4.2) time-reversal.

Notes, References and Bibliography

4.1 L.I. Schiff - Quantum Mechanics - McGraw-Hill Book Co., 2nd edition (1955) pages 36-40.

R.H. Dicke and J.P. Wittke - Introduction to Quantum Mechanics - Addison Wesley Publishing Co., Inc., Reading, Mass., U.S.A. (1960).

Many "modern" problems are formulated as one- dimensional problems and solved in:

H.J. Lipkin - Quantum Mechanics - North Holland Publishing Co., Amsterdam and London (1973).

4.2 E.P Wigner - Group Theory - Academic Press (1959) Chapter 26.

Chapter 4 Problems

4.1 An electron has an energy of 10 GeV = 10^{10} eV. For which of the following potentials will a classical approximation be valid?

a) A step function of height 10 GeV, .1 GeV, 10^{-3} GeV.

b) A potential $V = -V_o \exp - \frac{x^2}{a^2}$

$$V_o = 100 \text{ GeV}$$

$$a = 10^{-20} \text{ cm}, 10^{-15} \text{ cm}, 10^{-10} \text{ cm}, 10^{-5} \text{ cm}.$$

4.2 A beam of particles of mass m moving from left to right encounters a sharp drop in potential of amount V_o. Let E be the kinetic energy of the incoming particles and show that the fraction of particles reflected at the edge of the potential (located at x = 0) is given by

$$\left(\frac{\sqrt{E+V_o} - \sqrt{E}}{\sqrt{E+V_o} + \sqrt{E}}\right)^2 .$$

In view of this result, what will happen to a car moving at 10 km/hr if it meets the edge of a 200 m cliff? Is this answer reasonable? If not, why not?

4.3 A particle moving in one dimension interacts with a potential of the form

$$V(x) = \begin{cases} 0 & |x| > a \\ -\dfrac{V_0}{2a} & |x| < a \end{cases} .$$

Find the equation determining the energy eigenvalues of this system. Solve it approximately assuming a is very small. What happens in the limit $a \to 0$?

4.4 The wave-function of an electron in the ground state of a hydrogen-like atom is

$$\psi(r) = Ae^{-Zr/a} \quad , \quad a = \frac{\hbar^2}{me^2}$$

where Z is the charge on the nucleus.

a) Determine the constant A, so that the wave- function is normalized to unity.

b) At what distance from the origin is the probability of finding the electron a maximum?

c) Determine the average value of: the kinetic energy, the potential energy and the total energy.

Hint: Use the virial theorem.

4.5 Show that the wavefunction for a particle in a bound state may always be chosen to be real. By computing the current density give an explanation of the physical meaning of this result.

Chapter 5

More One-Dimensional Problems

We continue our familiarization with the Schrödinger equation in this chapter. To understand how boundary conditions lead to discrete eigenvalues we begin with a qualitative discussion of the bound state problem. We then consider tunneling through a square potential barrier.

One of the most important solvable problems in both classical and quantum physics is the simple harmonic oscillator. We introduce it and solve it, for the first of several times, at this stage.

In preparation for later needs we also introduce the concept of the delta function. To gain familiarity with it we solve two problems with it. These serve to illustrate that delta-function potentials are really a form of boundary condition.

Finally we again consider scattering from a potential well and construct parity invariant solutions. This allows us to introduce the concept of phase shifts.

5.1 General Considerations

Before proceeding with a study of several more specific cases we pause and consider the one-dimensional bound state problem quite generally.

The procedure will be heuristic but will hopefully give some insight. As always we start with the Schrödinger equation.

$$-\frac{\hbar^2}{2m}\frac{d^2\psi}{dx^2} + V\psi = E\psi$$

and rewrite it in the form

$$\frac{d^2\psi}{dx^2} + (\varepsilon - U)\psi = 0 \quad . \tag{5.1}$$

We are interested in the case where U is a potential well and $\varepsilon < 0$. Thus, we write

$$\frac{d^2\psi}{dx^2} = -(\varepsilon - U)\psi \ . \tag{5.2}$$

Furthermore, ψ is chosen to be real since we have real boundary conditions ($\psi \xrightarrow[x\to\pm\infty]{} 0$). Another property of the potential is that it is short range. This means $U \to 0$ for $|x| \to \infty$. Thus for large x we have

$$\frac{d^2\psi}{dx^2} \simeq \kappa^2\psi$$

where,

$$\kappa^2 = -\varepsilon$$

so that

$$\begin{aligned} \psi(x) &\sim e^{-\kappa x} \qquad x \text{ large} \\ \psi(x) &\sim e^{\kappa x} \qquad -x \text{ large} \ . \end{aligned}$$

It is precisely these condition that lead to the quantization of energy levels as we now show. In fact the condition that ψ be square

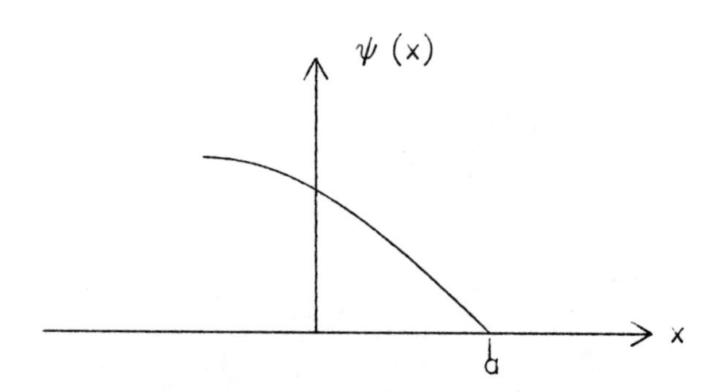

Fig. 5-1 Classically Permitted Region $\psi > 0$

integrable is what makes this an eigenvalue problem and gives a discrete set of eigenvalues.

Consider again equation 5.2

$$\frac{d^2\psi}{dx^2} = (U-\varepsilon)\psi \ , \tag{5.2}$$

and suppose we are in a classically permitted region

$$U - \varepsilon < 0 \ .$$

Thus,

$$\frac{d^2\psi}{dx^2} = -k^2\psi \tag{5.3}$$

in this region. Furthermore, if $\psi > 0$ then 5.3 states that $d^2\psi/dx^2 < 0$ so that ψ curves downwards as shown above (fig. 5.1). On the other hand, if $\psi < 0$ and $\frac{d^2\psi}{dx^2} > 0$ then ψ curves up as shown in (fig. 5.2). Thus, in the classically permitted region, ψ always bends towards the x axis and the behaviour may be described as oscillatory.

Suppose the classical bounds are given by $x = -a$, $x = b$. Thus,

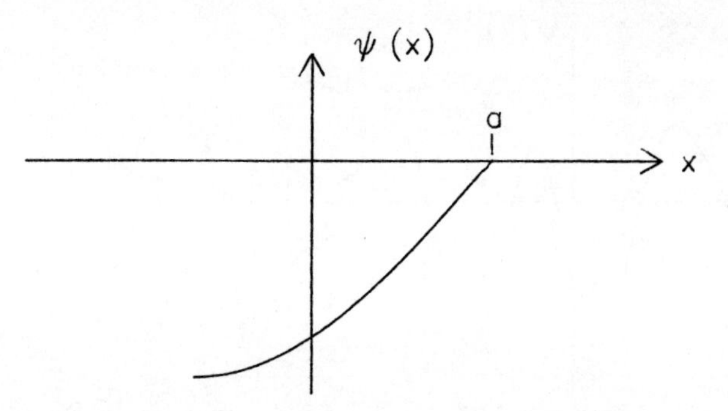

Fig. 5–2 Classically Permitted Region $\psi < 0$

$U = \varepsilon$ at $x = -a$ and $x = b$ as shown in figure 5.3. Then $-a \leq x \leq b$ is the classically permitted region and in this region ψ is "oscillatory".

In the classically forbidden regions ($x < -a$ and $x > b$) we have

$$\frac{d^2\psi}{dx^2} = k^2\psi$$

and ψ curves away from the axis. Now consider a solution as shown in

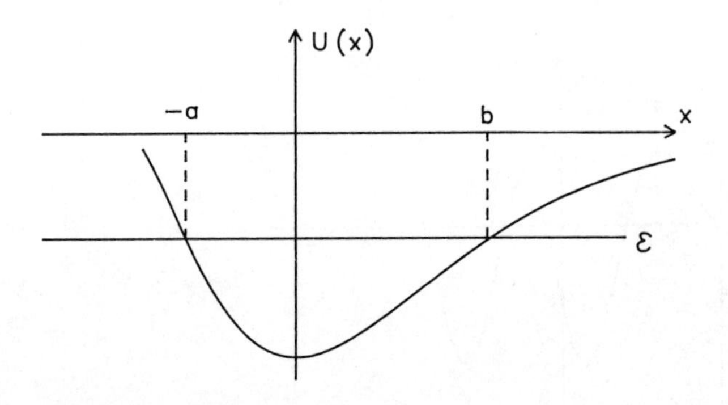

Fig. 5–3 Classical Turning Points

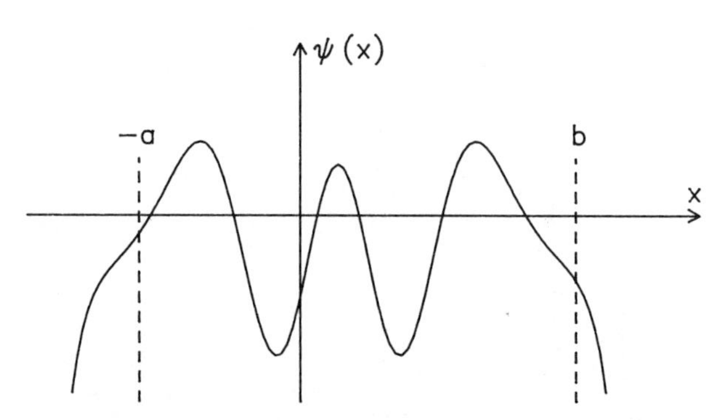

Fig. 5–4 Solution for Energy Slighty Too Large

(fig. 5.4). The solution is oscillatory between $-a \leq x \leq b$ and curves away from the x axis for $b < x < -a$, but goes to $-\infty$ as $|x|$ increases. This solution has exponential growth rather than decay and is not square integrable and hence not physically acceptable.

If k^2 is a little smaller, then ψ oscillates a little more slowly in the classically allowed region $-a \leq x \leq b$. In this case we may get a solution of the form shown in fig. 5.5. Again the growth for large $|x|$ is exponential and the solution is unacceptable. On the other hand, it is possible to choose a value of k^2 between the two values just considered so that ψ behaves as shown in fig. 5.6.

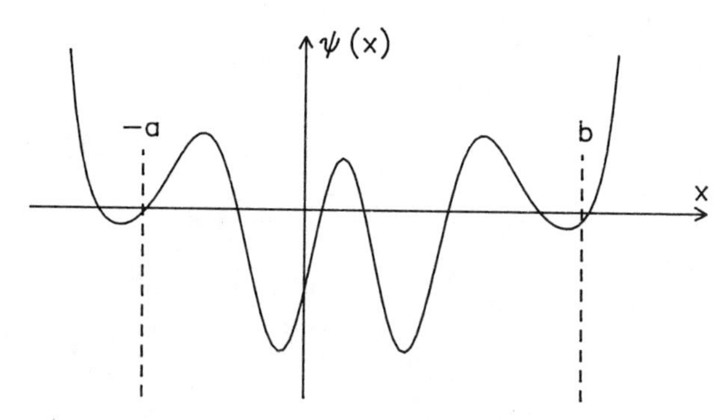

Fig. 5–5 Solution for Energy Slighty Too Small

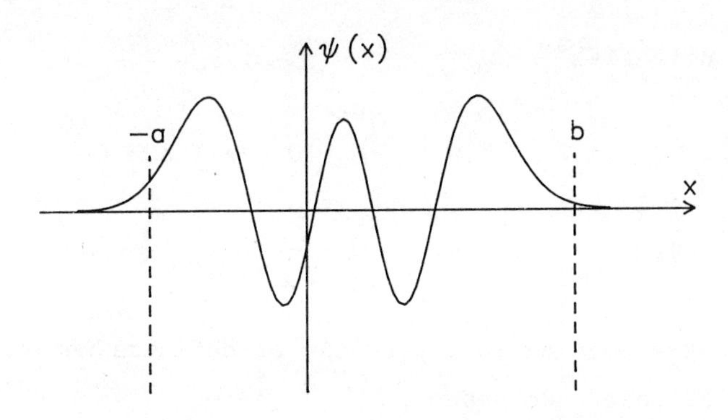

Fig. 5–6 Bound State Solution

In this case for large $|x|$ ψ is exponentially damped. Note that the value of k^2 so chosen means that ε is fixed. Thus for a given number of nodes in the classically permitted region we get a fixed value of ε and hence quantization.

We shall now study several problems which besides being soluble are of intrinsic physical interest. The first of these is tunneling.

5.2 Tunneling Through a Square Barrier

Classically a particle is never found in a region in which $V > E$, but as we saw previously, in the quantum mechanical case there is a non-zero probability of finding a particle in a classically forbidden region although the wave function decreases exponentially the further we penetrate into the forbidden region. If the forbidden region is not too broad, the tail of the wave function can penetrate part of it and thus, it is possible for a quantum particle to penetrate a potential barrier. This is known as tunneling. To illustrate this, consider the following simple problem.

We have a potential

$$V(x) = V_o > 0 \qquad 0 < x < a$$

$$V(x) = 0 \qquad a < x , \quad x < 0 \tag{5.4}$$

A wave e^{ikx} (corresponding to a particle of definite momentum $\hbar k$) is incident from the left. We assume

$$E = \frac{\hbar^2 k^2}{2m} < V_o \quad .$$

Thus, the region $0 < x < a$ is classically forbidden and a particle incident from the left would simply rebound elastically in the classical case. In the quantum mechanical case we must solve the Schrödinger equation which in this case can be written

$$\frac{d^2\psi}{dx^2} + k^2\psi = 0 \qquad a < x < 0$$

$$\frac{d^2\psi}{dx^2} - \kappa^2\psi = 0 \qquad 0 < x < a \tag{5.5}$$

where

$$\frac{2m}{\hbar^2}(V_o - E) = \kappa^2 \ . \tag{5.6}$$

The solution is

$$\psi = e^{ikx} + Re^{-ikx} \qquad x < 0 \ . \tag{5.7}$$

This is in fact the boundary condition. Also

$$\psi = Se^{ikx} \qquad x > a \ . \tag{5.8}$$

We dropped the term e^{-ikx} since there is no wave incident from the right and hence we only have a transmitted wave. The transmission coefficient T is given by

$$T = \frac{k}{k}|S|^2 = |S|^2 .$$

Also

$$\psi = A\,e^{\kappa x} + Be^{-\kappa x} \qquad 0 < x < a. \qquad 5.9$$

We now match the wave function and its derivative at $x = 0$ and $x = a$.

$$x = 0: \quad 1 + R = A + B \qquad 5.10$$

$$ik(1-R) = \kappa(A-B) \qquad 5.11$$

$$x = a: \quad Se^{ika} = Ae^{\kappa a} + Be^{-\kappa a} \qquad 5.12$$

$$ikSe^{ika} = \kappa(Ae^{\kappa a}-Be^{-\kappa a}) . \qquad 5.13$$

We can solve these four equations in four unknowns. However, only R and S are physically interesting. We get:

$$2ik = A(\kappa+ik) - B(\kappa-ik)$$

$$0 = A(\kappa-ik)e^{\kappa a} - B(\kappa+ik)e^{-\kappa a} .$$

Thus,

$$A = \frac{2ik(\kappa + ik)}{(\kappa+ik)^2 - (\kappa-ik)^2 e^{2\kappa a}} \qquad 5.14$$

$$B = \frac{2ik(\kappa - ik)e^{2\kappa a}}{(\kappa+ik)^2 - (\kappa-ik)^2 e^{2\kappa a}} . \qquad 5.15$$

Now,

$$R = \frac{1}{2ik}\left[-A(\kappa-ik) + B(\kappa+ik)\right]$$

$$S = \frac{e^{-ika}}{2ik}\left[A(\kappa+ik)e^{\kappa a} - B(\kappa-ik)e^{-\kappa a}\right] .$$

Substituting for A and B yields

$$R = \frac{(\kappa^2+k^2)(e^{\kappa a} - e^{-\kappa a})}{(\kappa+ik)^2 e^{-\kappa a} - (\kappa-ik)^2 e^{2\kappa a}}$$

or

$$R = \frac{(\kappa^2+k^2)\sinh \kappa a}{(k^2-\kappa^2)\sinh \kappa a + 2ik\kappa \cosh \kappa a} . \tag{5.16}$$

Similarly

$$S = \frac{e^{-ika}\, 2ik\kappa}{(k^2-\kappa^2)\sinh \kappa a + 2ik\kappa \cosh \kappa a} . \tag{5.17}$$

The transmission coefficient T is therefore given by

$$T = |S|^2 = \frac{4k^2\kappa^2}{(k^2-\kappa^2)^2 \sinh^2 \kappa a + 4k^2\kappa^2 \cosh^2 \kappa a}$$

using $\cosh^2 ka - \sinh^2 ka = 1$ we can rewrite this to read

$$T = \frac{4k^2\kappa^2}{(k^2+\kappa^2)^2 \sinh^2 \kappa a + 4k^2\kappa^2} . \tag{5.18}$$

In terms of the energy this reads

$$T = \frac{4E(V_o-E)}{V_o^2 \sinh^2 \kappa a + 4E(V_o-E)} \tag{5.19}$$

where as before

$$\kappa^2 a^2 = \frac{2ma^2}{\hbar^2}(V_o - E) \ .$$

Thus, the transmission coefficient decreases exponentially as the width of the barrier increases. Also, as $E \to \infty$ $T \to 1$ so that we get complete transmission. Again, as a check we have

$$T + R = 1 \qquad 5.20$$

where

$$R = |R|^2 \ . \qquad 5.21$$

Thus, the beam is either reflected or transmitted.

The reason for calling T and R transmission and reflection coefficients becomes obvious if we consider the probability current both to the left and right of the barrier. In section 3.2 we found that the conserved probability current is given by

$$j = -\frac{i\hbar}{2m}\left(\psi^* \frac{d\psi}{dx} - \frac{d\psi^*}{dx}\psi\right) \ . \qquad 5.22$$

Using the solutions 5.7 and 5.8 we get

$$j = \frac{\hbar k}{2m}[1 - |R|^2] = \frac{\hbar k}{m}(1-R) \qquad x < 0 \qquad 5.23$$

$$j = \frac{\hbar k}{m}|S|^2 = \frac{\hbar k}{m}T \qquad x > a \qquad 5.24$$

Since the current due to the incident beam is just $\frac{\hbar k}{m}$ our interpretation is vindicated.

5.3 The Simple Harmonic Oscillator

The Hamiltonian for this problem is

$$H = \frac{p^2}{2m} + \frac{1}{2} kx^2 . \qquad 5.25$$

Thus, the Schrödinger equation reads

$$i\hbar \frac{\partial \Psi}{\partial t} = H\Psi$$

or

$$i\hbar \frac{\partial \Psi}{\partial t} = (-\frac{\hbar^2}{2m} \frac{d^2}{dx^2} + \frac{1}{2} kx^2)\Psi . \qquad 5.26$$

For the stationary states

$$\Psi = e^{-iEt/\hbar} \psi \qquad 5.27$$

and then the Schrödinger equation reads

$$(-\frac{\hbar^2}{2m} \frac{d^2}{dx^2} + \frac{1}{2} kx^2)\psi = E\psi. \qquad 5.28$$

As in the case of the square well we introduce dimensionless variables.

Let $\omega = \sqrt{k/m}$ = classical frequency

$$\mu = \frac{2E}{\hbar\omega} \qquad y = (\frac{mk}{\hbar^2})^{1/4} x = (\frac{m\omega}{\hbar})^{1/2} x$$

In this case 5.28 becomes

$$\frac{d^2\psi}{dy^2} + (\mu - y^2)\psi = 0 . \qquad 5.29$$

Since we are looking for bound states the solutions must decrease at least exponentially for $y \to \pm\infty$. We therefore first look at the asymptotic behaviour of ψ. For $|y|$ large we can neglect μ. Then

$$\frac{d^2\psi}{dy^2} \simeq y^2\psi .$$

This yields that

$$\psi \simeq e^{\pm 1/2\, y^2}.$$

Obviously, for a square-integrable ψ, we must choose the asymptotic solution with the minus sign. To solve the problem we use the polynomial method of Sommerfeld (5.1). Thus we set

$$\psi = H(y)e^{-1/2\, y^2} \tag{5.30}$$

and look for the equation for H(y). When 5.30 is inserted into 5.29 the resultant equation is

$$\frac{d^2H}{dy^2} - 2y\frac{dH}{dy} + (\mu-1)H = 0 . \tag{5.31}$$

We solve this by the method of Frobenius, expanding H in a power series

$$H(y) = \sum_{n=0} a_n y^n . \tag{5.32}$$

Substituting into 5.31 gives

$$\sum_{n=0} [n(n-1)a_n y^{n-2} - 2n\, a_n y^n + (\mu-1)a_n y^n] = 0.$$

For $n \neq 0,1$ this can be written

$$\sum_{n'=0} [(n'+2)(n'+1)a_{n'+2} - 2n'a_{n'} + (\mu-1)a_{n'}]y^{n'} = 0.$$

Hence

$$a_{n+2} = \frac{2n + 1 - \mu}{(n+2)(n+1)}\, a_n \qquad n = 0,1,2,\ldots \tag{5.33}$$

This allows us to express

a_2 in terms of a_o and a_3 in terms of a_1

a_4 in terms of a_2 and a_5 in terms of a_3

and so forth. The two constants a_o and a_1 are arbitrary. Furthermore, the hamiltonian is invariant under the parity transformation that we introduced in section 4.3. As for the square well considered in that section the parity operator is also very useful here since we find that

$$PH = HP . \tag{5.34}$$

Therefore the eigenfunctions of the hamiltonian H may be chosen to be either even or odd functions. For $a_o = 0$, a_1 arbitrary we get odd solutions. For $a_1 = 0$, a_o arbitrary we get even solutions. Thus these are solutions of definite parity.

Now consider the recursion relation 5.23 for large n. In that case

$$a_{n+2} \simeq \frac{2}{n+2} a_n .$$

However, if we expand $e^{y^2} = \sum \frac{1}{n!} y^{2n}$, the coefficient b_n of y^n is

$$b_n = \frac{1}{(\frac{n}{2})!}$$

and

$$b_{n+2} = \frac{1}{(\frac{n}{2}+1)!} = \frac{1}{(\frac{n}{2}+1)} b_n = \frac{2}{n+2} b_n .$$

Thus, unless the series 5.32 terminates, H(y) behaves asymptotically as e^{y^2} so that ψ behaves like

$$e^{-y^2/2} e^{y^2} = e^{y^2/2} .$$

This behaviour is unacceptable. Thus the series 5.32 must terminate. Note that this is the condition for ϕ to be square integrable and leads to quantization. For the series to terminate requires that one of the a_n vanish. This will automatically happen if μ is an odd integer

$$\mu = 2N + 1 \qquad 5.35$$

Then a_{N+2}, a_{N+4}, a_{N+6}, etc. vanish. Thus, we have the quantization condition. Substituting back for μ we get the quantized energy levels

$$E_N = (N+\frac{1}{2})\,\hbar\omega \ . \qquad 5.36$$

Except for the additional $\frac{1}{2}\hbar\omega$ this is Planck's original assumption for the elementary oscillators of the electromagnetic field.

There is a simple way to generate all these polynomials satisfying equation 5.31. With their normalization chosen in this way they are called hermite polynomials. The formula is called a <u>Rodrigues' formula</u>.

Consider the expression

$$H_n(y) = (-1)^n e^{y^2} \frac{d^n}{dy^n} e^{-y^2} \qquad 5.37$$

and differentiate this expression to get

$$\frac{dH_n}{dy} = 2yH_n + (-1)^n e^{y^2} \frac{d^n}{dy^n} (-2ye^{-y^2}). \qquad 5.38$$

If we now use the Leibnitz formula

$$\frac{d^n}{dy^n}[F(y)G(y)] = \sum_{r=0}^{n} \binom{n}{r} \frac{d^rF}{dy^r} \frac{d^{n-r}G}{dy^{n-r}} \qquad 5.39$$

we find

$$\frac{d^n}{dy^n}(-2ye^{-y^2}) = -2y\frac{d^n}{dy^n}e^{-y^2} - 2n\frac{d^{n-1}}{dy^{n-1}}e^{-y^2}. \qquad 5.40$$

Replacing this result in equation 5.38 we then get

$$\frac{dH_n}{dy} = -2n(-1)^n e^{y^2}\frac{d^{n-1}}{dy^{n-1}}e^{-y^2} \qquad 5.41$$

Differentiating this expression once more we then obtain

$$\frac{d^2H_n}{dy^2} = -2n(-1)^n[2ye^{y^2}\frac{d^{n-1}}{dy^{n-1}}e^{-y^2} + e^{y^2}\frac{d^n}{dy^n}e^{-y^2}]$$

$$= 2y\frac{dH_n}{dy} - 2nH_n \qquad 5.42$$

where we have used 5.37 and 5.41. Equation 5.42 is precisely equation 5.31 with $\mu = 2n+1$. Thus the polynomials defined by equation 5.37 provide a solution to equation 5.31. They are called hermite polynomials.

The Rodrigues' formula (equation 5.37) is extremely useful. Thus 5.41 may be rewritten to read

$$\frac{dH_n}{dy} = 2nH_{n-1}(y). \qquad 5.43$$

This is an example of a recursion relation. We do not pursue this topic any further since in section 9.1 we shall again solve the harmonic oscillator problem in a completely different way and then the normalization of the wave-function as well as other properties will crop up quite naturally.

Why is the simple harmonic oscillator (S.H.O.) so important? Aside from the fact that it can be solved in closed form, it provides a possible means of quantizing fields, for example, the electromagnetic.

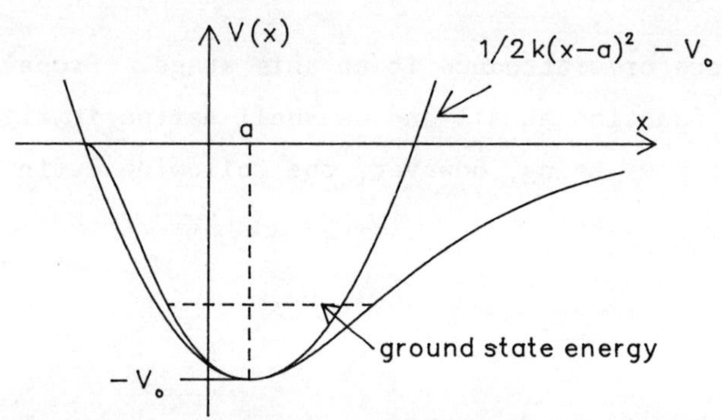

Fig. 5–7 S.H.O Approximation for Ground State

Furthermore, as we shall see later, the hermite functions $H(y)e^{-1/2\ y^2}$ are particularly convenient to work with; so they crop up in a large number of "practical" applications. Another use of the S.H.O. is for estimating the ground state (lowest) energy of a system. This works well if the potential is smooth. Consider the potential shown (fig. 5.7). Near the bottom, the potential is well approximated by a parabola. This is in fact what a S.H.O. is. Thus in the case shown the ground state energy is well approximated by $\frac{1}{2}\hbar\omega - V_o$. To get k we simply find the point $x = a$ where $V(x)$ is a minimum and evaluate

$$k = \frac{d^2V}{dx^2}\bigg|_{x=a} \quad . \qquad 5.44$$

This is an extremely simple method for obtaining an estimate of the energy for low lying levels in the case of a complicated but smooth potential.

5.4 The Delta Function

There is one more class of potentials that occurs in practice due to the extreme simplicity of the solutions of the Schrödinger equation, the so-called delta function. This function plays a large role in later discussions and we therefore introduce it at this stage. Properly speaking, it is not a function at all and we shall define it rigorously in chapter 7. For the time being, however, the following definitions will suffice.

Definition

$\delta(x) = 0 \qquad x \neq 0$

$\delta(x)$ is so singular at $x = 0$ that

$$\int_a^b f(x)\delta(x)dx = f(0) \tag{5.45}$$

whenever the interval (a,b) contains the origin.

$\delta(x)$ also has the following integral representation

$$\delta(x) = \frac{1}{2\pi}\int_{-\infty}^{\infty} e^{ikx}\, dx \tag{5.46}$$

where the integral is to be understood as

$$\delta(x) = \lim_{\varepsilon\to 0} \frac{1}{2\pi}\int_{-\infty}^{\infty} e^{ikx-\varepsilon k^2}\, dk. \tag{5.47}$$

Thus for example

$$\int_{-\infty}^{\infty} f(x)\delta(x)dx = \frac{1}{2\pi}\int_{-\infty}^{\infty}\int_{-\infty}^{\infty} e^{ikx}f(x)dxdk$$

$$= \frac{1}{2\pi}\int_{-\infty}^{\infty} \tilde{f}(k)\, dk$$

$$= f(0) \tag{5.48}$$

where

$$\tilde{f}(k) = \int_{-\infty}^{\infty} f(k)e^{ikx}dx \qquad 5.49$$

is the Fourier transform of $f(x)$. The inverse transform is given by

$$f(x) = \frac{1}{2\pi}\int_{-\infty}^{\infty} \tilde{f}(k)e^{-ikx}\, dk \qquad 5.50$$

so that clearly

$$f(0) = \frac{1}{2\pi}\int_{-\infty}^{\infty} \tilde{f}(k)dk. \qquad 5.51$$

Again a rigorous proof of all these relations will be delayed until Chapter 7.

Some of the more commonly used properties of the delta function (see problem 5.8) are listed below:

$$\int_{-\infty}^{\infty} f(x)\delta(x-a)\, dx = f(a) \qquad 5.52$$

$$\delta(ax) = \frac{1}{a}\,\delta(x) \qquad a > 0 \qquad 5.53$$

$$f(x)\,\delta(x-a) = f(a)\,\delta(x-a) \qquad 5.54$$

$$\int_{-\infty}^{\infty} \delta(x-y)\,\delta(y-z)\, dy = \delta(x-z) \qquad 5.55$$

$$\delta(x^2-a^2) = \frac{1}{2|a|}\,[\delta(x-a) + \delta(x+a)]\ . \qquad 5.56$$

With these preliminaries out of the way we are ready for some examples.

5.5 Attractive Delta Function Potential

We want to find the bound states for the Hamiltonian

$$H = \frac{p^2}{2m} - \Lambda\delta(x) \quad , \quad \Lambda > 0 \tag{5.57}$$

The Schrödinger equation is

$$H\psi = E\psi$$

with $E < 0$. Written out explicitly this reads

$$\left[-\frac{\hbar^2}{2m}\frac{d^2}{dx^2} - \Lambda\,\delta(x)\right]\psi(x) = E\psi(x)$$

or

$$\frac{d^2\psi}{dx^2} - \kappa^2\psi(x) = -\alpha^2\delta(x)\psi(x) \tag{5.58}$$

where

$$\kappa^2 = -\frac{2mE}{\hbar^2} > 0 \tag{5.59}$$

$$\alpha^2 = \frac{2m\Lambda}{\hbar^2} > 0 \; . \tag{5.60}$$

Thus for $x \neq 0$ we have the Schrödinger equation

$$\frac{d^2\psi}{dx^2} - \kappa^2\psi = 0 \tag{5.61}$$

with the solutions

$$\psi(x) = Ae^{-\kappa x} \qquad x > 0$$
$$\psi(x) = Be^{\kappa x} \qquad x < 0 \tag{5.62}$$

where we have already used the fact that $\psi(x)$ must be square integrable and therefore vanish at $x = \pm\infty$. We must choose ψ to be continuous at $x = 0$. This gives

$$A = B . \tag{5.63}$$

The derivative of ψ, say ψ', is however discontinuous at $x = 0$. This discontinuity can be computed directly from the Schrödinger equation 5.58. For this purpose we consider the expression

$$\lim_{\varepsilon\to 0+} \int_{-\varepsilon}^{\varepsilon} \frac{d^2\psi}{dx^2} dx = \psi'(0_+) - \psi'(0_-)$$

$$= \lim_{\varepsilon\to 0+} [\int_{-\varepsilon}^{\varepsilon} \kappa^2\psi(x)dx - \alpha^2\int_{-\varepsilon}^{\varepsilon} \psi(x)\delta(x)dx]$$

$$= -\alpha^2\psi(0)$$

where we have used the fact that $\psi(0_+) = \psi(0_-)$. Thus

$$\psi'(0_+) - \psi'(0_-) = -\alpha^2\psi(0). \tag{5.64}$$

Substituting the solution we get

$$-\kappa A - \kappa A = -\alpha^2 A$$

or

$$\kappa = \alpha^2/2 = \frac{m\Lambda}{\hbar^2} . \tag{5.65}$$

But since $E = -\hbar^2\kappa^2/2m$, we therefore get

$$E = -\frac{m\Lambda^2}{\hbar^2} \quad . \tag{5.66}$$

So the attractive delta function has one bound state with the above energy. The corresponding normalized wave function is

$$\psi = \sqrt{\kappa}\; e^{-\kappa|x|} \quad . \tag{5.67}$$

The main point of this problem was to show how to handle the delta-function. We now consider the corresponding scattering problem, but for variety we choose a repulsive delta function.

5.6 Repulsive Delta Function Potential

In this case the potential is

$$V(x) = \Lambda\delta(x) \qquad \Lambda > 0. \tag{5.68}$$

As before the wavefunction ψ is continuous at $x = 0$ and has a discontinuity in its first derivative at $x = 0$. The Schrödinger equation can be rewritten to read

$$\frac{d^2\psi}{dx^2} + k^2\psi(x) = \alpha^2\delta(x)\psi(x) \tag{5.69}$$

where $k^2 = 2mE/\hbar^2$ and $\alpha^2 = 2m\Lambda/\hbar^2$. If we call the solution for $x < 0$, $\psi_<$ and for $x > 0$, $\psi_>$ then the matching conditions at $x = 0$ are

$$\begin{aligned} \psi_<(0) &= \psi_>(0) \\ \psi'_<(0) &= \psi'_>(0) - \alpha^2\psi_>(0) \quad . \end{aligned} \tag{5.70}$$

Since for $x \neq 0$ the Schrödinger equation reduces to the equation for a free particle, the solutions are $e^{\pm ikx}$. If we start with a particle incident from the left we have

$$\psi_<(x) = e^{ikx} + Re^{-ikx} \tag{5.71}$$

$$\psi_>(x) = Te^{ikx} \tag{5.72}$$

where we have used our freedom in choosing the incoming wave as well as the physical requirements that for $x > 0$ we have only a transmitted wave. Using the matching condition at $x = 0$ we now get

$$1 + R = T \tag{5.73}$$

$$ik - ik\,R = (ik-\alpha^2)T. \tag{5.74}$$

The solutions of these equations are

$$R = \frac{\alpha^2}{-\alpha^2+2ik} \tag{5.75}$$

$$T = \frac{2ik}{-\alpha^2+2ik} \quad . \tag{5.76}$$

The transmission and reflection coefficients $\mathcal{T}$, $\mathcal{R}$ are then given by

$$\mathcal{T} = |T|^2 = \frac{4k^2}{4k^2+\alpha^4} \tag{5.77}$$

$$\mathcal{R} = |R|^2 = \frac{\alpha^4}{4k^2+\alpha^4} \quad . \tag{5.78}$$

5.7 Finite Square Well - Scattering and Phase Shifts

In section 4.2 we considered the finite square well

$$V(x) = \begin{cases} -U & |x| < a \\ 0 & |x| > a \end{cases} \tag{5.79}$$

and found all possible bound states. At the same time we saw that the invariance of this potential $V(x)$ under parity, that is

$$PV(x)P = V(-x) = V(x) \tag{5.80}$$

could be used to good advantage since the total hamiltonian was then also invariant under parity. This allowed us to consider separately the positive parity (even) eigenfunctions and negative parity (odd) eigenfunctions. Such a separation into positive and negative parity states is also possible for scattering solutions and frequently simplifies the computations. However these solutions of definite parity do not correspond directly to the scattering solutions with the physical boundary conditions that we have been considering so far. Corresponding to the potential above, physical boundary conditions are:

For $x < -a$ we have an incoming wave (normalization 1) plus a reflected wave, or

$$\psi(x) = e^{ikx} + Re^{-ikx} \quad \text{for} \quad x < -a. \tag{5.81}$$

For $x > a$ we have only a transmitted wave, or

$$\psi(x) = Te^{ikx} \qquad \text{for} \quad x > a. \tag{5.82}$$

To make contact between these solutions and those of definite parity is fortunately quite easy.

As in section 4.2 we now define

$$k^2 = \frac{2mE}{\hbar^2} \quad , \quad K^2 = \frac{2m(E+U)}{\hbar^2} . \tag{5.83}$$

The Schrödinger equation then reads:

$$\begin{aligned} &\frac{d^2\psi}{dx^2} + K^2\psi = 0 && |x| < a \\ &\frac{d^2\psi}{dx^2} + k^2\psi = 0 && |x| > a . \end{aligned} \tag{5.84}$$

Choosing the normalization in the regions $|x| > a$ to be unity, we can write the solution of definite parity as

$$\psi_+(x) = \begin{cases} \cos(kx-\delta_+) & x < -a \\ A \cos Kx & -a < x < a \\ \cos(kx+\delta_+) & x > a \end{cases} \tag{5.85}$$

and

$$\psi_-(x) = \begin{cases} i \sin(kx-\delta_-) & x < -a \\ B \sin Kx & -a < x < a \\ i \sin(kx+\delta_-) & x > a \end{cases} \tag{5.86}$$

where clearly we have

$$P\psi_{\pm}(x) = \psi_{\pm}(-x) = \pm\psi_{\pm}(x) \tag{5.87}$$

so that ψ_+ and ψ_- are respectively positive and negative parity solutions. The parameters $\delta_{\pm}$ are known as the <u>phase shifts</u> and reflect the presence of the potential. They contain all the information that the reflection and transmission amplitudes R and T contain. To see this we proceed as follows:

For $x < -a$ we consider

$$e^{i\delta_+}\psi_+(x) + e^{i\delta_-}\psi_-(x) = e^{ikx} + \frac{1}{2}(e^{2i\delta_+} - e^{2i\delta_-})e^{-ikx}. \tag{5.88}$$

Comparing this with the solution given by 5.81 for $x < -a$ we see that

$$R = \frac{1}{2}(e^{2i\delta_+} - e^{2i\delta_-}). \quad 5.89$$

Similarly for $x > a$ we consider

$$e^{i\delta_+}\psi_+(x) - e^{i\delta_-}\psi_-(x) = \frac{1}{2}(e^{2i\delta_+} + e^{2i\delta_-})e^{ikx} \quad 5.90$$

and comparing this with the solution given by 5.82 for $x > a$ we find

$$T = \frac{1}{2}(e^{2i\delta_+} + e^{2i\delta_-}) . \quad 5.91$$

Thus the problem is reduced to finding the phase shifts $\delta_\pm$.

To complete the solution and determine the phase shifts we have to match the wavefunctions and their slopes at $x = \pm a$. There is a simplification that can be used due to the fact that we are not interested in the constants A, B in equations 5.85 and 5.86. Thus instead of matching ψ and $d\psi/dx$ we match the so-called logarithmic derivative $\frac{1}{\psi}\frac{d\psi}{dx}$. This has the effect of cancelling the unwanted contants.

From 5.85 we therefore get:

$$k \tan(ka+\delta_+) = K \tan Ka \quad 5.92$$

and from 5.86 we get

$$k \cot(ka+\delta_-) = K \cot Ka . \quad 5.93$$

Thus,

$$\delta_+ = \arctan(\frac{K}{k} \tan Ka) - ka . \quad 5.94$$

$$\delta_- = \text{arccot}(\frac{K}{k} \cot Ka) - ka. \quad 5.95$$

Inserting these results in equations 5.89 and 5.91 we obtain after some simplifying algebra that

$$T = \frac{(k^2-K^2)e^{-2ika}}{(k^2+K^2) + ikK(\cot Ka - \tan Ka)} \tag{5.96}$$

$$R = \frac{ikK(\cot Ka + \tan Ka)e^{-2ika}}{(k^2+K^2) + ikK(\cot Ka - \tan Ka)} \quad . \tag{5.97}$$

So as always $|T|^2 + |R|^2 = 1$. This result is in fact obvious from equation 5.89 and 5.91 if $\delta_{\pm}$ are real.

The expressions 5.96 and 5.97 for T and R could have been obtained directly by solving the Schrödinger equation 5.84 as we did in the previous examples. From the way we introduced the phase shifts, it would appear that they are only a computational tool. This is probably true for one- dimensional scattering problems. However in three- dimensional scattering problems (as we shall later see) the phase shifts have intrinsic physical interest, corresponding to quantities directly measurable in the laboratory. Thus in anticipation of their future utility we introduced them now.

So far we have been solving very concretely specified problems. The purpose of this has been two-fold: to develop our technical abilities and to build up our intuition. We now leave the discussion of specific problems for a while to embark on a consideration of the formal aspects of quantum mechanics. In this process we shall develop a certain mathematical language which may, at first, seem artifical but is, in fact, very important if not essential for a clear formulation of quantum mechanics.

Notes, References and Bibliography

All the references for chapter 4 apply. In addition, it may be useful to consult:

S. Flügge - Practical Quantum Mechanics - Springer-Verlag, New York, Heidelberg, Berlin (1974).

This book is a collection of very many worked examples together with a discussion of the physics of these examples.

5.1 A. Sommerfeld - Wave Mechanics - Dutton, New York (1929).

Chapter 5 Problems

5.1 A wave function

$$\psi(x,t) = \int_{-\infty}^{\infty} f(x)e^{i[kx-\omega(k)t]}dk$$

is normalized such that

$$\int_{-\infty} |\psi(x,t)|^2 \, dx = 1.$$

Assume $f(k)$ is a smooth function vanishing rapidly at infinity. Show that the velocity of the center of mass $\bar{x}$ defined by

$$\bar{x} = \int_{-\infty}^{\infty} x|\psi(x,t)|^2 dx$$

is given by

$$\frac{d\bar{x}}{dt} = 2\pi \int_{-\infty}^{\infty} \frac{d\omega}{dk} |f(k)|^2 dk.$$

Hint: Use the fact that $\int_{-\infty}^{\infty} e^{i(k-q).x}dx = 2\pi\delta(k-q)$.

Show furthermore if $\omega(k)$ is adequately approximated by

$$\omega(k) \simeq \omega_o + (k-k_o)\frac{d\omega}{dk}\Big|_{k_o} \qquad \text{(eqn. 3.17)}$$

then $\dfrac{d\bar{x}}{dt} \simeq \dfrac{d\omega}{dk}\Big|_{k_o}$ (the group velocity).

5.2 Compute the probability density ρ and the current j for the wavefunction in problem 3.3. What are the possible forms of $\omega(k)$ such that the equation of continuity (eqn. 3.8) is satisfied? Explain this result recalling that $E = \hbar\omega$ and $p = \hbar k$.

5.3 A potential sometimes used in molecular physics is the so-called 6-12 potential

$$V(x) = \varepsilon_o[(\frac{\sigma}{x})^{12} - 2(\frac{\sigma}{x})^6].$$

a) Sketch this potential and indicate the values of the most interesting points in the diagram.

b) Estimate the ground state (lowest) energy eigenvalue for a particle of mass m bound in this potential.

5.4 A particle is in the potential

$$V = V_o \exp 1/2\ ax^2 \qquad a > 0.$$

Estimate the energy of the 2 lowest eigenstates. What are the parities of these states?

5.5 Find the wave-function for a particle at rest at the origin of a coordinate system fixed in space.

5.6 Calculate the transmission probability for a particle of mass m incident on a potential

$$V(x) = \Lambda_o[\delta(x+a) + \delta(x-a)].$$

Compute also the phase shifts.

5.7 Complete all the steps in going from 5.89, 5.91, 5.94, 5.95 to 5.96, and 5.97.

5.8 Using the representation

$$\delta(x) = \lim_{\varepsilon\to 0} h_\varepsilon(x)$$

where

$$h_\varepsilon(x) = \begin{cases} \frac{1}{2\varepsilon} & -\varepsilon < x < \varepsilon \\ 0 & -\varepsilon > x > \varepsilon \end{cases}$$

verify equation 5.44 - 5.48.

Hint: For 5.45 to 5.48 integrate both sides of the equation with a well-behaved function $f(x)$.

5.9 Repeat problem 5.8 using the representation

$$\delta(x) = \lim_{\varepsilon \to 0} \frac{1}{2\pi} \int_{-\infty}^{\infty} e^{ikx-\varepsilon k^2} dk.$$

Chapter 6

Mathematical Foundations of Quantum Mechanics

In this chapter we present a mathematical interlude to provide a more formal language. The first two sections providing a discussion of Hilbert space should be read by all those not familiar with this concept. The remaining sections are somewhat more mathematical in nature and are provided for those who desire more rigour.

In section 3 we discuss linear operators in Hilbert space and introduce the concept of self-adjointness. The Cayley transform is introduced in section 4 and used to classify all self-adjoint extensions of a symmetric operator. Section 5 is devoted to some examples illustrating the results of section 4. More examples are also provided in section 6.

6.1 Geometry of Hilbert Space

The language we have used so far is one of wave-functions or state vectors and operators on these wave-functions. There is a ready-made mathematical language for this. This is the language of Hilbert space. Thus, quantum mechanics is naturally formulated in Hilbert space. Actually, for practical purposes Hilbert space is too small and the appropriate generalization is to a so-called rigged Hilbert space. We shall ignore this for the time being and discuss it briefly in Chapter 8, where we shall also give references for those who are interested in more of the details.

What is Hilbert space? First of all it is a vector space analogous to the usual Euclidean spaces E_3 or E_n, however unlike E_3 which is 3-dimensional and E_n which is n-dimensional, Hilbert space is ∞ dimensional. As a conceptual model of Hilbert space H one can think of taking some representation of a vector in E_n, say $(a_1,\ldots,a_n)$ and writing it

$$(a_1,\ldots,a_n, 0,\ldots,0,\ldots \quad)$$

Then letting n increase without limit we arrive at the notion of H. However, unlike all ordered n-tuples which automatically can be considered in E_n, not all infinite ordered sequences can be considered to belong to H. The reason for this is that not all of them have finite "length" and we wish to include in H only elements of finite length. This is tantamount to saying in the language of wavefunctions that we want the wavefunction to be square integrable. We start by listing those properties of E_n which remain true in the transition to H. The elements of the space are vectors f on which certain operations are defined.

1) Scalar multiplication:

If $f \in H$ and λ is a complex number

then $\lambda f \in H$

2) Addition:

If f_1 and $f_2 \in H$

then $f_1 + f_2 \in H$

Thus, combining 1) and 2) we see that all finite linear combinations of elements in H belong to H.

3) In H there is defined an inner product

$$(f,g) \quad \text{where} \quad f,g \in H$$

This inner product maps elements of H into complex numbers and satisfies the following conditions:

a) $(f,g) = \overline{(g,f)}$ 6.1

where the bar means complex conjugation.

b) $(\lambda_1 f, \lambda_2 g) = \lambda_1^* \lambda_2 (f,g)$ 6.2

where λ_1 and λ_2 are numbers and * also means complex conjugation. We shall use either notation as the occasion arises.

c) $(f_1+f_2,g) = (f_1,g) + (f_2+g)$

6.3

and $(f,g_1+g_2) = (f,g_1) + (f,g_2)$.

d) $|(f,g)|^2 \leq (f,f)(g,g)$ 6.4

This last is the Schwarz inequality. We shall later derive this inequality.

In terms of the inner product we define the "length" $\|.\|$ or norm of a vector in the usual way by

$$\|f\|^2 = (f,f) . \tag{6.5}$$

It is also possible to define orthogonality using the inner product.

Thus f is orthogonal to g if and only if (abbreviated iff)

$$(f,g) = 0.$$

A set of vectors $\{f_i\}$ is orthonormal that is orthogonal and normal iff

$$(f_i,f_j) = \delta_{ij} .$$

We now come to some of the differences. For this we need two definitions.

Definition

A set $\{f_i\}$ of vectors is complete iff any vector in H can be written as a linear combination of vectors from the set $\{f_i\}$. A complete set of orthonormal vectors forms a basis.

Example

Consider the Euclidean space E_3 and choose three orthonormal vectors $\{\hat{e}_1,\hat{e}_2,\hat{e}_3\}$. Thus

$$(\hat{e}_i,\hat{e}_j) = \delta_{ij} \ .$$

Then any vector $f \in E_3$ can be written

$$f = \sum_{i=1}^{3} \lambda_i \hat{e}_i$$

and in fact

$$\lambda_i = (\hat{e}_i,f) \ .$$

The $\hat{e}_i$ are obviously complete and form a basis in E_3. On the other hand, if we choose just two of these vectors, say $\hat{e}_1$ and $\hat{e}_2$, they do not form a complete set since for example any vector with a component along $\hat{e}_3$ cannot be expressed in terms of just the first two.

The λ_i are usually called the components of the vector. If we then agree to keep the basis fixed, we can suppress the basis vectors and write

$$f = (\lambda_1,\lambda_2,\lambda_3) \ .$$

In this manner we establish a one-one correspondence between vectors in

E_3 and ordered triplets. The norm of f is given by

$$\|f\|^2 = \sum_{i,j=1}^{3} \lambda_i^* \lambda_j (\hat{e}_i, \hat{e}_j)$$

$$= \sum_{i,j=1}^{3} \lambda_i^* \lambda_j \delta_{ij}$$

$$= \sum_{i=1}^{3} |\lambda_i|^2$$

This formalism above is all exceedingly trivial and you may wonder why bother. The reason is to establish a precise formalism so that when the situation becomes complicated we can rely on the formalism and not just our intuition.

Another use for the word <u>complete</u> is in the description of a vector space. This concept again is trivial for E_n but is non-trivial for H. Consider a sequence of vectors $f_1, f_2, \ldots$. Furthermore, suppose that for every $\varepsilon > 0$ we can find an ℓ such that for any finite m

$$\|f_{\ell+m} - f_\ell\| < \varepsilon \; . \tag{6.6}$$

This is just a statement of the Cauchy criterion for convergence using the norm $\|.\|$ rather than the absolute value as is the usual case for numerical sequences. We shall call such a sequence a Cauchy sequence. Now if H in this case is finite dimensional say E_n then it is trivial to show that the limit of the sequence exists and is a vector in E_n . This property that all Cauchy sequences have a limit in E_n is stated by saying that E_n is complete. In fact all finite dimensional vector spaces are complete. That this is also true for ∞-dimensional Hilbert spaces is a deep theorem of analysis known as the Riesz-Fischer Theorem. We shall now look at a specific model of a Hilbert space which is called L_2 by mathematicians.

6.2 L_2 - A Model Hilbert Space

The elements of L_2 are square-integrable complex- valued functions $f(x)$ of a real variable x. More generally x is a vector in some real finite vector space so that f is a function of n real variables. This generalization has no effect whatever on the ensuing statement and so we ignore it. The norm $\|.\|$ in L_2 is defined by

$$\|f\|^2 = (f,f) = \int f^*(x)f(x)dx \ . \qquad 6.7$$

The range of integration in 6.7 is over the full range of the variable x. Thus, if x is unrestricted, the integral runs from $-\infty$ to $+\infty$.

It is trivial to check our first 2 conditions for elements in L_2. Thus

$$f \in L_2 \qquad \Rightarrow \lambda f \in L_2$$

$$f_1, f_2 \quad L_2 \quad \Rightarrow \quad f_1+f_2 \quad L_2 \ .$$

Furthermore, as defined by 6.7 the inner product obviously satisfies conditions a), b) and c).

a) $\quad (f,g) = \int f^* g dx = (\int fg^* \, dx)^* = (g,f)^*$

and

b) $\quad (\lambda_1 f,\lambda_2 g) = \int(\lambda_1 f)^*(\lambda_2 g)dx = \lambda_1^*\lambda_2 \int f^* g \, dx = \lambda_1^*\lambda_2(f,g)$

c) $$(f_1+f_2,g) = \int(f_1+f_2)^* gdx = \int f_1^* gdx + \int f_2^* gdx$$
$$= (f_1,g)+(f_2,g)$$

and

$$(f,g_1+g_2) = \int f^*(g_1+g_2)dx = \int f^* g_1 dx + \int f^* g_2 dx$$
$$= (f,g_1)+(f,g_2)$$

The only condition left to verify on the inner product is the Schwarz inequality. To do this consider the vector

$$h = f + \lambda(g,f)g \qquad 6.8$$

where $f,g \in L_2$ and λ is a real number. Then

$$(h,h) = \|h\|^2 \geq 0.$$

$$\therefore \quad 0 \leq (f + \lambda(g,f)g,\ f + \lambda(g,f),g)$$

$$= (f,f) + 2\lambda|(f,g)|^2 + \lambda^2|(f,g)|^2(g,g) .$$

Therefore, the quadratic polynomial in λ cannot have two real distinct zeroes and hence the discriminant is negative, giving

$$|(f,g)|^4 - |(f,g)|^2(f,f)(g,g) \leq 0.$$

The equality sign obviously applies when $(f,g) = 0$ or $f = \mu g$. Thus even if $(f,g) = 0$ we get

$$|(f,g)|^2 \leq (f,f)(g,g) \qquad 6.4$$

as required.

Orthogonality is still given by: f is orthogonal to g iff

$$(f,g) = 0.$$

An example of 2 orthogonal vectors in L_2 is

$$f = \frac{1}{\pi^{1/4}} e^{-x^2/2} \qquad g = \frac{2x}{\pi^{1/4}} e^{-x^2/2}$$

where the range of intergration is $(-\infty,\infty)$. These are the first 2 hermite functions. As we shall see later, it is a general fact that eigenfunctions corresponding to different eigenvalues are orthogonal. In fact for physical hamiltonians, the eigenfunctions properly normalized can be taken as a basis. This is an important fact since it implies that these eigenvectors form a complete set.

Although we will not prove the Riesz-Fischer Theorem we restate it here. Consider a sequence of functions $f_1, f_2, \ldots\ldots$ all of which belong to L_2. Furthermore, let this be a Cauchy sequence. This means that given any $\varepsilon > 0$ we can find an $\ell > 0$ such that

$$\left[\int |f_{\ell+m} - f_\ell|^2 dx\right]^{1/2} < \varepsilon .$$

Then the Riesz-Fischer Theorem asserts that

a) $\lim_{m\to\infty} f_m(x) = f(x)$

exists and

b) $f(x) \in L_2$.

That is

$$\int |f(x)|^2 dx < \infty .$$

This guarantees that we can take limits of sequences in L_2 and more generally in any Hilbert space. Clearly L_2 is the model for the quantum mechanical Hilbert spaces. We shall see that the inner product plays an exceedingly important role in the physical interpretation of quantum mechanics. There is one more technical point, namely that for any set of linearly independent vectors $\{f_j\}$ it is possible to construct an orthonormal set of vectors $\{e_j\}$ which span the same space. The orthogonalization process is called the Schmidt Orthogonalization Procedure. The proof is by construction. Choose one of the f_j say f_1. Then

$$e_1 = f_1/\|f_1\| .$$

Now form

$$g_2 = f_2 - (e_1,f_2)e_1$$

and

$$e_2 = g_2/\|g_2\| .$$

Clearly, e_1, e_2 are orthonormal. Now form

$$g_3 = f_3 - (e_1,f_3)e_1 - (e_2,f_3)e_2$$

and

$$e_3 = g_3/\|g_3\| .$$

The process is now obvious.

We now turn to another aspect of Hilbert space, namely operators.

6.3 Operators on Hilbert Space - Mainly Definitions

An operator on Hilbert space is a mapping which maps certain elements of $\mathcal{H}$ into $\mathcal{H}$. Thus, if A is an operator with domain $D_A \subset \mathcal{H}$ then for all $f \in D_A$

$$g = Af \in \mathcal{H} .$$

The domain D_A consists simply of all those vectors in $\mathcal{H}$ such that the result of operating with A on a vector in D_A is a vector in $\mathcal{H}$. Thus, for example, if $\mathcal{H}$ is the space L_2 and A is the operator x^2 (multiplication by x^2) then

$$(Af)(x) = x^2 f(x) .$$

Clearly even if $f \in L_2$ not all functions $x^2 f(x)$ are in L_2. For example

$$f(x) = (x^2 + a^2)^{-\mu}$$

is in L_2 for $\text{Re}\,\mu > \frac{1}{4}$. But

$$x^2 f(x) = x^2 (x^2 + a^2)^{-\mu}$$

is not in L_2 unless $\text{Re}\,\mu > \frac{9}{4}$.

Again we shall only be interested in linear operators. Thus

$$A(\lambda_1 f_1 + \lambda_2 f_2) = \lambda_1 A f_1 + \lambda_2 A f_2 \ . \qquad 6.9$$

For example, the operator x^2 defined above is linear and so are the operators

$$pf = \frac{\hbar}{i} \frac{df}{dx} \qquad 6.10$$

and

$$(Kf)(x) = \int_{-\infty}^{\infty} K(x,y) f(y) dy \ . \qquad 6.11$$

On the other hand, log f and $\sqrt{f}$ are definitely not linear operators.

Linear operators are also familiar in finite dimensional vector spaces. They are usually represented by matrices in this case. There is an analogous representation for operators in Hilbert space. Formulated in this way quantum mechanics is called matrix mechanics to distinguish it from the Schrödinger formulation or wave mechanics. Both formulations are just two different mathematical ways of looking at the same thing. We shall examine matrix mechanics after we have developed all the necessary mathematical machinery.

To illustrate the matrix operator formalism we first derive the form of the most general linear operator on a finite vector space say E_n. Let A be such an operator. Call

$$g = Af \qquad 6.12$$

and consider taking for f different elements of a basis set $\{e_i\}$. Thus

$$g_i = Ae_i \; . \qquad 6.13$$

Then writing

$$f = \sum \lambda_i e_i \qquad 6.14$$

$$g = \sum \mu_i e_i \qquad 6.15$$

we get by linearity:

$$\sum \mu_i e_i = \sum \lambda_i A e_i = \sum \lambda_i g_i \; . \qquad 6.16$$

Now using

$$(e_i, e_j) = \delta_{ij} \qquad 6.17$$

and taking inner products in 6.16 we get

$$\mu_j = \sum_i \lambda_i (e_j, Ae_i) = \sum_i \lambda_i (e_j, g_i) \; . \qquad 6.18$$

Thus, the operator A is completely determined in this basis by the matrix of numbers

$$A_{ji} = (e_j, Ae_i) \; . \qquad 6.19$$

Conversely if we are given a matrix $(n \times n)$ then it can always be used to define a linear operator according to 6.19. Thus, as stated previously the most general linear operator on E_n can be considered to be an $n \times n$ matrix. With only some attention to details the same argument will go through for an ∞ dimensional Hilbert space.

Just as a matrix algebra is possible, so an algebra of linear operators is generally possible. It is only necessary to pay due attention to such things as domains of the operators.

Let A be an operator on H with domain D_A. Then λA is an operator on H with domain D_A and acts as follows:

$$(\lambda A)f = \lambda(Af) . \qquad 6.20$$

This is almost too obvious. If A, B are operators with domains D_A and D_B respectively, then $A + B$ is an operator with domain $D_A \cap D_B$ defined by

$$(A + B)f = Af + Bf . \qquad 6.21$$

The range R_A of an operator A is defined as the set of all vectors obtained by operating with A on elements in D_A. Symbolically

$$R_A = AD_A . \qquad 6.22$$

Then if A,B are operators with domains D_A and D_B and $D_B \supset R_A$ we can define the product operator BA according to

$$(BA)f = B(Af) . \qquad 6.23$$

This is well defined since by assumption

$$f \in D_A$$
$$\therefore \quad Af \in R_A \subset D_B$$

and hence

$$Af \in D_B .$$

Conversely if $D_A \supset R_B$ we can define the product

$$(AB)f = A(Bf) \; . \tag{6.24}$$

This points out the interesting possibility that although BA may exist as an operator AB might not, and conversely.

Another property which many operators of physical interest possess is hermiticity. Actually the interesting property is self-adjointness and we shall examine these two properties in some detail to bring out the difference. First we need some definitions.

Let A be an operator on $\mathcal{H}$ and $f \in D_A$; then consider the expression (g,Af).
If for some $g \in \mathcal{H}$ we find that

$$(g,Af) = (h,f) \tag{6.25}$$

for all $f \in D_A$ then we define the adjoint operator $A^\dagger$ of A by

$$h = A^\dagger g \tag{6.26}$$

and domain $D_{A^\dagger}$ the set of all g for which 6.25 holds.

In that case we can write

$$(g,Af) = (A^\dagger g,f) \; . \tag{6.27}$$

Note that h in 6.25 is defined uniquely by g if the domain D_A contains sufficiently many vectors. The precise statment of this is that D_A is <u>dense</u> in $\mathcal{H}$. For our purposes a set in $\mathcal{H}$ is dense if any element in $\mathcal{H}$ can be approximated arbitrarily closely by an element from this set. Thus, D_A is dense if for any $f \in \mathcal{H}$ there exists a $g \in D_A$ such that given $\varepsilon > 0$

$$\|f - g\| < \varepsilon \; .$$

In this case the proof that h is unique is trivial. For, assume there is another such vector h'. Then

$$(g, Af) = (h', f)$$

as well. Combining this with 6.25 we get

$$(h-h', f) = 0.$$

Thus, h-h' is orthogonal to every vector in D_A. But D_A is dense in H so that for any vector $\phi \in H$

$$|(h-h', \phi)| < \varepsilon \quad .$$

This is possible only if h = h'. The adjoint operator is also a linear operator as is immediately obvious.

Now again let A be an operator in H with domain D_A then A is <u>hermitian</u> if for all $f, g \in D_A$

$$(Af, g) = (f, Ag) \quad . \tag{6.28}$$

An operator A is <u>symmetric</u> if it is hermitian and its domain of definition D_A is dense in H.

From the definition of $A^\dagger$ it then follows that for symmetric A

$$D_A \subset D_{A^\dagger} \tag{6.29}$$

as we show for an example. If in addition

$$D_A = D_{A^\dagger} \tag{6.30}$$

or as this implies

$$A = A^\dagger \tag{6.31}$$

Then A is self-adjoint.

To make this less abstract consider the momentum operator

$$p = \frac{\hbar}{i}\frac{d}{dx} \tag{6.32}$$

defined on the Hilbert space $L^2(a,b)$ of functions square- integrable on the interval (a,b). As domain of this operator we choose

$$D_p = \{f \in L^2(a,b) \mid \frac{df}{dx} \text{ is bounded on } (a,b) \text{ and } f(a) = f(b) = 0\}. \tag{6.33}$$

With this definition it is easy to see that p is hermitian and in fact symmetric. The domain D_p is dense in $L^2(a,b)$. Thus if p is hermitian, it is symmetric. To see hermiticity let

$$f \in D_p \quad , \quad g \in D_p \ .$$

Then

$$\begin{aligned}(f,pg) &= \int_a^b f^*(x)\, \frac{\hbar}{i}\frac{dg}{dx}\, dx \\ &= \frac{\hbar}{i} f^*(x) g(x) \Big|_a^b + \int_a^b \left(\frac{\hbar}{i}\frac{df}{dx}\right)^* g(x) dx\end{aligned} \tag{6.34}$$

Since $f^*(a) = f^*(b) = g(a) = g(b) = 0$, the term obtained from integration by parts vanishes and so we have

$$(f,pg) = (pf,g) \ . \tag{6.35}$$

Thus p is hermitian (symmetric) as claimed. On the other hand $p \neq p^\dagger$ since, as we now show, the domain $D_{p^\dagger}$ of $p^\dagger$ is much larger than the domain D_p of p, ie. $D_p \subset D_{p^\dagger}$. To see this consider $g \in D_p$ and let f be any function whose derivative is bounded over (a,b) and such that

$$f(b) = e^{i\theta} f(a) \tag{6.36}$$

where θ is a constant. Then by a computation, identical to the one above, we again find

$$(f,pg) = (pf,g) .$$

Thus as a differential operator

$$p^\dagger = \frac{\hbar}{i}\frac{d}{dx} \tag{6.37}$$

but the domain of $p^\dagger$ is larger than the domain of p. It is furthermore easy to check that if we define

$$D_{p^\dagger} = \{f \in L^2(a,b) \mid \frac{df}{dx} \text{ is bounded on } (a,b) \text{ and }$$

$$f(b) = e^{i\theta} f(a)\}$$

then $p^\dagger$ is also symmetric. Thus we say that $p^\dagger$ is a symmetric extension of p. In fact one can check that $p^\dagger$ is self-adjoint because $D_{p^\dagger} = D_{p^{\dagger\dagger}}$. Thus we have a self- adjoint extension of p.

A symmetric operator A is <u>essentially self-adjoint</u> if $A^{\dagger\dagger}$ is self-adjoint. What this means is that although A itself is not self-adjoint there is a unique way to extend it to a self-adjoint operator. That $A^{\dagger\dagger}$ is an extension of A follows from

$$D_A \subset D_{A^\dagger} \subset D_{A^{\dagger\dagger}} . \tag{6.38}$$

The operator p discussed above is not essentially self-adjoint because, for each value of the parameter θ used to define $D_{p^\dagger}$, we get a <u>different</u> self-adjoint extension.

6.4 The Cayley Transform and Self-Adjoint Operators

We now examine under what conditions a general symmetric operator possesses self-adjoint extensions and how many. To do this we need some more machinery. The operation analogous to a rotation in a Euclidean space E_n is a unitary transformation in H. The characteristic property of a rotation in E_n is that it preserves length and angles or more succintly, it preserves the inner product. This is also its characteristic in H.

<u>Definition</u>

U is unitary iff $D_U = R_U = H$ and

$$(Uf, Uf) = (f,f) . \qquad 6.39$$

From this we immediately get

$$U^\dagger U = 1 \qquad 6.40$$

And since $D_U = R_U = H$ we also get

$$U U^\dagger = 1 . \qquad 6.41$$

Note, unlike the case for finite vector spaces, 6.40 does not imply 6.41 without the additional assumptions on U. We now show that A is self-adjoint iff the operator

$$U = (A - i\,1)(A + i\,1)^{-1} \qquad 6.42$$

called the Cayley transform of A is unitary.

Proof:

Suppose A is self-adjoint in $\mathcal{H}$ and $f \in D_A$. Then

$$\|Af \pm if\|^2 = (Af,Af) \pm i\,(Af,f) \mp i(f,Af) + (f,f)$$
$$= \|Af\|^2 + \|f\|^2 \,. \qquad 6.43$$

Therefore, $(A \pm i\,1)f = 0$ is only possible if $f = 0$. Thus, the operators $(A \pm i\,1)^{-1}$ and hence U exist. Furthermore, as we now show, the ranges R_{A+i1} and R_{A-i1} are dense in $\mathcal{H}$. For suppose g is orthogonal to all vectors in $R_{A\pm i1}$. Then for $f \in R_{A\pm i1}$ or equivalently, $f = (A\pm i1)h$ we have

$$0 = (g,f) = (g,Ah \pm ih) = (g,Ah) \mp (ig,h).$$

Thus

$$(g,Ah) = \pm(ig,h) \,.$$

So

$$g \in D_{A^\dagger} = D_A$$

and

$$A^\dagger g = Ag = \pm ig.$$

But as we have seen this is not possible unless $g = 0$. Thus the ranges $R_{A\pm i1}$ are dense in $\mathcal{H}$. We now prove that in fact

$$R_{A\pm i1} = \mathcal{H}. \qquad 6.44$$

Let $g \in H$, then since $R_{A \pm i1}$ is dense in H the limit $g_n = Af_n \pm if_n \to g$ exists. Also, using 6.43

$$\|g_n - g_m\|^2 = \|A(f_n - f_m) \pm i(f_n - f_m)\|^2$$

$$= \|A(f_n - f_m)\|^2 + \|f_n - f_m\|^2$$

and thus the f_n and Af_n converge to some vectors f and h respectively. Furthermore, because A is self- adjoint

$$f \in D_A$$

and

$$h = Af.$$

Hence by definition of g as the limit of g_n it follows that

$$g = Af \pm if \in R_{A \pm i1} .$$

Thus, the limit g of the approximating vector g_n itself belongs to $R_{A \pm i1}$. However, this limit may be any element in H. Hence

$$R_{A \pm i1} = H. \qquad 6.44$$

Thus, we have that

$$D_U = R_U = H. \qquad 6.45$$

Now choose any element f, then $f \in D_U$ and hence $f \in D_{(A+i1)^{-1}}$. Thus we can write

$$f = (A+i1)g$$

and

$$\begin{aligned} Uf &= (A-i1)(A+i1)^{-1}(A+i1)g \\ &= (A-i1)g. \end{aligned} \qquad 6.46$$

Theretore

$$\begin{aligned} \|Uf\|^2 &= \|(A-i1)g\|^2 = \|Ag\|^2 + \|g\|^2 \\ &= \|(A+i1)g\|^2 \\ &= \|f\|^2 \qquad \text{by } 6.47. \end{aligned}$$

Thus assuming $A = A^\dagger$ we conclude that $UU^\dagger = U^\dagger U = 1$. It is always possible to recover A from U according to

$$A = i(1 - U)^{-1}(1 + U) = i(1 + U)(1 - U)^{-1} . \qquad 6.48$$

We now prove the converse, that if U is unitary, A is self-adjoint.

Let $g \in D_{A^\dagger}$ and define $g^* = A^\dagger g$. Then for any $f \in D_A$

$$(g,Af) = (g^*,f) \qquad 6.49$$

But since $A = i(1 + U)(1 - U)^{-1}$, all $f \in D_A$ are of the form

$$f = (1 - U)h \qquad 6.50$$

where

$$h \in D_U = \mathcal{H}.$$

Therefore 6.49 reads

$$(g,i(1 + U)h) = (g^*,(1 - U)h) \qquad 6.51$$

for any $h \in \mathcal{H}$. Now since U is unitary and therefore defined everywhere and conserves inner products, we can replace (h,g) by (Uh,Ug) and (h,g^*) by (Uh,Ug^*) to get from 6.51

$$(Ug,iUh) + (g,iUh) - (Ug^*,Uh) + (g^*,Uh) = 0$$

or

$$(-iUg - ig - Ug^* + g^*,Uh) = 0.$$

Thus $-iUg - ig - Ug^* + g^*$ is orthogonal to all elements of $\mathcal{H}$ and hence vanishes

$$-iUg - ig - Ug^* + g^* = 0 . \qquad 6.52$$

From this we get

$$g = -ig^* - U(g - ig^*) . \qquad 6.53$$

We now perform some algebra. Thus,

$$g = \frac{g - ig^*}{2} - \frac{g + ig^*}{2} - U(g - ig^*).$$

Using 6.52 again this becomes

$$g = \frac{g - ig^*}{2} + \frac{1}{2}U(g - ig^*) - U(g - ig^*)$$

and hence

$$g = (1 - U)\frac{g - ig^*}{2} . \qquad 6.54$$

Similarly we get

$$g^* = i(1 + U)\,\frac{g - ig^*}{2}\,. \tag{6.55}$$

This proves to two things: If $g \in D_{A^\dagger}$ then

a) $g \in D_A$ according to 6.54, ie. it is in $D_{(1-U)^{-1}}$.

b) $Ag = g^* = A^\dagger g$ (6.56)

since

$$Ag = i(1 + U)(1 - U)^{-1}(1 - U)\,\frac{g - ig^*}{2}$$

$$= i(1 + U)\,\frac{g - ig^*}{2} = g^*\,.$$

This proves that $A^\dagger = A$ and hence A is self-adjoint. Before proceeding let us examine the reasons for our interest in self-adjointness. The examination will be, of necessity, somewhat cursory.

6.5. Some Properties of Self-Adjoint Operators

To begin, consider the operator A which means multiplying by a. Self-adjointness implies that for $f,g \in \mathcal{H}$

$$(f,ag) = (af,g). \tag{6.57}$$

But according to the definition of the inner product we have

$$(f,ag) = (a^*f,g)\,.$$

Thus

$$a = a^* \tag{6.58}$$

and hence a must be real. This is not a coincidence. In fact self-adjoint operators, in some sense which will become precise, correspond to real numbers. To make this precise we now discuss the eigenvalue problem for self-adjoint operators.

Again, let A be a self-adjoint operator; then there are certain vectors belonging to the domain of A on which operations by A are particularly simple. The operation involves multiplicaton by a number. We have already encountered this in our solution of the Schrödinger equation

$$H\psi_E = E\psi_E \ .$$

Here, operating with H on the vector ψ_E involves multiplying ψ_E by E. The vector ψ_E is called an <u>eigenfunction</u> of H belonging to the <u>eigenvalue</u> E.

More generally, f_j is an eigenfunction of the operator A belonging to the eigenvalue a_j if

$$Af_j = a_j f_j \ . \qquad 6.59$$

The <u>imporant properties</u> of self-adjoint operators are that

a) All eigenvalues of a self-adjoint operator are real.

b) Eigenvectors belonging to different eigenvalues are orthogonal.

c) The eigenvectors form a complete set.

We now prove a) and b). The eigenvalues of a self-adjoint operator are real and the eigenvectors belonging to different eigenvalues are orthogonal.

<u>Proof</u>.

Let f_1, f_2 be eigenfunctions of A belonging to the eigenvalues a_1 and a_2 respectively. Thus,

$$Af_1 = a_1 f_1 \qquad f_1 \in D_A \qquad 6.60$$

$$Af_2 = a_2 f_2 \ . \qquad f_2 \in D_A \qquad 6.61$$

From 6.59 we get

$$(f_1,Af_1) = (Af_1,f_1) = (f_1,a_1f_1) = (a_1f_1,f_1)$$

$$\therefore \quad a_1 = a_1^*$$

as required. Note the self-adjointness was necessary for otherwise we do not know that $f_1 \in D_{A^\dagger}$. Using 6.59 or 6.60 we get

$$(f_1,Af_2) = (Af_1,f_2)$$

$$\therefore \quad a_2(f_1,f_2) = a_1(f_1,f_2)$$

$$\therefore \quad (a_2-a_1)(f_1,f_2) = 0 .$$

Thus, if $a_1 \neq a_2$, then

$$(f_1,f_2) = 0 .$$

Thus, we have established the results.

The proof of completeness of the eigenfunctions of a self-adjoint operator is beyond the scope of this book. Consequently we only show a sort of converse which makes the result appear plausible. The general theorem is known as the "Spectral Theorem" and is discussed in detail in reference 6.1.

Let A be a linear operator with a <u>complete</u> orthonormal set of eigenvectors $\{f_n\}$ and corresponding set of <u>real</u> eigenvalues $\{a_n\}$, then A is self-adjoint. Thus, we have that if

$$Af_n = a_nf_n$$

$$(f_n,f_m) = \delta_{nm}$$

$\{a_n\}$ are real $\{f_n\}$ complete. Then $A = A^\dagger$.

Proof:

We must show that $D_A = D_{A^\dagger}$ and for $f,g \in D_A$

$$(Af,g) = (f,Ag).$$

The proof is based on knowing A on a basis (the eigenfunctions). Suppose $f,g \in D_A$. Then because $\{f_n\}$ are complete we have

$$f = \sum \alpha_n f_n$$

$$g = \sum \beta_n f_n$$

where

$$\alpha_n = (f_n, f)$$

$$\beta_n = (f_n, g) \ .$$

Suppose $f \in D_A$. Then

$$(f,Af) = \sum_{m,n} (\alpha_m f_m, A\alpha_n f_n)$$

$$= \sum_n a_n \alpha_n^* \alpha_n$$

$$= \sum_{m,n} (a_m \alpha_m f_m, \alpha_n f_n)$$

$$= \sum_{m,n} (A\alpha_m f_m, \alpha_n f_n)$$

$$= (Af,f) = (A^\dagger f,f) \qquad 6.62$$

Thus $f \in D_{A^\dagger}$ and in a similar manner we get $(g,Af) = (Ag,f)$. Thus, A is self-adjoint.

6.6 Classification of Symmetric Operators

We now complete the classification of symmetric operators. For the purposes of physics there is no need to distinguish between self-adjoint and essentially self- adjoint operators since the latter always have a unique and obvious extension to self-adjoint operators. We are mainly concerned in determining which symmetric operators have several self-adjoint extensions. Our main tool in this investigation will be the Cayley transform that we discussed previously.

Suppose A is an arbitrary symmetric operator. In that case U need not be unitary and the domain D_U and the range $R_U = UD_U$ need not coincide with the whole Hilbert space $\mathcal{H}$. If we consider the sets of vectors $D_U^\perp$ and $R_U^\perp$ orthogonal to D_U and R_U, their "size" gives us an indication of the extent to which U is not unitary and A is not self-adjoint. We call these subspaces the deficiency subspaces of A and their dimensions the deficiency indices. Thus, the deficiency indices of A are

$$(m,n) = (\dim D_U^\perp, \dim R_U^\perp) \; . \qquad 6.63$$

Now all elements in D_U are of the form

$$f = (A + i1)g.$$

So $\dim D_U^\perp$ is given by the number of linearly independent solutions of

$$Ag = -ig \qquad 6.64a)$$

belonging to $\mathcal{H}$.

Similarly all vectors in R_U are of the form $(A - i1)g$ and hence $\dim R_U^\perp$ is given by the number of linearly independent solutions of

$$Ag = +ig \qquad 6.64b)$$

belonging to H.

Now from our previous results we know that A is self-adjoint iff U is unitary and hence iff $R_U = D_U = H$. Thus, A is self-adjoint iff the deficiency indices are (0,0).

If the deficiency indices are (m,n), (m,n≠0) it is possible to extend the operator A as follows. Let two solutions of 6.64a and b be g_- and g_+ respectively. Then for $g \in D_A$ define the operator A' by

$$A'[g+\theta(g_+ + g_-)] = Ag + i\theta(g_+ - g_-) \qquad 6.65$$

Clearly A' is an extension of A since now $g_+ + g_-$ belongs to $D_{A'}$. Furthermore, the deficiency indices for A' are (m-1,n-1). One can proceed in this manner until one gets deficiency indices (r,0) or (0,r). In this case no further extension is possible. Such an operator has no self-ajoint extensions. If the deficiency indices are (1,1) we get a one-parameter family of self-adjoint extensions, and for deficiency indices (n,n) we get an n-parameter family of self-adjoint extensions.

Mathematically this is as far as one can go. To pick the "correct" extension in these cases depends on the physical situation and cannot be decided by mathematics. To illustrate these points we shall now discuss some examples.

Consider the momentum operator $p = \frac{\hbar}{i}\frac{d}{dx}$ on L^2. Let p be defined on the interval $(-\infty,\infty)$. In this case the equations for the deficiency indices read

$$\frac{\hbar}{i}\frac{df}{dx} = \pm\, if \qquad 6.66$$

or

$$\frac{df}{dx} = \mp \frac{1}{\hbar} f \,. \qquad 6.67$$

The solutions are

$$f = Ae^{\mp x/\hbar} \tag{6.68}$$

but neither of these solutions is square integrable on $(-\infty,\infty)$ and hence the deficiency indices are $(0,0)$ and defined over $-\infty < x < \infty$ p is self-adjoint.

Consider p defined on the interval $(0,\infty)$. As before the solutions are

$$f = Ae^{\mp x/\hbar} .$$

This time $Ae^{-x/\hbar}$ is square integrable. Thus, the deficiency indices are $(1,0)$ and defined over $0 \leq x < \infty$ p has no self-adjoint extensions. The reason for this is easy to see.

Consider

$$(f,pg) = \int_0^\infty f^* \frac{\hbar}{i}\frac{dg}{dx}\,dx$$

$$= \frac{\hbar}{i} f^* g\Big|_0^\infty + \int_0^\infty \left(\frac{\hbar}{i}\frac{df}{dx}\right)^* g\,dx$$

$$= \frac{\hbar}{i} f^* g\Big|_0^\infty + (pf,g) \; . \tag{6.69}$$

This requires that

$$f^* g\Big|_0^\infty = 0. \tag{6.70}$$

Now we know that both f^* and g vanish at $x \to \infty$. Therefore we require

$$f(0) = 0 \quad \text{or} \quad g(0) = 0. \tag{6.71}$$

In fact to make p self-adjoint requires $D_p = D_{p^\dagger}$ and hence both

$f(0) = g(0) = 0$. On the other hand, the "eigenfunctions" of p are

$$Ae^{ikx}$$

and only vanish at $x = 0$ if $A = 0$. Thus, p would have no eigenfunction if it were to be self-adjoint.

Finally, consider p defined on the interval $a \leq x \leq b$. In this case, both $Ae^{\mp x/\hbar}$ are square integrable and the deficiency indices are (1,1). The self-adjoint extensions depend on precisely one parameter as we already saw. In this case,

$$(f,pg) = \int_a^b f^* \frac{\hbar}{i} \frac{dg}{dx}\, dx$$

$$= \frac{\hbar}{i} f^* g\Big|_a^b + (pf,g) \quad . \qquad 6.72$$

Therefore for self-adjointness we require

$$f^*(b)g(b) = f^*(a)g(a) \qquad 6.73$$

$$(f(b)/f(a))^* = (g(a)/g(b))$$

$$f(b)/f(a) = g(b)/g(a) = e^{i\theta} \quad . \qquad 6.74$$

So θ is the parameter determining the different extensions. Actually, this is a specification of boundary conditions. Thus, if we choose $\theta = 0$ we have periodic boundary conditions. These are the most common. Note that once we have specified that the domain of p is the set of all square-integrable functions for $a \leq x \leq b$ (abbreviated $L^2(a,b)$) such that

$$f(b) = e^{i\theta} f(a); \qquad 6.75$$

then the deficiency indices become (0,0) and p is self- adjoint.

To give some idea of the physics behind this we state a theorem without proof and then use it.

Stone's Theorem

If A is a self-adjoint operator, then

$$U(\lambda) = e^{i\lambda A} \qquad 6.76$$

is a unitary operator for every real number λ. Furthermore

$$U(\lambda)U(\mu) = U(\lambda+\mu) \qquad 6.77$$

$$U(-\lambda) = U(\lambda)^{-1} = U(\lambda)^{\dagger} \qquad 6.78$$

Conversely given a set of continuous unitary operators satisfying 6.77 and 6.78 then there is a self-adjoint operator A such that 6.76 holds and A is given by

$$iA = \lim_{\varepsilon \to 0} \frac{1}{\varepsilon} [U(\varepsilon) - 1] \ . \qquad 6.79$$

Operators satisfying 6.77 and 6.78 form an algebraic structure called a group. We now construct such a unitary set of operators and use them for interpreting what we did.

Consider the operator

$$U(\lambda)f(x) = f(x+\lambda) \ . \qquad 6.80$$

Then,

$$U(0)f(x) = f(x) \qquad 6.81$$

and

$$U(\mu)U(\lambda)f(x) = U(\mu)f(x+\lambda) = f(x+\lambda+\mu) \ . \qquad 6.82$$

Thus, the operators defined by 6.80 satisfy 6.77 and 6.78 as we see by setting $\mu = -\lambda$ for a left inverse and $\lambda = -\mu$ for a right inverse. On the other hand, if $f(z)$ is analytic for $a \leq \mathrm{Re}\, z \leq b$ it has the Taylor expansion

$$f(x+\lambda) = \sum_{n=0}^{\infty} \frac{\lambda^n}{n!} \frac{d^n}{dx^n} f(x)$$

$$= \sum_{n=0}^{\infty} \left(\frac{i\lambda p}{\hbar}\right)^n \frac{1}{n!} f(x)$$

$$= e^{i\lambda p/\hbar} f(x) \qquad 6.83$$

Thus,

$$U(\lambda) = e^{i\lambda p/\hbar} \qquad 6.84$$

and we have succeeded in expressing $U(\lambda)$ in the form 6.76. It makes sense to call $U(\lambda)$ the translation operator since it "translates" functions by an amount λ. Furthermore we call p the generator of translations since for infinitesimal translations

$$U(\lambda) \underset{\lambda/\hbar \to 0}{\to} 1 + i \frac{\lambda}{\hbar} p \; . \qquad 6.85$$

In terms of these considerations we can understand why p has a one-parameter family of self-adjoint extensions on $L^2(a,b)$. Actually, our considerations depend on the fact that f is analytic for $a \leq \mathrm{Re}\, z \leq b$. This, however, is not a restriction since the functions square-integrable and analytic over $a \leq \mathrm{Re}\, z \leq b$ form a dense set in $L^2(a,b)$.

Suppose $f(x)$ is an infinitely differentiable function which is non-zero only on an interval completely contained in $a \leq x \leq b$. An

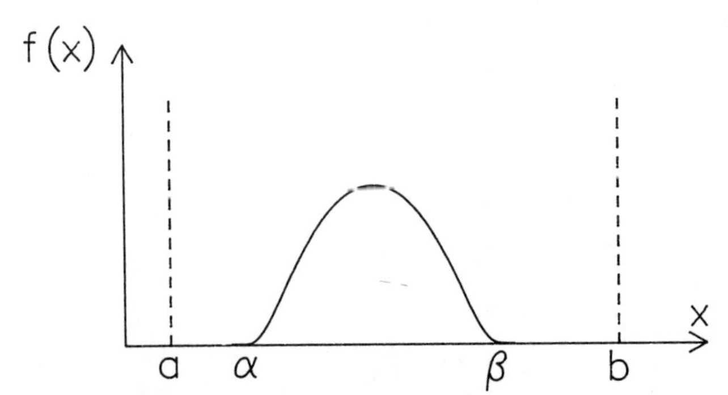

Fig. 6–1 A Wave–Packet of Compact Support

example of such a function is

$$f(x) = \begin{cases} 0 & x \leq \alpha \text{ and } x \geq \beta \\ \exp\left(-\frac{1}{x-\alpha} - \frac{1}{\beta-x}\right) & \alpha \leq x \leq \beta \end{cases} \qquad 6.86$$

where $a < \alpha < \beta < b$.

A picture of such a function is as shown in fig. 6.1. Now consider the probability density associated with f, ie. $\int_a^b |f(x)|^2 \, dx$.

We want the translation operator $U(\lambda)$ to be unitary. Thus, we need

$$\int_a^b |U^*(\lambda)f(x)|^2 \, dx = \int_a^b |f(x)|^2 dx \quad .$$

But,

$$U(\lambda)f(x) = f(x+\lambda)$$

and if $\lambda > b-\beta$ part of the wavefunction "disappears" past the right end point. To conserve the integral above requires that what disappears at the right must reappear from the left. Of course the phase of the function can be shifted in reappearing from the left. Furthermore, all functions will experience the same phase shift. Thus, if f_1 and f_2 were two such functions and if their phase shifts were θ_1 and θ_2, then translation of the function $f = f_1 + f_2$ would eventually give $f' = e^{i\theta_1} f_1 + e^{i\theta_2} f_2$. But

$$\int_a^b |f'|^2 \, dx \neq \int_a^b |f|^2 dx$$

unless $\theta_1 = \theta_2$. So the superposition principle limits the number of phase shift parameters to one.

Why does p then not have any self-ajoint extensions on $L^2(0,\infty)$? The answer is as follows. Translating x to the right will never take the function past the right end point. On the other hand, by translating to the left we can always bring the function past the left end point (the origin). In this case there is not anywhere from where the function can reappear to conserve probability and hence p cannot be made self-adjoint. This also explains why p is already self-adjoint on $L^2(-\infty,\infty)$.

Another extremely simple problem is the case of a particle in a strongly repulsive potential such as a quartic or cubic potential

$$V = -gx^n \qquad g > 0 , \quad n = 3,4 \quad . \tag{6.87}$$

In this case the hamiltonian is

$$H = \frac{p^2}{2m} + V = -\frac{\hbar^2}{2m}\frac{d^2}{dx^2} - gx^n . \tag{6.88}$$

By redefining the variables we can bring this to the form

$$H = -\frac{d^2}{dx^2} - ax^n \qquad a > 0 .$$

We want to examine this hamiltonian on $L^2(-\infty,\infty)$. To get a feel for the physical situation consider the problem classically. In the previous example we saw that the existence of different self-adjoint extensions depended on the fact that the particle can reach a boundary (a or b) and have to be transmitted or reflected.

In this case the boundaries are at $\pm\infty$. So we must examine if the particle can in fact reach these boundaries. Now suppose the particle starts at $x = 0$ with energy $E > 0$. Then classically its velocity v is given by

$$\frac{1}{2} mv^2 - gx^n = E. \tag{6.89}$$

or

$$v = \left(\frac{2E}{m} + \frac{2}{m} gx^n\right)^{1/2} .$$

Therefore the time to reach ∞ is

$$t = \int_0^\infty \frac{dx}{v} = \int_0^\infty \frac{dx}{\left(\frac{2E}{m} + \frac{2}{m} gx^n\right)^{1/2}} . \tag{6.90}$$

And for $n = 3$ or 4, $t < \infty$. Thus the particle reaches $+\infty$ in a finite time. To conserve probability it must be reflected and return to the origin in a finite time. Thus the time-averaged particle position is near the origin. We therefore expect to find that all eigenfunctions of this hamiltonian are square integrable and that the spectra of the self-adjoint extensions of H are discrete. This is in fact the case and this hamiltonian is analogous to a free particle hamiltonian on a finite interval (a,b), see problem 6.7. For the repulsive quartic potential the points $\pm\infty$ behave like the end points (a,b). So it is not surprising that the deficiency indicies are (2,2) for both cases.

For the repulsive cubic potential the situation is different. The particle can again reach $x = +\infty$ in a finite time but it can never reach $x = -\infty$. Thus we need only specify boundary conditions at $x = -\infty$. In this case the deficiency indices turn out to be (1,1).

The main point of this discussion is that whenever an operator, which is a candidate for representing an observable, is not self-adjoint but has self-adjoint extensions, then there are good physical reasons for this.

6.7 Spontaneously Broken Symmetry

The concept of spontaneously broken symmetries plays a very important role in some field theories of elementary particles. Since it fits naturally into the topics we have just discussed, we shall start by defining the concept and then proceed to illustrate it with an example.

Suppose we have some observable Q that commutes with the hamiltonian H. Then either Q corresponds to a discrete symmetry operation Q_D, such as parity, or else we can use Q to define a unitary operator

$$U(\alpha) = e^{i\alpha Q} . \qquad 6.91$$

It then follows from

$$[Q,H] = 0 \qquad 6.92$$

that

$$e^{i\alpha Q} H e^{-i\alpha Q} = H \qquad 6.93$$

or else for a discrete symmetry

$$Q_D H Q_D^{-1} = H . \qquad 6.94$$

We also require that the ground state of the hamiltonian ϕ_o should be invariant under either $U(\alpha)$ or Q_D. This means

$$U(\alpha)\phi_o = \phi_o \qquad 6.95$$

or

$$Q_D \phi_o = \phi_o \ . \qquad 6.96$$

<u>Definition</u>

A symmetry corresponding to an observable $Q(Q_D)$ is spontaneously broken if all the above statements except 6.95 (6.96) hold.

For this to occur requires that the ground state be degenerate, in itself an unusual phenomenon.

To illustrate this phenomenon we consider the hamiltonian

$$H = \frac{p^2}{2m} \qquad 6.97$$

defined on the interval $-a \leq x \leq a$. We furthermore pick for $p = \frac{\hbar}{i}\frac{d}{dx}$ the self-adjoint extension corresponding to the domain

$$D_p = \{f(x) \in C^1 | \ f(a) = -f(a)\} \ . \qquad 6.98$$

Thus, instead of periodic we pick "anti-periodic" boundary conditions.

The complete set of normalized eigenfunctions of this momentum operator are given by

$$f_n(x) = (2a)^{-1/2} e^{i\pi(n+1/2)\frac{x}{a}} \qquad n = 0,\pm1,\pm2,\ldots \qquad 6.99$$

with corresponding eigenvalues $\frac{\pi\hbar}{a}(n+1/2)$.

These wavefunctions have the following symmetry properties

$$f_n(x) = f_{-(n+1)}(-x) \qquad 6.100$$

$$f_n^*(x) = f_{-(n+1)}(x) \ . \qquad 6.101$$

Thus the parity operator P and the time-reversal operator T have the following action on them

$$(Pf_n)(x) = f_{-(n+1)}(x) \tag{6.102}$$

$$(Tf_n)(x) = f_{-(n+1)}(x) \quad . \tag{6.103}$$

The set of functions $\{f_n\}$ are also eigenfunctions of the hamitonian. In fact

$$Hf_n = \frac{\pi^2\hbar^2}{2ma^2}(n+ 1/2)^2 f_n \tag{6.104}$$

$$Hf_{-(n+1)} = \frac{\pi^2\hbar^2}{2ma^2}(n+ 1/2)^2 f_{-(n+1)} \quad . \tag{6.105}$$

Thus all eigenvalues including the ground state eigenvalue $\pi^2\hbar^2/8ma^2$ are <u>doubly degenerate</u>. We further see that although the hamiltonian H, the parity operator P, and the time-reversal operator T commute, the two ground states f_o and f_{-1} are not eigenstates of either the parity operator or the time-reversal operator.

$$(Tf_o)(x) = (Pf_o)(x) = f_{-1}(x) \tag{6.106}$$

$$(Tf_{-1})(x) = (Pf_{-1})(x) = f_o(x) \; . \tag{6.107}$$

Thus parity and time-reversal are spontaneously broken symmetries.

It is possible to restore these symmetries by defining states

$$g_n^+(x) = 2^{-1/2}[f_n(x) + f_{-(n-1)}(x)] = a^{-1/2}\cos(n+ \tfrac{1}{2})\frac{\pi x}{a} \tag{6.108}$$

$$g_n^-(x) = 2^{-1/2}i[-f_n(x) + f_{-(n-1)}(x)] = a^{-1/2}\sin(n+ \tfrac{1}{2})\frac{\pi X}{a}. \tag{6.109}$$

These are now simultaneous eigenstates of H, P and T

$$(Pg_n^{\pm})(x) = \pm g_n^{\pm}(x) \qquad 6.110$$

$$(Tg_n^{\pm})(x) = g_n^{\pm}(x) \qquad 6.111$$

$$Hg_n^{\pm} = \frac{\pi^2\hbar^2}{2ma^2}(n+1/2)^2 g_n^{\pm} \; . \qquad 6.112$$

Thus parity and time reversal are no longer broken symmetries. In this case, however, we have an even more surprising symmetry breaking, for although the momentum operator p and the hamiltonian $p^2/2m$ commute, these eigenstates of the hamiltonian are not eigenstates of the momentum operator. In fact,

$$pg_n^{\pm} = \mp \frac{\pi\hbar}{a}(n+1/2)\, g_n^{\mp} \qquad 6.113$$

so that in particular

$$pg_o^{+} = -\frac{\pi\hbar}{2a} g_o^{-} \qquad 6.114$$

$$pg_o^{-} = \frac{\pi\hbar}{2a} g_o^{+} \; . \qquad 6.115$$

In this case we therefore have translational symmetry spontaneously broken since the translation operator

$$U(\lambda) = e^{i\lambda p/\hbar} \qquad 6.116$$

does not leave the ground states $g_o^{\pm}$ invariant. In fact, by expanding

$$U(\lambda) = \sum_{n=0}^{\infty} \frac{1}{n!}\left(\frac{i\lambda p}{\hbar}\right)^n \qquad 6.117$$

and repeatedly applying 6.115 and 6.116 we get

$$U(\lambda)g_o^{\pm} = \cos(\frac{\pi\lambda}{2a})g_o^{\pm} \mp i \sin(\frac{\pi\lambda}{2a})\, g_o^{\mp} . \qquad 6.118$$

This demonstrates conclusively that the translational symmetry is broken.

We now relate the mathematical model we have displayed, to a definite physical system. If one considers a one-dimensional crystal consisting of only one type of atom, then the boundary condition in going from nearest neighbour to nearest neighbour is periodic. The situation repeats itself. Similarly for a one-dimensional crystal with alternating atoms (ABAB....), as in an antiferromagnet, the boundary condition from an atom to its next nearest neighbour is periodic, and hence from nearest neighbour to nearest neighbour anti-periodic.

We can now visualize the physical situation corresponding to our model and get a clearer understanding of the cause of the broken symmetry. If we consider such an antiferromagnetic crystal and consider the interval between nearest neighbours as fundamental, we must impose antiperiodic boundary conditions. Furthermore, since the end points correspond physically to different situations (atoms) it makes a difference whether a particle travels freely from left to right or right to left. The situations are not mirror images of each other and hence not eigenstates of the parity operator. Since time-reversal reverses the direction of travel, these states are also not eigenstates of the time-reversal operator.

One can take a superposition of states of particles travelling to the left and right, as we did, to get standing waves which are then automatically time-reversal as well as parity invariant. In this case, however, conservation of probability brings about a loss of translation invariance. It is clear now that this "unusual" self- adjoint extension of the momentum operator has just as physical an interpretation as the usual one with periodic boundary conditions.

It is perhaps also worth while to notice that the commutation relation

$$[x,p] = i\hbar \qquad 6.119$$

is not valid in this representation since for $f \in D_p$, $xf \notin D_p$ in general. In fact, in this case, $xf \in D_p$ only if $f(a) = f(-a) = 0$. Nevertheless, it is true that

$$[x^{2n},p] = i2n\hbar x^{2n-1} \qquad n = 0,1,2,\ldots \quad . \qquad 6.120$$

This concludes our mathematical treatment of self- adjointness. We now turn to a systematic analysis of the physical interpretation of quantum mechanics.

Notes, References and Bibliography

G.F. Simmons - Introduction to Topology and Modern Analysis - McGraw-Hill Book Co. (1963) Chapter 10.

A.S. Wightman - Chapter 8 in Cargèse Lecture in Theoretical Physics, Edited by M. Lévy - Gordon and Breach Inc. (1967).

A.H. Zemanian - Distribution Theory and Transform Analysis - McGraw-Hill Book Co. (1965) page 174.

J. von Neumann - Mathematical Foundations of Quantum Mechanics - Princeton University Press (1955).

T. Kato - Perturbation Theory for Linear Operators - Springer-Verlag, New York Inc. (1966).

This book gives a very concise and complete summary of useful Hilbert space techniques.

One of the standard treatises on functional analysis is:

6.1 F. Riesz and B.S.Z. Nagy - Functional Analysis, Frederick Unger Publishing Co., N.Y. (1955).

Chapter 6 Problems

6.1 Consider the set of functions $\{f_k(x) = e^{ikx}f(x)\ ,\ f(x) \in L^2\}$. Show that $\lim_{k\to\infty} (g,f_k) = 0 \qquad g \in L^2$ whereas $\|f_k\|^2 = \|f\|^2 \neq 0$. The above type of convergence of $f_k \to 0$ is called weak convergence as opposed to the notion of strong convergence defined in the text. Hint: Use the Riemann-Lebesgue Theorem - ref. 6.3.

6.2 Show that every Cauchy sequence in a finite dimensional vector space converges strongly (see problem 6.1).

6.3 Consider the operator $A = px^{2n+1} + x^{2n+1}p$ where $p = \frac{\hbar}{i}\frac{d}{dx}$ and $n = 1,2,\ldots$. Find the eigenvalues and eigenfunctions of A. What are the deficiency indices of A? The Hilbert space in this case is $L^2(-\infty,\infty)$.

6.4 A projection operator is a self-ajoint, non-negative operator P satisfying $P^2 = P$. Let ϕ_n be a normalized eigenfunction of a self-adjoint operator A with only discrete eigenvalues λ_n.

a) Show that the operator $P_n\psi = \phi_n(\phi_n,\psi)$ is a projection operator.

b) Show that A can be written

$$A\psi = \sum_n \int \lambda_n \phi_n(x)\phi_n^*(y)\psi(y)dy = \sum_n \lambda_n P_n \psi$$

This is called the spectral resolution of the operator A. Hint: Assume completeness of the eigenfunction.

6.5 Find the spectral resolution (see problem 6.4) of the operator

$$A = \begin{vmatrix} a_3 & a_1+ia_2 \\ a_1-ia_2 & -a_3 \end{vmatrix}$$

6.6 For any operator A the corresponding operator $R(z) = (A- z1)^{-1}$ is called the resolvent operator. Show that for any square matrix, A, $R(z)$ is analytic in z with poles at the eigenvalues of A.

6.7 Find the deficiency indices and hence all self- adjoint extensions of the hamiltonian

$$H = -\frac{\hbar^2}{2m}\frac{d^2}{dx^2}$$

defined on the interval (a,b).

Hint: It may be useful to express the boundary conditions on a function $f \in D_H$ in terms of 2-component quantities

$$F(a) = \begin{pmatrix} f(a) \\ f'(a) \end{pmatrix} \quad \text{and} \quad F(b) = \begin{pmatrix} f(b) \\ f'(b) \end{pmatrix}$$

and assume that $F(b) = UF(a)$ where U is a unitary 2×2 matrix.

6.8 Find the spectrum (all eigenvalues) of each of the self-adjoint extensions as well as the corresponding eigenfunctions for the Hamiltonian above.

Chapter 7

Physical Interpretation of Quantum Mechanics

At this stage we have developed all the formal mathematical machinery that we need. This does not mean we have developed all the techniques needed or useful in solving concrete problems. What we have developed is a machinery that allows us to form the physical interpretation of the theory in a precise and economical manner. We shall do this by displaying classical and quantum mechanics side by side.

In classical mechanics (C.M.) the state of a physical system is described by a point in phase space. Thus, specifying x and p for each one of a system of point particles specifies the system completely. For simplicity we shall consider systems consisting of only one point particle. Thus, in C.M. there is no distinction between specifying the values of certain observables (x,p) and specifying the state of a system. In quantum mechanics (Q.M.) the situation is quite different. Here, there is a definite distinction. We now state this in terms of a series of assumptions or axioms.

7.1 Assumption (A1) - Physical States

The state of a physical system is completely specified by a ray in

Hilbert space. A ray is a constant multiple of a vector. Conversely to every vector in Hilbert space corresponds the state of a physical system.

This last assumption needs to be modified in certain cases, called super-selection rules, with which we are not concerned. Also, these modifications do not occur for the physics we shall discuss so we shall ignore them. Notice, that A1 is a very strong assumption in that it states that the physical state is completely specified by the Hilbert space ray. Thus, if $f \in \mathcal{H}$, then f and cf both describe the same state and furthermore, everything that can in principle be determined about the state is contained in f. Furthermore, A1 says nothing about observables. This is contained in our next assumption, but first we must define observables. In C.M. an observable is any dynamical variable or any function of dynamical variables. In Q.M. the situation is again quite different. We define observables as follows.

An observable is any physical quantity whose "value" is obtained by a definite physical operation. Thus, the physical operation or method of measuring defines the observable. The measurement need not be performed but it must be possible in principle to perform the operation yielding the measurement. Furthermore, we shall assume that the measurement operation is ideal in the sense that all experimental errors are zero. It will also turn out that there is not a one-one correspondence between observables in C.M. and observables in Q.M. To be explicit we shall denote observables by fancy letters $\mathcal{A}$, $\mathcal{B}$, $\mathcal{C}$ etc.

Another point is worth mentioning. In C.M. we do not distinguish between the mathematical representation of an observable and the value of the observable. Thus $x(t)$ represents both the function and the value of the position at time t. In Q.M. the situation is quite different.

7.2 Assumption (A2) - Observables

Any physical observable $\mathcal{A}$ is represented in Q.M. by a self-adjoint operator A in the Hilbert space of physical states. Furthermore, since A is self-adjoint it possesses a complete set of eigenfunctions with a corresponding set of real eigenvalues

$$Af_j = a_j f_j \tag{7.1}$$

Also any measurement of $\mathcal{A}$ can yield as a <u>value</u> only one of the eigenvalues $\{a_j\}$ and no other number. Conversely to every self-adjoint operator A there corresponds a physical observable $\mathcal{A}$.

For simplicity we assume that all the eigenvalues a_j are distinct. In that case the f_j are orthogonal and we may as well consider them normalized. Thus, the f_j form a basis set in H called the <u>eigenbasis</u> of A. This means that any physical state f can be written as a linear superposition of the f_j

$$f = \sum (f_j, f) f_j \, . \tag{7.2}$$

This has important implications. Notice also that A being self-adjoint guarantees that the value of $\mathcal{A}$ is a real number, namely one of the a_j.

We now know that the possible outcome of a measurement is one of the a_j but we do not know which one or, if this cannot be stated, what is the probability for a given one. This is in fact the content of our next assumption.

7.3. Assumption (A3) - Probabilities

For an observable $\mathcal{A}$ as specified in (A2) and any physical state f (assumed normalized) the most detailed statement one can make regarding a measurement of $\mathcal{A}$ is that any one of the eigenvalues of A may occur and that the probability P_j that a given eigenvalue a_j occurs is given by

$$P_j = |(f_j, f)|^2 \, . \tag{7.3}$$

Notice that (f_j,f) is the coefficient of f_j in the expansion of f as a linear superposition of the $\{f_j\}$. Thus, writing

$$f = \sum \alpha_j f_j \qquad 7.4$$

we have

$$\alpha_j = (f_j,f) \; . \qquad 7.5$$

It is therefore appropriate to call α_j the probablity amplitude for observing the value a_j since this probability is given by

$$P_j = |\alpha_j|^2 \; . \qquad 7.6$$

This is of course the point at which Q.M. differs most radically from C.M. Thus, although we may know a state completely, we will nevertheless be unable, in general, to predict the outcome of an experiment to measure a given observable with certainty. Even worse if we make two separate but identical measurements on a system taking due care that in both cases the system is in exactly the same state before each measurement, the results of the two measurements will generally differ. The point is that states and observables are defined differently in Q.M. whereas a state in C.M. is defined in terms of observables.

We must now check that our interpretation is consistent. This requires some elementary considerations. In order that P_j be a probability we need

1) $0 \leq P_j \leq 1$
2) $\sum P_j = 1.$

The first follows trivially from Schwarz's inequality since

$$0 \leq |(f_j,f)|^2 \leq \|f_j\|^2 \|f\|^2 = 1.1 = 1.$$

The second follows from 7.2

$$f = \sum (f_j, f) f_j$$

$$\therefore \quad (f,f) = 1 = \sum (f_j, f)(f, f_j)$$

$$= \sum |(f_j, f)|^2 \ .$$

Although in general we can only make predictions of probabilities, there are occasions when predictions can be made with absolute certainty. For example, if in 7.6

$$\alpha_j = 0 \qquad j \neq k$$

and

$$\alpha_j = 1 \qquad j = k \ ,$$

we get that

$$P_k = 1$$

and

$$P_j = 0 \qquad j \neq k \ .$$

Thus, in this case the result of a measurement is certain to be a_k. Furthermore, using 7.4 we see that the state on which the measurement is performed is $f = f_k$.

Thus, if the system is in an eigenstate of an observable, then a measurement of that observable is certain to yield the eigenvalue corresponding to this eigenstate. This in no way implies that the states which are eigenstates of some observable $\mathcal{A}$ are more precisely specified than states which are not eigenstates of $\mathcal{A}$, because if $\mathcal{B}$ is another observable, the eigenstates of A (the operator corresponding to $\mathcal{A}$) are generally not eigenstates of B as well. Thus, although the value of $\mathcal{A}$ will be well-defined in this state the value of $\mathcal{B}$ will not. In fact we shall agree that an observable has a value only for states which are eigenstates of this observable. Of course if the state under consideration is "almost" an eigenstate then we may say that the observable has approximately the value corresponding to the predominant eigenstate. To make this more precise we introduce the concepts of average or expectation value of an observable and the root-mean-square (RMS) deviations of an observable.

The expectation or average value $\langle A\rangle$ of an observable A is given by summing all possible values multiplied by their probabilities. Thus, using assumption A2 we get

$$\langle A\rangle = \sum P_j a_j \tag{7.7}$$

and using 7.3 this becomes

$$\begin{aligned}\langle A\rangle &= \sum |(f_j,f)|^2 a_j \\ &= \sum (f,f_j)(f_j,f)a_j \\ &= \sum (f,Af_j)(f_j,f)\ .\end{aligned}$$

Since A is self-adjoint we can write this

$$\begin{aligned}\langle A\rangle &= \sum (Af,f_j)(f_j,f) \\ &= (Af,f) = (f,Af)\end{aligned}$$

where we have used the fact that the $\{f_j\}$ form a complete set. Thus

$$\langle A\rangle = (f,Af). \tag{7.8}$$

The R.M.S. deviation ΔA is defined by

$$(\Delta A)^2 = \langle (A - \langle A \rangle)^2 \rangle \tag{7.9}$$

$$= \langle A^2 - 2A\langle A \rangle + \langle A \rangle^2 \rangle$$

$$= \langle A^2 \rangle - 2\langle A \rangle\langle A \rangle + \langle A \rangle^2$$

$$\therefore \quad (\Delta A)^2 = \langle A^2 \rangle - \langle A \rangle^2 , \tag{7.10}$$

ΔA provides a measure of how spread out the "value" of $\mathcal{A}$ is. Thus, if f is an eigenstate say f_j

$$\langle A \rangle = (f_j, Af_j) = (f_j, a_j f_j)$$

$$= a_j(f_j, f_j)$$

$$= a_j .$$

Also as is obvious from the definition $\Delta A = 0$.

$$(\Delta A)^2 = \langle A^2 \rangle - \langle A \rangle^2$$

$$= \langle f_j, A^2 f_j \rangle - a_j^2$$

$$= a_j^2 - a_j^2 = 0 .$$

Thus, in an eigenstate the RMS deviation from the expectation value is zero. This means that $\mathcal{A}$ has a sharp value, and in this case the expectation value will coincide with the observed value. Thus, the expectation value corresponds to the average of a large number of identical measurements on systems in the same state prior to the measurement. It may be worth pointing out that the expectation value will not usually coincide with any of the actually measured values.

So far we have talked about performing measurements on systems in the same state. This implies that we have somehow "prepared" the systems in this state. We now show how to do this. For although two separate measurements of an observable made on systems in the same state do not generally yield the same result, we do not have complete chaos. In fact if we make a measurement on a system and then immediately repeat it before the system has a chance to evolve, the results of the two measurements are identical. This is in fact the content of our next assumption.

7.4 Assumption (A4) - Reduction of the Wave Packet

If the system is in a state specified by the wavefunction ψ and a measurement of an observable $\mathcal{A}$ is made yielding the value a_j, then immediately after the measurement the state of the system will be specified by the wavefunction f_j where f_j is an eigenfunction of the operator A corresponding to the eigenvalue a_j. Also A is the operator corresponding to the observable $\mathcal{A}$. Thus,

$$Af_j = a_j f_j \ .$$

The wording above was deliberate when we stated that f_j is an eigenfunction of A. This is because it is possible for an operator to have several eigenfunctions corresponding to one eigenvalue. In that case we say that the eigenvalue is degenerate and any linear combination of the eigenfunctions corresponding to this eigenvalue is also an eigenfunction. Thus, the wavefunction f_j in the assumption above may be any such linear combination if a_j is degenerate.

Suppose a_j is non-degenerate, then we know the state precisely and thus performing a measurement in quantum mechanics is the same as preparing a state. In fact, that is how we prepare states quantum mechanically. Thus, unless ψ is an eigenstate of A it is impossible to predict the result of a measurement of $\mathcal{A}$ prior to the measurement. So performing the measurement causes an uncontrollable change in the wavefunction. Thus, the single measurement cannot tell us anything about what the state of the system was prior to the measurement, it only tells us the state immediately after the measurement. This is quite different from what a measurement does in C.M.

In C.M. an observable always has a value and in principle we can measure its value without disturbing the system. Thus, in C.M. a measurement tells us both what the state of the system was before as well as what the state of the system is after the measurement.

7.5 Example

Consider the operator

$$\sigma_3 = \begin{pmatrix} 1 & 0 \\ 0 & -1 \end{pmatrix}$$

corresponding to an "observable" in a two-dimensional Hilbert space. Clearly the eigenvalues are ± 1 and the corresponding eigenvectors are

$$f_+ = \begin{pmatrix} 1 \\ 0 \end{pmatrix} \quad \text{and} \quad f_- = \begin{pmatrix} 0 \\ 1 \end{pmatrix} \ .$$

Suppose we are given a wave-function

$$\psi = \begin{pmatrix} \cos\theta \\ \sin\theta \end{pmatrix}$$

Then $(\psi,\psi) = \cos^2\theta + \sin^2\theta = 1$. So ψ is normalized. Also,

$$\psi = \cos\theta\, f_+ + \sin\theta\, f_- \ .$$

A measurement of the observable corresponding to σ_3 must yield either $+1$ or -1. The corresponding probabilities are $\cos^2\theta$ and $\sin^2\theta$. The expectation value is

$$\langle\sigma_3\rangle = (\psi,\sigma_3\psi) = \cos^2\theta - \sin^2\theta = \cos 2\theta$$

and can lie anywhere between -1 and $+1$ depending on θ. Thus if $\theta = 0$ or $\psi = f_+$, then

$$\langle\sigma_3\rangle = +1 .$$

If $\theta = \pi/2$ or $\psi = f_-$

$$\langle\sigma_3\rangle = -1 .$$

The R.M.S. deviation is given by

$$(\Delta\sigma_3)^2 = \langle\sigma_3{}^2\rangle - \langle\sigma_3\rangle^2 .$$

But

$$\sigma_3{}^2 = \begin{pmatrix} 1 & 0 \\ 0 & 1 \end{pmatrix} \quad \text{so} \quad \langle\sigma_3{}^2\rangle = 1 .$$

$$\therefore \quad (\Delta\sigma_3)^2 = 1 - \cos^2 2\theta$$

$$= \sin^2 2\theta .$$

Thus for $\theta = 0$ or $\pi/2$,

$$(\Delta\sigma_3)^2 = 0 .$$

This means that if ψ is an eigenstate, $\Delta\sigma_3 = 0$. On the other hand, for any value $0 < \theta < \pi/2$, σ_3 does not have a definite value but only a probability for obtaining $+1$ or -1 can be given. For example if $\theta = \pi/4$ we have a 50% chance of obtaining either $+1$ or -1 in a measurement. Incidentally σ_3 does correspond to a physical observable and we shall encounter this operator again later.

So far we have only considered measurements on a single observable and have found a profound difference between C.M. and Q.M. We now consider measurements of several different observables. This will tend to accentuate the difference between C.M. and Q.M. As stated before, in C.M. every dynamical variable has a definite value in every conceivable state of the system. On the other hand, in Q.M. a dynamical variable or observable has a definite value only if the system is an eigenstate of the corresponding operator. If we now consider two observables, they can both have sharp values iff the system is a simultaneous eigenstate of both of them. What does this entail?

Let $\mathcal{A}$ and $\mathcal{B}$ be two observable with the corresponding self-adjoint operators A and B having a common dense domain. Then if a measurement of $\mathcal{A}$ yielding a_n, is immediately followed by a measurement of $\mathcal{B}$ and a second measurement of $\mathcal{A}$, we say that $\mathcal{A}$ and $\mathcal{B}$ are compatible if the second value obtained for $\mathcal{A}$ is always the same as the first (namely a_n in this case). We saw an example of compatible operators in the case of a hamiltonian H with an even potential

$$V(x) = V(-x).$$

In that case the corresponding compatible observables were H and the parity operator P. In fact the following three statements are equivalent.

7.6. Compatibility Theorem and Uncertainty Principle

The following statements are equivalent for a pair of observables $\mathcal{A}$, $\mathcal{B}$.

1) $\mathcal{A}$ and $\mathcal{B}$ are compatible.
2) A and B possess a common eigenbasis.
3) A and B commute, that is $[A,B] \equiv AB - BA = 0$.

An example of two observables that are always compatible is $\mathcal{A}$ and $f(\mathcal{A})$. This is reasonable in view of how $f(\mathcal{A})$ is obtained.

We are now ready to discuss the - Heisenberg Uncertainty Principle

Suppose $\mathcal{A}$ and $\mathcal{B}$ are not compatible, then $[A,B] \neq 0$. This means that there exists at least one f belonging to the common domain of A and B such that

$$(AB - BA)f \neq 0 \ .$$

Now there are many pairs of observables (in fact all classically conjugate variables) for which the commutator is particularly simple, namely just a number. For example, using

$$p = \frac{\hbar}{i}\frac{d}{dx}$$

we have

$$[p,x]f = \frac{\hbar}{i}\frac{d}{dx}(xf) - \frac{\hbar}{i}x\frac{df}{dx}$$

$$= -i\hbar f$$

so that

$$[p,x] = -i\hbar \ . \qquad 7.11$$

For such observables their lack of compatibility is expressed by the Heisenberg uncertainty principle which states:

Let the state of a system be described by the wavefunction ψ and let $\mathcal{A}$ and $\mathcal{B}$ be two observables. Then the uncertainties (RMS deviations from the mean) in $\mathcal{A}$ and $\mathcal{B}$ satisfy the inequality

$$\Delta\mathcal{A}\,\Delta\mathcal{B} \geq \frac{1}{2}\left|(\psi,[A,B]\psi)\right| \ . \qquad 7.12$$

This is a rather formidable looking expression but if [A,B] is just a c-number (ordinary number rather than an operator which is called a q-number), then the expression simplifies. Thus if

$$[A,B] = c \ , \qquad 7.13$$

then 7.12 reads for a normalized ψ

$$\Delta A \Delta B \geq \frac{1}{2} |c| \ . \tag{7.14}$$

We shall derive 7.12 shortly but first let us consider its implications.

Classically observables are not operators and therefore the commutator is always 0, implying that all observables are compatible. Now for conjugate variables such as p and x in the example above, $c = i\hbar$ so that

$$\Delta p \Delta x \geq \hbar/2 \ . \tag{7.15}$$

Thus we see that in some sense classical mechanics corresponds to the limit $\hbar = 0$. Of course we know this already from the way Planck introduced $\hbar$ as the smallest lump of action possible.

In Q.M. according to 7.14 we see, however, that if we increase the precision with which a given observable is known, the corresponding non-compatible observable becomes more uncertain. As we shall see, for p and x this is strictly a property of Fourier transforms. However, it is one of the most outstanding features of Q.M. We now give a physical derivation of 7.15 using the so-called Heisenberg microscope and then we proceed to the mathematical derivation.

7.7 The Heisenberg Microscope

The experiment considered here is not an actual but rather a "Gedanken" or thought experiment: We have a microscope as shown below (figure 7.1) illuminated by light from the left. Suppose we are examining an electron represented by the little sphere. Also assume that the electron is at rest. We now try to determine its position as precisely as possible. We already know that the momentum is zero. Now in order to "observe" the electron, we must scatter at least one photon off it. This will impart some momentum to it and since the aperture of the

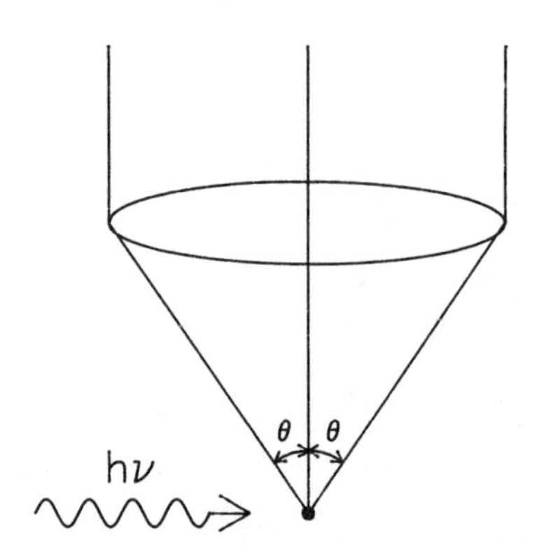

Fig. 7–1 Heisenberg Microscope

microscope is of finite width we do not know how much momentum was imparted. In fact the reflected photon can travel anywhere in the cone designated by the angle θ. If the wavelength of the photon is λ then the resolving power of the microscope is

$$\Delta x = \lambda/\sin\theta. \tag{7.16}$$

The momentum transferred to the electron (the uncertainty in the momentum) is given by

$$\Delta p \simeq p \sin\theta \tag{7.17}$$

where p is the momentum of the photon. But

$$p = h/\lambda$$

$$\therefore \quad \Delta x \Delta p \simeq h\ . \tag{7.18}$$

Notice that this "derivation" implies that this inaccuracy is intrinsic and cannot be decreased by getting better instruments. We now turn to the mathematical proof. It is based on the following properties:

1) A and B are self-adjoint

2) The expressions for ΔA and ΔB are

$$(\Delta A)^2 = \langle(A - \langle A\rangle)^2\rangle$$

$$(\Delta B)^2 = \langle(B - \langle B\rangle)^2\rangle$$

3) The Schwarz inequality,

$$|(f,g)|^2 \leq (f,f)(g,g)$$

The statement to be proven is:

$$\Delta A \Delta B \geq \frac{1}{2} |(\psi,[A,B]\psi)| \quad . \tag{7.12}$$

Define first the operators

$$A' = A - \langle A\rangle \tag{7.19}$$

$$B' = B - \langle B\rangle \tag{7.20}$$

These are still self-adjoint. Furthermore,

$$[A',B'] = [A,B] \quad .$$

All we have done is to subtract the mean values. Furthermore, we now have

$$(\Delta A)^2 = (A'\psi,A'\psi) \tag{7.21}$$

since

$$(A'\psi, A'\psi) = (\psi, A'^2\psi)$$

$$= (\psi, (A - \langle A\rangle^2\psi)$$

$$= \langle (A - \langle A\rangle)^2\rangle \ .$$

To obtain 7.12 we now just apply the Schwarz inequality after the following short computation

$$(\psi, [A,B]\psi) = (\psi, [A',B']\psi)$$

$$= (\psi, A'B'\psi) - (\psi, B'A'\psi)$$

$$= (A'\psi, B'\psi) - (B'\psi, A'\psi)$$

$$= (A'\psi, B'\psi) - (A'\psi, B'\psi)^*$$

$$= 2i \ \mathrm{Im}(A'\psi, B'\psi) \ .$$

Thus,

$$|(\psi, [A,B]\psi)| = 2|\mathrm{Im}(A'\psi, B'\psi)|$$

$$\leq 2|(A'\psi, B'\psi)|$$

and by Schwarz's inequality

$$\leq 2(A'\psi, A'\psi)^{1/2}(B'\psi, B'\psi)^{1/2}$$

$$= 2\Delta A \Delta B \ .$$

Thus,

$$\Delta A \Delta B \geq \frac{1}{2} \left|(\psi, [A,B]\psi)\right| \qquad 7.12$$

as stated.

We now proceed to discuss the time-evolution of a quantum state. In C.M. the evolution of a state is given by a set of first order differential equations, Hamilton's equations, and is therefore unique. The corresponding thing in Q.M. is Schrödinger's equation which is also first order in time and hence predicts the evolution of a state uniquely. Furthermore, the evolution of a state $\Psi(t,x)$ gives the time evolution of any observable's expectation value according to

$$\langle A \rangle_t = (\Psi(t,x), A\Psi(t,x)). \qquad 7.22$$

On the other hand, the "value" of the observable does not evolve in a completely predictable fashion. Just as in C.M. the hamiltonian plays a special role in determining the evolution so does the hamiltonian operator in Q.M. We now state this as our fifth and final assumption.

7.8 Assumption (A5) - The Schrödinger Equation

For every physical system there exists an observable, the total energy, to which there corresponds a self-adjoint operator H called the hamiltonian. The hamiltonian determines the time evolution of the system according to the time-dependent Schrödinger equation

$$i\hbar \frac{\partial \Psi(t,x)}{\partial t} = H\Psi(t,x) \qquad 7.23$$

provided the system is not disturbed.

Note, the last proviso states that if a measurement is made, 7.23 ceases to hold. This is due to the fact that making a measurement involves disturbing the system in an <u>essentially unpredictable</u> manner as we saw with the Heisenberg microscope. If the disturbance were predictable we could include it as some new, perhaps time-dependent, "potential" in the hamiltonian and thus in principle predict the outcome of all experiments. This is not possible, however.

In order that our interpretation of Q.M. as given by A1, A2, A3 and A4 be consistent with A5 we need that if Ψ is normalized at some instant, that it remain so. This implies that $(\Psi(t,x),\Psi(t,x))$ must be time-independent. But

$$\frac{d}{dt}(\Psi(t,x),\Psi(t,x)) = \left(\frac{\partial\Psi(t,x)}{\partial t},\Psi(t,x)\right) + \left((\Psi(t,x), \frac{\partial\Psi(t,x)}{\partial t}\right)$$

and using 7.23

$$= (-\frac{iH}{\hbar}\Psi(t,x), \Psi(t,x)) + (\Psi(t,x), -\frac{iH}{\hbar}\Psi(x,t))$$

$$= \frac{i}{\hbar}[(H\Psi,\Psi) - (\Psi,H\Psi)]$$

$$= \frac{i}{\hbar}[(\Psi,H\Psi) - (\Psi,H\Psi)] = 0$$

where we have used the fact that H is self-adjoint.

The way we have formulated the t-dependence via the Schrödinger equation allows us to consider t as a parameter labelling different states on $\mathcal{H}$. All that the Schrödinger equation does is to determine how one vector in $\mathcal{H}$ evolves into another. Thus, it is natural to look for an operator that performs this evolution. In fact, to preserve the inner product, the evolution operator must be unitary. Also if we compare 7.20 with our statement of Stone's Theorem we see that H is the generator of the evolution operator. Rather than use Stone's Theorem we shall rederive this result. Thus, assume that there exists a unitary operator $U(t,t_o)$ such that

$$\Psi(t,x) = U(t,t_o)\Psi(t_o,x) \tag{7.24}$$

where

$$U(t_o,t_o) = 1. \tag{7.25}$$

Substituting into 7.23 we get

$$[i\hbar \frac{\partial U(t,t_o)}{\partial t} - HU(t,t_o)]\Psi(t_o,x) = 0.$$

This must be true for all possible initial states $\Psi(t_o,x)$, and hence

$$i\hbar \frac{\partial U(t,t_o)}{\partial t} = HU(t,t_o) . \tag{7.26}$$

$$U(t,t_o) = \exp - \frac{iH}{\hbar}(t-t_o) \tag{7.27}$$

or

$$U(t,t_o) = 1 + \sum_{n=1}^{\infty} \frac{1}{n!} [- \frac{iH(t-t_o)}{\hbar}]^n \tag{7.28}$$

where 7.28 gives a definition of the function exp. Notice that $U(t,t_o)$ is <u>not</u> an observable since it is not self-adjoint, nevertheless it is a very physical object. We shall encounter this operator again when we examine the so-called transformation theory of Q.M.

7.9 Time Evolution of Expectation Values and Constants of the Motion

The expectation value of some observable A is given by

$$\langle A\rangle_t = (\Psi(t,x),A\Psi(t,x)) . \tag{7.29}$$

Hence

$$\frac{d}{dt}\langle A\rangle_t = \frac{d}{dt}(\Psi,A\Psi) = (\frac{\partial\Psi}{\partial t},A\Psi) + (\Psi,A\frac{\partial\Psi}{\partial t}) .$$

Using 7.23 this becomes

$$\frac{d}{dt}\langle A\rangle_t = \frac{i}{\hbar}[(H\Psi,A\Psi) - (\Psi,AH\Psi)]$$

$$= \frac{i}{\hbar}(\Psi,(HA-AH)\Psi)$$

where we have used the self-adjointness of H. Thus,

$$\frac{d}{dt}\langle A\rangle_t = \frac{i}{\hbar}(\Psi,[H,A]\Psi) \ . \qquad 7.30$$

This gives the evolution of the "mean value" of an observable.

Now suppose $[H,A] = 0$. Then 7.30 gives

$$\frac{d}{dt}\langle A\rangle_t = 0.$$

Thus the expectation value of A does not change in time and we can state that A is a constant of the motion. In a similar manner we can show that $\frac{d}{dt}(\Delta A)_t = 0$ (Problem 7.8). An important special case of this is the hamiltonian itself. Since $[H,H] = 0$ we immediately get that the hamiltonian is a constant of the motion. But the hamiltonian represents the total energy of the undisturbed system. Thus we have shown that for an undisturbed system, the total energy does not change with time.

7.10 Time-Energy Uncertainty Relation

The inequality 7.12 states that for classically conjugate variables (i.e, those satisfying $[p,q] = -i\hbar$) we have the Heisenberg uncertainty or indeterminacy relation

$$\Delta p \Delta q \geq \hbar/2. \qquad 7.22$$

Such a relation holds also for time and energy even though they do not satisfy a commutation relation

$$[E,t] = i\hbar. \tag{7.31}$$

This commutator might be conjectured from the fact that in the time-dependent Schrödinger equation the "energy operator" appears as $E = i\hbar \frac{\partial}{\partial t}$. As we now show equation 7.31 is wrong for any physical system. The reason for this is that the energy for a physical system must have a finite lower bound. Otherwise the system would wind up in the lowest state of negative energy (by radiating its energy away) and stay there forever. No perturbation would be strong enough to excite the system from $E = -\infty$ to a finite value. This being the case we must have $E \geq E_o$ and thus there can be no self-adjoint extensions for the operator t satisfying the commutator equation 7.31 [see problem 7.10]*. Hence time could not be an observable if equation 7.31 were to hold.

We now show directly that if there exists a finite number E_o (the lowest energy eigenvalue of H) such that $E \geq E_o$ then we obtain a contradiction from equation 7.31 if the Hamiltonian (total energy) is to be an observable.

Let ϕ_o be the energy eigenfunction corresponding to E_o

$$H\phi_o = E_o\phi_o . \tag{7.32}$$

Now pick any positive frequency ω and define the operator

$$b = i\omega t. \tag{7.33}$$

---

* If the energy is also bounded above, then t has a one-parameter family of self-adjoint extensions. In this case, however, the spectrum of the t-operator would be discrete.

Using equation 7.31 we find

$$Hb = bH - \hbar\omega \ . \tag{7.34}$$

Next we take expectation values with ϕ_o of both sides of equation 7.34 and use 7.32 as well as the fact that H is self-adjoint

$$(\phi_o, Hb\phi_o) = (\phi_o, bH\phi_o) - \hbar\omega(\phi_o, \phi_o) \tag{7.35}$$

$$(H\phi_o, b\phi_o) = E_o(\phi_o, b\phi_o) - \hbar\omega(\phi_o, \phi_o) \tag{7.36}$$

or

$$E_o(\phi_o, b\phi_o) = E_o(\phi_o, b\phi_o) - \hbar\omega(\phi_o, \phi_o) \ . \tag{7.37}$$

This implies $(\phi_o, \phi_o) = 0$. Thus either H does not have a lowest eigenvalue E_o and corresponding eigenfunction ϕ_o, or H is not self-adjoint, or equation 7.34 is false.

Since we want H to be self-adjoint and the energy bounded below we are forced to reject the commutator equation 7.34 and also 7.31.

In spite of the absence of a "time-energy commutation relation", a time-energy uncertainty relation holds. This relation is extremely useful and it is important to understand what it means. It is for this reason we have given such a lengthy discussion above. We now proceed to the derivation of this relation.

The "time" involved in the time-energy uncertainty relation is <u>not</u> the time parameter in the Schrödinger equation, instead it is the "time of a process" associated with an observable A. We therefore define the evolution time T_a of A by

$$T_a = \frac{(\Delta A)_t}{\left| \frac{d\langle A\rangle_t}{dt} \right|} \ . \tag{7.38}$$

To see what this means let us consider the change in the expectation value $\langle A\rangle_t$ of the observable A in a time interval Δt. This is

$$|\Delta\langle A\rangle| = |\langle A\rangle_{t+\Delta t} - \langle A\rangle_t|$$

$$\simeq \left|\frac{d\langle A\rangle}{dt}\right| \Delta t \ . \qquad 7.39$$

Now for the change $\Delta\langle A\rangle$ to be measurable requires that $\Delta\langle A\rangle$ be at least as large as the uncertainty $(\Delta A)_t$ in the observable A. This gives us the length of time Δt which we must wait to be able to "see" any change in $\langle A\rangle_t$. Equating $\Delta\langle A\rangle$ with $(\Delta A)_t$ we find

$$\Delta t = \frac{(\Delta A)_t}{\left|\frac{d\langle A\rangle}{dt}\right|} = T_a \ . \qquad 7.40$$

But this is just the time T_a. So T_a is the time required for the expectation value of A to change by an amount equal to the uncertainty in A.

Returning to the time-energy uncertainty relation we find according to 7.30 that

$$\left|\frac{d}{dt}\langle A\rangle_t\right| = \frac{1}{\hbar}|(\Psi,[H,A]\Psi)|$$

$$= \frac{1}{\hbar}|(\Psi,[H',A']\Psi)| \qquad 7.41$$

where as before

$$H' = H - \langle H\rangle$$

$$A' = A - \langle A\rangle \qquad 7.42$$

Then by the same sequence of steps used to obtain 7.12, we get:

$$\left| \frac{d\langle A \rangle_t}{dt} \right| \leq \frac{2}{\hbar} \Delta H (\Delta A)_t \qquad 7.43$$

Writing ΔE for ΔH and solving for T_a we find

$$\Delta E T_a \geq \hbar/2 \qquad 7.44$$

or writing Δt for T_a we have

$$\Delta E \Delta t \geq \hbar/2. \qquad 7.45$$

This is the famous time-energy uncertainty relation. Its meaning is clear from the derivation. Thus let ΔE be the uncertainty (RMS deviation from the mean) in the total energy and let Δt be the minimum time required for a measurable change to occur in a given observable evolving according to the Hamiltonian describing the total energy then the relation

$$\Delta E \Delta t \geq \hbar/2 \qquad 7.45$$

holds.

7.11 Time Evolution of Probability Amplitudes

We shall now derive an equation that will tell us how the probability amplitudes evolve in time. Since H is self-adjoint it has a complete set of eigenvectors $u_k(x)$. Assume these are normalized and therefore form a basis. Thus,

$$Hu_k = E_k u_k \qquad 7.46$$

where E_k are the eigenvalues of H and thus the allowed values of the total energy of the system. A given eigenstate then evolves according to

$$\Psi_k(t,x) = e^{-i\frac{E_k}{\hbar}t} u_k(x) \ . \tag{7.47}$$

Now suppose an observable A has an eigenvalue α_n and eigenstate $\phi_n(x)$. If the state of the system is $\Psi(t,x)$ we can write the probability amplitude for observing α_n as

$$a_n(t) = (\phi_n(x),\Psi(t,x)) \ . \tag{7.48}$$

Thus,

$$\Psi(t,x) = \sum_n a_n(t)\phi_n(x) \ . \tag{7.49}$$

7.47 is obviously a special case of this. From 7.48 we get

$$\begin{aligned}\frac{da_n(t)}{dt} &= (\phi_n \ , \frac{\partial\Psi}{\partial t}) \\ &= -\frac{i}{\hbar}(\phi_n,H\Psi) \\ &= -\frac{i}{\hbar}(\phi_n, H\sum_m a_m(t)\phi_m) \\ &= -\frac{i}{\hbar}\sum_m a_m(t)(\phi_n,H\phi_m) \ .\end{aligned}$$

Thus,

$$\frac{da_n(t)}{dt} = -\frac{i}{\hbar}\sum_m (\phi_n,H\phi_m)a_m(t) \ . \tag{7.50}$$

For an arbitrary observable A this is as far as we can go. If, however, A is the hamiltonian H so that $\phi_m = u_m$, then we get

$$\frac{d}{dt} a_n(t) = -\frac{i}{\hbar} \sum_m E_m \delta_{nm} a_m(t)$$

$$= -\frac{i}{\hbar} E_n a_n(t) .$$

So

$$a_n(t) = a_n(0)\, e^{-\frac{i}{\hbar} E_n(t)} . \qquad 7.51$$

This result follows of course immediately from 7.47 by recalling that $a_n(t) = (\phi_n, \Psi)$. In fact this is precisely how we previously introduced the concept of stationary state in connection with the Schrödinger equation.

We now use these results to examine the time development of observables that are constants of the motion. Thus, if A is a constant of the motion then

$$[A,H] = 0. \qquad 7.52$$

We have already found that this implies that $\langle A \rangle_t$ does not change in time and similarly $(\Delta A)_t$ is constant in time. Now since 7.52 is time independent and A and H are both self-adjoint with an assumed common dense domain, they have a common eigenbasis. Thus both

$$Hu_k = E_k u_k$$

and

$$A\phi_n = \alpha_n \phi_n .$$

We can by relabelling (or rearranging) arrange that

$$\phi_n = u_n$$

so that as stated the operators have a common eigenbasis. The probability of measuring α_n for a system in the state $\Psi(t,x)$ is as before given by

$$|a_n(t)|^2 = |(\phi_n,\Psi)|^2 = |(u_n,\Psi)|^2 \ . \tag{7.53}$$

Thus

$$\begin{aligned} |a_n(t)|^2 &= |(u_n,\Psi(0,x)e^{-iE_nt/\hbar})|^2 \\ &= |(u_n,\Psi(0,x))|^2 \\ &= |(\phi_n,\Psi(0,x))|^2 \end{aligned}$$

and is independent of t. This means that all measurements we can perform on A do not change with time and so we are truly justified in calling A a constant of the motion whenever $[A,H] = 0$.

We have now formulated non-relativistic quantum mechanics as a theory of linear operators on a Hilbert space. A predominant role is played by the hamiltonian operator. Thus a rule for obtaining this operator is desirable. Unfortunately aside from the considerations we gave in Chapter 3 not much more can be done in general. There are more formal ways to approaching this problem of making a classical observable into a self-adjoint operator on the Hilbert space of physical states. Basically the problem is as follows.

Find pairs of classically conjugate variables p_r, q_r. Then impose the formal algebraic relation

$$[p_r,q_s] = -i\hbar\ \delta_{rs} \ . \tag{7.54}$$

Now look for operators on a Hilbert space H such that these operators satisfy the relation 7.54 and are self-adjoint. Worded in a slightly different way this problem was considered by von Neumann who found the following result.

Except for unitary transformations (which amount to rotation of the Hilbert space and are thus of no physical consequence) the Schrödinger representation of the commutation relations 7.54 is unique. The Schrödinger representation is

$$p_r = \frac{\hbar}{i}\frac{\partial}{\partial x_r} \qquad 7.55$$

$$q_r = x_r \, . \qquad 7.56$$

Actually von Neumann showed that there is a slight generalization of 7.55 and 7.56 called a direct sum, but this is also of no importance to us. It essentially amounts to writing

$$p_r = \begin{pmatrix} \frac{\hbar}{i}\frac{\partial}{\partial x_r} & 0 \\ 0 & \frac{\hbar}{i}\frac{\partial}{\partial y_r} \end{pmatrix}$$

$$q_r = \begin{pmatrix} x_r & 0 \\ 0 & y_r \end{pmatrix}$$

or even larger diagonal matrices with different independent variables x,y,z etc. in each new row. We shall have no need for this. Now once we have the representation 7.55 and 7.56 we form operators as stated in Chapter 3, paying due attention to the possible ambiguities there considered.

Notes, References and Bibliography

D.T. Gillespie - A Quantum Mechanics Primer - International Textbook Co., Scranton, Pa. (1970).
This is a very readable little book describing the orthodox interpretation of quantum mechanics.

One of the best books is still:

P.A.M. Dirac - The Principles of Quantum Mechanics - Oxford at the Clarendon Press, 4th edition (1958).

A book containing numerous physical examples and written by one the the founders of the subject is:

W. Heisenberg - Physical Principles of the Quantum Theory - Dover Publishing, Inc., New York.

Chapter 7 Problems

7.1 A particle is in a state given at $t = 0$ by

$$\psi = \frac{1}{3}u_o(x) + \frac{i\sqrt{2}}{3}u_1(x) - \frac{\sqrt{6}}{3}u_2(x)$$

where u_o, u_1, u_2 are simple harmonic oscillator eigenfunctions corresponding to the energies $\frac{1}{2}\hbar\omega$, $\frac{3}{2}\hbar\omega$, and $\frac{5}{2}\hbar\omega$ respectively.

a) What is the most likely value of the energy that will be found in a single observation on this system? What is the probability of finding this value?

b) What is the average of the energy that would be obtained if the experiment in part a) could be repeated many times? What is the probability of getting this value?

c) A measurement of the energy yields a value $\frac{3}{2}\hbar\omega$. The measurement is immediately repeated. What is the resultant value of the energy? What is the wave-function immediately after the second measurement?

d) What is the wavefunction of the undisturbed system after a time t has elapsed?

7.2 A free particle is located at $x = a$ at $t = 0$ ie. its wavefunction at $t = 0$ is given by

$$\Psi(0,x) = \delta(x-a).$$

Find the wavefunction for $t > 0$. This solution is called the free particle propagator. Hint: To evaluate an integral of the form

$\int_{-\infty}^{\infty} e^{i\lambda x^2+i\beta x}dx$ pretend that λ is $\lambda+i\varepsilon$ with $\varepsilon > 0$ so that the integral is convergent. Then complete the square in the exponent of the exponential and change variables. The $i\varepsilon$ in your answer will allow you to decide whether to take the positive or negative square root. Finally let $\varepsilon \to 0$.

7.3 Minimum Uncertainty Wavefunction

We have seen (eqn. 7.15) that $\Delta p\Delta x \geq \hbar/2$. Assume $\langle x\rangle = \langle p\rangle = 0$. Now use the Schwarz inequality (eqn. 6.4)

$$\|f\|^2\|g\|^2 \geq |(f,g)|^2$$

and put $f = x\psi$, $g = p\psi$. Show that the inequality in the uncertainty will hold only if

$$p\psi = \lambda x\psi \quad \text{with } \lambda \text{ a constant}$$

and

$$(\psi,(xp+px)\psi) = 0.$$

Hence derive an equation for ψ and solve it explicitly.

7.4 A <u>free</u> particle of mass m is at $t = 0$ in a state described by

$$\psi(x,0) = [2\pi L^2]^{-1/4}e^{-(x/2L)^2}.$$

What is the wavefunction for an arbitrary time $t > 0$? Compute the uncertainties Δx and Δp as functions of time. This illustrates the "spreading" of a wave packet.

7.5 A free particle is, at $t = 0$, in a state described by the wave-function

$$\Psi(0,x) = A\sin^2\frac{\pi x}{a} \qquad |x| < a$$

$$\Psi(0,x) = 0 \qquad |x| > a$$

Find for $t > 0$ the following expectation values $\langle p\rangle_t$, $\langle x\rangle_t$, $\langle p^2\rangle_t$, $\frac{d}{dt}\langle xp+px\rangle_t$ and finally $\langle x^2\rangle_t$.

Hint: Use equation 7.30. If you try to evaluate these results using the time-dependent wave-function $\Psi(t,x)$ you will get some impossible integrals.

7.6 Consider a particle under the influence of a hamiltonian

$$H = \frac{p^2}{2m} + V(x)$$

so that

$$i\hbar \frac{\partial \Psi}{\partial t} = H\Psi \; .$$

Show that if $\langle x \rangle$ is the "center of mass" of the wave packet and $\langle p \rangle$ the average momentum of the particle then

$$\frac{d\langle x \rangle}{dt} = \frac{1}{m} \langle p \rangle$$

and

$$\frac{d}{dt} \langle p \rangle = \langle F(x) \rangle = -\langle \frac{dV}{dx} \rangle \; .$$

These are known as the Ehrenfest Equations. To be equivalent to Newton's equations requires that $\langle F(x) \rangle = \langle F(\langle x \rangle) \rangle$. Discuss under what circumstances this condition is approximately valid.

7.7 Prove the compatibility theorem.

7.8 Show that if $[H,A] = 0$ then ΔA does not change in time.

7.9 Use the solution to problem 3.8 to answer the following questions for a free electron described by a Gaussian wavepacket

a) What is $\langle x \rangle$ at any time?

b) What is Δx as a function of time? This is known as the spreading of a wave packet. If a packet corresponding to an electron with 20 keV energy has a width of 100 Å at $t = 0$, what is its width after travelling 100 m?

7.10 Consider the space of functions of E belonging to $L_2(0,\infty)$. The relationship

$$[E,t] = i\hbar \qquad 7.31$$

can be represented on this space by

$$Ef(E) = Ef(E)$$

$$tf(E) = -i\hbar \frac{\partial f(E)}{\partial E}$$

Show that the operator t so defined can not be an observable, ie. that it has no self-adjoint extensions. This proves that if the energy has a lower bound, a relationship such as 7.31 can not hold if time is to be an observable.

Chapter 8

Distributions, Fourier Transforms, and Rigged Hilbert Spaces

In this chapter we develop some more mathematical tools. Again these tools are not important for computational purposes, but they are important as a justification for the calculations performed in practice. Furthermore in areas such as quantum field theory where it is not known whether the difficulties encountered are due to bad mathematics, or bad physics, or both it is important to ensure that the mathematics, at least, is correct. To this end, we give here a brief introduction to some of the results of modern analysis. The presentation, although at a submathematical level is intended for the more mathematically inclined student. We give definitions and theorems, but the proofs for the theorems are only sketched, or omitted altogether. To compensate for this we list some relevant references.

8.1 Functionals

A function may be considered as a mapping from a certain well-defined set of numbers called the domain into another set of numbers called the range. Thus if f denotes a function then $f(x)$ denotes the value of the function at the point x. This distinction is not always made but clearly there is such a distinction. We shall now consider a mapping whose domain is a set of functions called test functions and whose range is the set of real numbers. Such a mapping is called a functional. Thus if T is a functional then $T(f)$ is the <u>value</u> of the functional at the function f. Thus the arguments of functionals are functions. From the class of all possible functionals we pick out a particularly simple class, namely the <u>linear functionals</u>. Thus T is linear if for f and g belonging to the domain of T and a,b two numbers

$$T(af + bg) = aT(f) + bT(g). \tag{8.1}$$

An example of such a functional is

$$T(f) = \int_{-\infty}^{\infty} t(x)f(x)dx \tag{8.2}$$

where $t(x)$ is a fixed function and $f(x)$ is in the domain of T if the right hand side is convergent. Furthermore a functional T is bounded if for all f in a given space $T(f) \leq c\|f\|$ where c is a constant. There is a remarkable theorem for bounded linear functionals on a Hilbert space.

<u>Riesz Representation Theorem</u>

Let $\mathcal{H}$ be a Hilbert space and T a <u>bounded linear functional</u> on $\mathcal{H}$. Then there exists a uniquely determined vector f_T of $\mathcal{H}$ such that

$$T(g) = (f_T, g) \tag{8.3}$$

for all $g \in \mathcal{H}$. Conversely of course any vector $f \in \mathcal{H}$ defines a bounded linear functional T_f by

$$T_f(g) = (f,g) \ .$$

Proof

The proof is rather straightforward and is a proof by construction. Uniqueness is obvious. For suppose f' is another vector besides f_T satifying 8.3, then

$$(f' - f_T, g) = 0$$

for all $g \in H$. Thus $f' - f_T = 0$ as desired. To prove that f_T exists consider the null space N_T of T where

$$N_T = \{g \in H \; ; \; T(g) = 0\} \; . \qquad 8.4$$

If $N_T = H$ take $f_T = 0$. This is the trivial case. Now assume $N_T \neq H$. Then there exists at least one vector $f_o \neq 0$ belonging to $N_T^{\perp}$, the orthogonal complement of N_T. In this case define

$$f_T = (T(f_o)^* / \|f_o\|^2) f_o \; . \qquad 8.5$$

This is the desired f_T as we now prove. Suppose $g \in N_T$. Then

$$T(g) = 0 = (f_T, g) \; .$$

Next if g is of the form

$$g = \alpha f_o$$

then we have

$$(f_T, g) = (f_T, \alpha f_o) = \alpha T(f_o) = T(\alpha f_o) = T(g)$$

as required. We now show that any $g \in H$ can be written

$$g = \alpha f_o + \beta f_1 \qquad 8.6$$

where $f_1 \in N$. To prove this recall that

$$T(f_T) \neq 0 .$$

Thus we have the identity

$$g = \left(g - \frac{T(g)}{T(f_T)} f_T\right) + \frac{T(g)}{T(f_T)} f_T \qquad 8.7$$

which is of the form 8.6. Thus since T is linear we have completed our proof and shown that

$$T(g) = (f_T, g)$$

for all $g \in \mathcal{H}$. This shows that on a Hilbert space the only linear functionals are those given by inner products. We want to extend this notion somewhat. Thus it is natural that we must go beyond the concept of Hilbert space.

In general to define a space we must have a criterion for deciding when two points of the space are "close". This criterion defines the <u>topology</u> of the space. Thus in the finite dimensional vector spaces E_n we use the Euclidean norm $(x_1^2 + x_2^2 + \ldots + x_n^2)^{1/2}$ to measure closeness. In Hilbert space we use the norm

$$\|f\| = (f,f)^{1/2}$$

to measure closeness. For functions one also frequently uses point-wise estimates of the form $|f(x)-g(x)|$. All of these criteria are useful. For functionals one also has an estimate which is derived by analogy with 8.2. Thus if T and S are bounded linear functionals, meaning that there are constants c, c' such that

$$T(f) \leq c\|f\|$$

$$S(f) \leq c'\|f\|$$

for all f in a given space X, then T and S are "close" if

$$T(f) - S(f).$$

is small. Here $\|f\|$ denotes the appropriate norm in X. Thus the notion of "close" (or topology) of the linear functionals on X is derived from the topology of X itself.

Dual Space

Let X be a space of functions with a given topology. Now consider the set X' of all bounded linear functionals on X. Then X' is itself a linear vector space with the topology of X' determined by the topology of X. We call X' the dual space of X.

An example of these concepts is Hilbert space itself. Thus the dual of $\mathcal{H}$ is $\mathcal{H}$ itself. In fact it is logically correct to consider the inner product on a Hilbert space as being formed by elements from two spaces, the Hilbert space $\mathcal{H}$ and its dual , which is of course a copy of $\mathcal{H}$.

The point of all this is that one can take linear functionals that are as singular or pathological as one wishes if it is possible to find a space of functions sufficiently nice to compensate for these pathologies. The space of nice functions is called the space of test-functions. There are many test-function spaces. One of the most useful of these is the Schwartz space $\mathcal{S}$. Its dual space is called $\mathcal{S}'$, the space of tempered distributions and is sufficiently general to encompass almost any kind of "function" we shall encounter. To describe these spaces we need some more terminology.

A function with continuous derivatives up to and including the n'th is called C^n. Thus continuous functions are called C^o. If a function is C^n for all n it is called C^∞ . Using this terminology we can define $\mathcal{S}$ as the space of all C^∞ functions which together with their derivatives vanish at infinity faster than the inverse of any polynomial. To make this more explicit we define the sequence of norms

$$\|f\|_{r,n} = \sup_{x} \left| x^r \frac{d^n f}{dx^n} \right| \tag{8.8}$$

where sup means "least upper bound". In that case f belongs to iff

$$\|f\|_{r,n} < \infty$$

for all integers r,n. This specifies the topology or notion of $\mathcal{S}$ closeness in $\mathcal{S}$. Thus for example a sequence $\{f_j\}$ of functions in converges to f if for each r and n

$$\lim_{j\to\infty} \|f_j - f\|_{r,n} = 0 \ .$$

In terms of this the <u>tempered distributions</u> also have a topology whose definition can be made very similar to the ε,δ definition for ordinary functions. Thus T is continuous at f_o if given an $\varepsilon > 0$ there exist integers r,n and a $\delta > 0$ such that for

$$\|f - f_o\|_{r,n} < \delta \tag{8.9}$$

we have

$$|T(f) - T(f_o)| < \varepsilon \ . \tag{8.10}$$

One way to ensure that 8.10 follows from 8.9 is to insist that for all r,n there exists a constant c such that

$$|T(f)| \leq c\|f\|_{r,n} \tag{8.11}$$

since then

$$|T(f) - T(f_o)| = |T(f-f_o)| \leq c\|f-f_o\|_{r,n} \ .$$

As a matter of fact, there is a theorem that states that every continuous linear functional T on $\mathcal{S}$ satisfies 8.11. Thus one can use 8.11 to define the topology on $\mathcal{S}'$.

To prepare us for future applications we introduce two more notations for distributions. To specify the value of T at f we have used T(f). We can also write this (T,f). This does not mean we have an inner product, it is simply another way of writing T(f). As a matter of fact physicists carry this even a step further and write this as

$$\int_{-\infty}^{\infty} T(x)f(x)dx \; . \qquad 8.12$$

Again this is a purely symbolic way of writing T(f) and does not imply that any integral such as 8.12 exists in any of the usual senses of integral. Nevertheless, the notation 8.12 is extremely suggestive and thus if applied with due caution one may treat this as an integral.

The most common of the distributions so treated is the so-called δ function. It is defined by

$$\delta(f) = f(0) \; . \qquad 8.13$$

On the other hand we frequently write this as

$$\int_{-\infty}^{\infty} \delta(x)f(x)dx = f(0) \; . \qquad 8.14$$

It is an easy matter to prove that no function δ with the property 8.14 can exist, see von Neumann's book (ref. 6.4) "Mathematical Foundations of Quantum Mechanics, pages 23-25". However, if we realize that 8.14 does not imply a genuine integral or function and is just another way of writing 8.13 then all objections to 8.14 are removed. That δ is not a function can also be seen from the fact that although functions may be multiplied by functions to give functions, distributions cannot generally be multiplied by distributions or functions. Thus if we consider the product of 1/x and $\delta(x)$ this is not defined in general. Nevertheless there is a smaller domain for which this product makes sense. An even more acute example is the product

$$\delta'(x)\delta(x)$$

where $\delta'(x)$ is the derivative of $\delta(x)$.

We now define the differentiation of distributions. In fact the definition is given by analogy with integration by parts using 8.12. Thus we define

$$\frac{d^nT}{dx^n}(f) = (-1)^n T\left(\frac{d^nf}{dx^n}\right) . \qquad 8.15$$

This expression is obviously well-defined for all $T \in \mathcal{S}'$ since if $f \in \mathcal{S}$ so is $\frac{d^nf}{dx^n} \in \mathcal{S}$.

In the notation 8.12 the definition of the derivative reads

$$\int_{-\infty}^{\infty} \frac{d^nT}{dx^n} f(x)dx = (-1)^n \int_{-\infty}^{\infty} T(x) \frac{d^nf}{dx^n} dx . \qquad 8.16$$

It is a simple matter to generalize these results to test functions of several variables and distributions over these variables. Thus if $f(x_1,x_2,\ldots,x_k)$ is an element of $\mathcal{S}^{(k)}$ in each variable, then we may have a distribution T in the dual space $\mathcal{S}^{(k)'}$ such that $T(f)$ is well defined. Again another possible symbolic notation for $T(f)$ would be

$$T(f) \int T(x_1,x_2,\ldots,x_k)f(x_1,x_2,\ldots,x_k)dx_1\ldots dx_k . \qquad 8.17$$

We emphasize once more that although 8.17 looks like an integral it is not. This is simply a symbolic way of writing $T(f)$. Nevertheless we shall use this way of writing almost all the time since it is the standard notation for physicists.

8.2 Fourier Transforms

Consider the linear transformations F and $\bar{F}$ defined on S according to

$$(Ff)(p) = \frac{1}{\sqrt{2\pi}}\int_{-\infty}^{\infty} e^{-ip.x}f(x)dx \equiv F(p) \qquad 8.18$$

$$(\bar{F}F)(x) = \frac{1}{\sqrt{2\pi}}\int_{-\infty}^{\infty} e^{ip.x}F(p)dp \qquad 8.19$$

Clearly 8.18 defines a uniformly and absolutely convergent integral since $e^{-ip.x}$ can only improve the convergence. We shall now prove that F and $\bar{F}$ map S onto S in a continuous one to one manner. The proof will give us as a side benefit the formal result

$$\frac{1}{(2\pi)}\int_{-\infty}^{\infty} e^{ip(x-y)}dp = \delta(x-y) . \qquad 8.20$$

Consider the expression

$$\lim_{\varepsilon\to 0+} \frac{1}{2\pi}\int_{-\infty}^{\infty} e^{ip.x}\, dp\, e^{-\varepsilon p^2}\int_{-\infty}^{\infty} dy\, e^{-ip.y}f(y). \qquad 8.21$$

Now for $\varepsilon > 0$ both integrals exist and we may interchange their order. Furthermore

$$\int_{-\infty}^{\infty} e^{ip.x-\varepsilon p^2}\, dp = \sqrt{\frac{\pi}{\varepsilon}}\, e^{-x^2/4\varepsilon} .$$

Thus we get for 8.21

$$\lim_{\varepsilon\to 0} \frac{1}{\sqrt{4\pi\varepsilon}}\int_{-\infty}^{\infty} e^{-(x-y)^2/4\varepsilon}\, f(y)dy . \qquad 8.22$$

Now consider a circle $(x-y)^2 = R^2$. Clearly due to the factor

$$e^{-(x-y)^2/4\varepsilon}$$

any contribution to the integral 8.22 from points outside the circle vanishes in the limit as $\varepsilon \to 0$. Thus we can estimate the difference between $f(x)$ and 8.22 by

$$\left| \frac{1}{\sqrt{4\pi\varepsilon}} \int_{|x-y|\leq R} e^{-(x-y)^2/4\varepsilon} f(y)-f(x)dy \right|$$

$$\leq \sup_{|x-y|\leq R} |f(x)-f(y)| \underset{R\to 0}{\longrightarrow} 0.$$

This justifies 8.20 and shows that $\bar{\mathcal{F}}\mathcal{F} = 1$. Using 8.20 we now also get $\mathcal{F}\bar{\mathcal{F}} = 1$. Also we have that $\mathcal{F}$, $\bar{\mathcal{F}}$ map $\mathcal{S}$ onto $\mathcal{S}$ as stated.

Now, in mapping $\mathcal{S}$ onto a copy of itself using $\mathcal{F}$ what happens to $\mathcal{S}'$? In order to keep things well-defined, $\mathcal{S}'$ must be mapped onto $\mathcal{S}'$. Using the symbolic notation 8.12 this is trivial. Since $\mathcal{F}$ is a unitary operator on Hilbert space we have for $f,g \in \mathcal{H}$ that

$$(\mathcal{F}f, \mathcal{F}g) = (f,g) \quad . \tag{8.23}$$

This is known as Parseval's theorem and written out reads

$$\int_{-\infty}^{\infty} F^*(p)G(p)dp = \int_{-\infty}^{\infty} f^*(x)g(x)dx \tag{8.24}$$

or

$$\frac{1}{\sqrt{2\pi}} \int_{-\infty}^{\infty} F^*(p)dp \int_{-\infty}^{\infty} e^{-ip.x} g(x)dx$$

$$= \frac{1}{\sqrt{2\pi}} \int_{-\infty}^{\infty} g(x)dx \int_{-\infty}^{\infty} e^{-ip.x} F^*(p)dp$$

where we have used the formulae defining F and G in terms of f and g and vice-versa. 8.24 is already in the desired form to define the Fourier transform of distributions. Thus suppose $g \in \mathcal{S}'$ then 8.24 reads

$$(\mathcal{F}g)(F^*) = g((\mathcal{F}F)^*) . \qquad 8.25$$

Thus we define the Fourier transform of distributions in $\mathcal{S}'$ using 8.25. In other words if $T \in \mathcal{S}'$ then the Fourier transform $\mathcal{F}T$ is defined by

$$(\mathcal{F}T)(f) = T(\mathcal{F}f) \qquad 8.26$$

where $f \in \mathcal{S}$.

It is now a simple matter to use 8.26 to show that the Fourier transform maps $\mathcal{S}'$ onto $\mathcal{S}'$.

8.3 Rigged Hilbert Space

To motivate the use and definition of rigged Hilbert spaces, we begin by considering the following eigenvalue problem on $L_2(-\infty,\infty) = \mathcal{H}$.

$$pu_k(x) = \hbar k u_k(x) . \qquad 8.27$$

Since

$$p = \frac{\hbar}{i}\frac{d}{dx}$$

we get

$$u_k(x) = \frac{1}{\sqrt{2\pi}} e^{ikx} . \qquad 8.28$$

Now the operator p corresponds to the physical observable, momentum and hence the eigenvalue problem 8.27 has a definite physical meaning. It tells us what the possible results of measurements of p are and is also supposed to give the probability amplitude for obtaining a given measurement. Nevertheless the "eigenfunctions" $u_k(x)$ are not square-integrable and hence do not belong to our Hilbert space. This is an undesirable situation. It can of course be obviated by forming wave packets. However the plane waves 8.28 are particularly convenient for practical calculations and we would be loath to have to give up using them. Thus we are tempted to enlarge our state vector space beyond Hilbert space. Actually this also provides many simplifications in the analysis of operators. However we shall not study that aspect.

To show one possible extension we first note that $u_k(x) \in \mathcal{S}'$ if we define them as follows

$$u_k(f) = \int_{-\infty}^{\infty} u_k(x)f(x)dx \quad . \tag{8.29}$$

Why did we choose $\mathcal{S}'$? The reasons are mainly technical. Thus $\mathcal{F}\mathcal{S}'$ is again $\mathcal{S}'$ and this is desirable. Actually other spaces of distributions may be used, but for the sake of concreteness we concentrate only on $\mathcal{S}'$. Now how does considering u_k as elements of $\mathcal{S}'$ help? To answer this we start with a definition.

Let A be a linear operator in $\mathcal{S}$. This means that A is also a linear operator in $\mathcal{H}$. In fact if A has a dense domain in $\mathcal{S}$ it has a dense domain in $\mathcal{H}$ since $\mathcal{S}$ is dense in $\mathcal{H}$. To see this consider the hermite functions $H_n(x)e^{-x^2/2}$. All of these are in $\mathcal{S}$ and any element in $L_2(-\infty,\infty)$ can be approximated by linear combinations of these functions. Thus $\mathcal{S}$ is dense in $\mathcal{H}$.

Now given such an operator A then $T \in \mathcal{S}'$ is called a <u>generalized eigenvector</u> of A corresponding to the eigenvalue λ if

$$T(Af) = \lambda T(f) \tag{8.30}$$

for all $f \in \mathcal{S}$.

Notice that by definition $u_k(x)$ is a generalized eigenvector of the momentum operator $p = \frac{\hbar}{i}\frac{d}{dx}$ since in the notation 8.12 we have

$$\int_{-\infty}^{\infty} \frac{1}{\sqrt{2\pi}} e^{ikx} \frac{\hbar}{i}\frac{df}{dx}\, dx = -\hbar k \int_{-\infty}^{\infty} \frac{1}{\sqrt{2\pi}} e^{ikx} f(x)dx \qquad 8.31$$

for $f \in \mathcal{S}$. We have simply "integrated" by parts. Thus we can now legitimately consider functions such as $u_k(x)$ as generalized eigenvectors. We still have to tie this together with the concept of Hilbert space. One more example is in order first.

Formally, the eigenvalue problem

$$xg_a(x) = ag_a(x) \qquad 8.32$$

has the solution

$$g_a(x) = \delta(x - a) \ . \qquad 8.33$$

Clearly $\delta(x-a)$ is not square integrable and hence is not in $\mathcal{H}$. But for $f \in \mathcal{S}$ we have

$$\int_{-\infty}^{\infty} \delta(x-a)xf(x)dx = a \int_{-\infty}^{\infty} \delta(x-a)f(x)dx.$$

Thus $\delta(x-a)$ is a generalized eigenvector of the position operator x.

We now define our rigged Hilbert space.

We are given the space $\mathcal{S}$. On $\mathcal{S}$ are defined a countable sequence of norms $\|f\|_{n,r}$. We now also define on $\mathcal{S}$ an inner product which coincides with the L^2 inner product

$$(f,g) = \int_{-\infty}^{\infty} f^*(x)g(x)dx \ .$$

Now as stated $\mathcal{S}$ is dense in L^2 and we identify $\mathcal{H}$ with L^2. Thus $\mathcal{S}$ is identified as a subset of $\mathcal{H}$. Together with $\mathcal{S}$ and $\mathcal{H}$ we consider the space S'. The triplet of spaces

$$S \, , \; H \, , \; S'$$

form a rigged Hilbert space. It is usually denoted by

$$S \subset H \subset S' \; .$$

The advantage of the symbolic notation 8.6 is now obvious. Thus "inner products" exist between elements of H and H and elements of S and S'. We do not form inner products between elements of S' and S'.

This is all about rigged Hilbert spaces that we shall need. It is sufficient to provide a justification of most of the manipulations that we shall carry out. Further details are readily available in the references.

From now on we shall proceed as if "functions" like $\delta(x)$ and e^{ikx} were elements of H. To justify our manipulations we can always fall back on the concept of rigged Hilbert spaces, but we shall not explicitly do so. As stated at the beginning, this chapter was simply to show that our formal manipulations can be fully justified.

Notes, References and Bibliography

The standard reference on Rigged Hilbert Spaces is:

I.M. Gel'fand, M.I. Graev and N.Ya. Vilenkin - Generalized Functions Vol. 4 - Academic Press, N.Y. (1964) Chapters I.3 and I.4

There are numerous other books on distribution theory of which we list several below:

I.M. Gel'fand and N.Ya. Vilenkin - Generalized Functions Vol. 1 - Academic Press, N.Y. (1964).

A.H. Zemanian - Distribution Theory and Transform Analysis - McGraw-Hill Book Co. (1965).

H. Bremmerman - Distributions, Complex Variables, and Fourier Transforms - Addison-Wesley Publishing Co., Inc., Reading Mass. (1965).

The original reference for the subject is:

L. Schwarz - Théorie des Distributions, Hermann, Paris, Part I (1957), Part II (1959).

Chapter 8 Problems

8.1 Show that T is a tempered distribution if T is defined by

$$T(f) = \sum_{k=0}^{m} \int F_k(x) \frac{d^k f(x)}{dx^k} dx$$

where F_k are continuous functions bounded by

$$|F_k(x)| \le C_k(1+ |x|^j)$$

for some C_k and j depending on k.

As a matter of fact every tempered distribution can be written in this form. Symbolically one then writes

$$T = \sum_{k=0}^{m} (-1)^k \frac{d^k}{dx^k} F_k(x) .$$

This formula cannot be taken literally however since the $F_k(x)$ need not be differentiable. It arises from a formal integration by parts of the first equation above.

8.2 The test function space $\mathcal{D}$ consists of the space of C^∞ functions of bounded support. The support of a function f, (supp f) is the complement of the largest open set on which the function vanishes. Show that if

$$\tilde{f} \in F\mathcal{D}$$

then $\tilde{f}$ is an entire function.

8.3 Prove the Theorem: The Fourier transform of a tempered distribution is a C^∞ function bounded by a polynomial. HINT: To prove that it is bounded by a polynomial use the result of problem 8.1.

8.4 Let $f(z)$ be an entire function vanishing rapidly at large $|\mathrm{Re}\, z|$. Show that

$$\lim_{\varepsilon\to 0+} \frac{1}{2}\int_{-\infty}^{\infty} \left(\frac{1}{x-a+i\varepsilon} + \frac{1}{x-a-i\varepsilon}\right) f(x)dx$$

$$= P\int_{-\infty}^{\infty} \frac{f(x)}{x-a}\, dx$$

where the principle value integral is defined by

$$P\int_{-\infty}^{\infty} \frac{f(x)}{x-a}\, dx = \lim_{\varepsilon\to 0+} \left[\int_{-\infty}^{a-\varepsilon} \frac{f(x)}{x-a}\, dx + \int_{a+\varepsilon}^{\infty} \frac{f(x)}{x-a}\, dx\right].$$

Furthermore show that

$$\lim_{\varepsilon\to 0} \frac{\varepsilon/\pi}{x^2+\varepsilon^2} = \delta(x) \ .$$

Hence conclude that considered as distributions

$$\lim_{\varepsilon\to 0+} \frac{1}{x-a\pm i\varepsilon} = P\,\frac{1}{x-a} \mp i\pi\delta(x-a) \ .$$

i.e.

$$\lim_{\varepsilon\to 0+} \int_{-\infty}^{\infty} \frac{f(x)dx}{x-a\pm i\varepsilon} = P\int_{-\infty}^{\infty} \frac{f(x)dx}{x-a} \mp i\pi f(a) \ .$$

8.5 Using the result of problem 8.4 and defining

$$\frac{1}{2\pi}\int_{-\infty}^{\infty} e^{ikx}\, dk \equiv \lim_{\varepsilon\to 0+} \left[\frac{1}{2\pi}\int_{0}^{\infty} e^{ik(x+i\varepsilon)}dk + \frac{1}{2\pi}\int_{-\infty}^{0} e^{ik(x-i\varepsilon)}\, dk\right] .$$

Prove that $\frac{1}{2\pi}\int_{-\infty}^{\infty} e^{ikx}\, dk = \delta(x)$.

8.6 Let $f(k)$ be a C^∞ function bounded by a polynomial. Show that

$$F(z) = \int_0^\infty f(k)e^{ikz}dk$$

is an entire function for $\text{Im } z > 0$. Using this and the result of the Theorem proved in problem 8.3 show that every tempered distribution is the boundary value of an analytic function.

Chapter 9

Algebraic Approach to Time-Independent Problems

Having developed all the tools we need we now turn to "practical" applications of these concepts. The first problem we reconsider is the simple harmonic oscillator.

We then move on to consider the rigid rotator in one and three dimensions. This allows us to introduce the concept of angular momentum. The angular momentum eigenvalue problem is solved analytically as well as algebraically. We then discuss rotational invariance.

The algebraic solution of the angular momentum eigenvalue problem yields half odd-integral as well as integral eigenvalues. These half odd-integral eigenvalues correspond to a new degree of freedom of a particle, called spin. This is the final topic of discussion in this chapter.

9.1 Simple Harmonic Oscillator

The hamiltonian is

$$H = \frac{p^2}{2m} + \frac{1}{2} kx^2 \ . \tag{9.1}$$

Set

$$\omega = \sqrt{k/m} \ . \tag{9.2}$$

As always we have

$$[x,p] = i\hbar \ . \tag{9.3}$$

Defining,

$$P = \frac{1}{\sqrt{m\hbar\omega}} \, p \tag{9.4}$$

$$Q = \sqrt{\frac{m\omega}{\hbar}} \, x \tag{9.5}$$

we get

$$H = \frac{1}{2} \hbar\omega (P^2 + Q^2) \ . \tag{9.6}$$

Then the commutator between P and Q is

$$[P,Q] = -i. \tag{9.7}$$

So far as we haven't done anything. Unlike the previous time we solved this problem, we shall now solve it using operator algebra techniques. To this end we introduce two non-hermitian operators

$$a = \frac{1}{\sqrt{2}} (Q + iP) \tag{9.8}$$

$$a^\dagger = \frac{1}{\sqrt{2}} (Q - iP). \tag{9.9}$$

Recall that Q and P are both self-adjoint on our Hilbert space which is $L_2(-\infty,\infty)$. Now consider the commutator

$$[a,a^\dagger] = \frac{1}{2}[Q + iP, Q - iP]$$

$$= \frac{1}{2}\{[Q,Q] + i[P,Q] - i[Q,P] + [P,P]\}$$

$$= \frac{1}{2}(0 + 1 + 1 + 0) = 1 .$$

Thus,

$$[a,a^\dagger] = 1. \qquad 9.10$$

Solving 9.8 and 9.9 we get

$$Q = \frac{1}{2}(a + a^\dagger) \qquad 9.11$$

and

$$P = \frac{i}{\sqrt{2}}(a^\dagger - a). \qquad 9.12$$

$$\therefore \quad H = \frac{1}{2}\hbar\omega(P^2 + Q^2)$$

$$= \frac{1}{4}\hbar\omega[(a + a^\dagger)^2 - (a^\dagger - a)^2]$$

$$= \frac{1}{4}\hbar\omega[a^2 + aa^\dagger + a^\dagger a + a^{\dagger 2} - a^{\dagger 2} + a^\dagger a + aa^\dagger - a^2]$$

or

$$H = \frac{1}{2}\hbar\omega(a^\dagger a + aa^\dagger) . \qquad 9.13$$

Using 9.10 we can rewrite H to read

$$H = \hbar\omega(a^\dagger a + \tfrac{1}{2}) = \hbar\omega(aa^\dagger - \tfrac{1}{2}) \ . \tag{9.14}$$

Furthermore

$$[H,a] = -\hbar\omega a \tag{9.15}$$

and

$$[H,a^\dagger] = \hbar\omega a^\dagger \ . \tag{9.16}$$

Now since P and Q are self-adjoint it follows from 9.6 that H is a non-negative self-adjoint operator. Thus all the eigenvalues of H are ≥ 0. In fact they are positive.

Thus let ψ_o be the lowest eigenstate of H. That is, ψ_o is the state corresponding to the smallest eigenvalue. Let this eigenvalue be E_o. Then,

$$H\psi_o = E_o\psi_o \ . \tag{9.17}$$

And hence

$$aH\psi_o = E_o a\psi_o \ .$$

But

$$aH = Ha + [a,H]$$

$$= Ha + \hbar\omega a \ .$$

$$\therefore \quad (Ha + \hbar\omega a)\psi_o = E_o a\psi_o$$

and hence

$$H(a\psi_o) = (E_o - \hbar\omega)a\psi_o \ . \tag{9.18}$$

Thus clearly we have that either

$$a\psi_o = 0 \qquad 9.19$$

or $a\psi_o$ is another eigenfunction of H corresponding to the eigenvalue $E_o - \hbar\omega$. On the other hand, E_o is the lowest eigenvalue by assumption. Thus $a\psi_o = 0$. From 9.19 we also get, by multiplying by $h\omega a^\dagger$ from the left, the equation

$$\hbar\omega a^\dagger a\psi_o = 0 \ .$$

But,

$$\hbar\omega a^\dagger a = H - \frac{1}{2}\hbar\omega \ .$$

$$\therefore \quad (H - \frac{1}{2}\hbar\omega)\psi_o = 0$$

or

$$H\psi_o = \frac{1}{2}\hbar\omega\psi_o \ . \qquad 9.20$$

Thus,

$$E_o = \frac{1}{2}\hbar\omega \ .$$

Now operate on 9.20 once more with $a^\dagger$ to get

$$a^\dagger H\psi_o = \frac{1}{2}\hbar\omega a^\dagger\psi_o \ .$$

Using 9.16 this gives

$$(Ha^\dagger - \hbar\omega a^\dagger)\psi_o = \frac{1}{2}\hbar\omega a^\dagger\psi_o \ .$$

Thus,

$$H(a^\dagger\psi_o) = (1 + \frac{1}{2})\hbar\omega(a^\dagger\psi_o) \ . \qquad 9.21$$

So $a^\dagger\psi_o$ is another eigenfunction corresponding to the eigenvalue $(1 + \frac{1}{2})\hbar\omega$. This shows that $a^\dagger$ is a "raising operator"; it raises the eigenvalue of the hamiltonian by one unit. Hence it is called a raising or step-up operator. We also call it a "creation operator" since it creates a quantum of energy $\hbar\omega$. If we repeat this procedure n times we get

$$H((a^\dagger)^n\psi_o) = (n + \tfrac{1}{2})\hbar\omega((a^\dagger)^n\psi_o) \ . \tag{9.22}$$

Thus in this manner we get a whole sequence of eigenfunctions. We now find $\psi_o(x)$ explicitly. We have

$$a\psi_o(x) = 0 \ .$$

Writing this out we get

$$(Q + iP)\psi_o(x) = 0$$

or

$$\left(\sqrt{\frac{m\omega}{\hbar}}\, x + \sqrt{\frac{\hbar}{m\omega}}\, \frac{d}{dx}\right) \psi_o = 0 \ .$$

This can be written

$$\left(\frac{d}{dx} + \frac{k}{\hbar\omega}\, x\right)\psi_o(x) = 0 \ . \tag{9.23}$$

The solution of this equation is

$$\psi_o(x) = Ae^{-kx^2/2\hbar\omega} \ . \tag{9.24}$$

To normalize ψ_o we compute

$$(\psi_o,\psi_o) = 1 = |A|^2 \int_{-\infty}^{\infty} e^{-kx^2/\hbar\omega}\, dx$$

$$= |A|^2\sqrt{\frac{\pi\hbar\omega}{k}} \ .$$

Thus choosing the phase of A we get

$$A = \left(\frac{k}{\pi\hbar\omega}\right)^{1/4}$$

and hence

$$\psi_o(x) = \left(\frac{k}{\pi\hbar\omega}\right)^{1/4} e^{-kx^2/2\hbar\omega} \; . \qquad 9.25$$

Futhermore, note that the solution is <u>unique</u>. Why do we emphasize this? The reason is that it allows us later to conclude that we have found <u>all</u> the eigenfunctions. First, however, we shall find the proper normalization for all the $\psi_n(x)$. Thus we have

$$\psi_n(x) = c_n(a^\dagger)^n\psi_o(x) \; . \qquad 9.26$$

The c_n is included for normalization, that is, to make

$$(\psi_n,\psi_n) = 1.$$

Then,

$$\begin{aligned}(\psi_n,\psi_n) = 1 &= \left|\frac{c_n}{c_{n-1}}\right|^2 (a^\dagger\psi_{n-1},a^\dagger\psi_{n-1}) \\ &= \left|\frac{c_n}{c_{n-1}}\right|^2 (\psi_{n-1},aa^\dagger\psi_{n-1}) \\ &= \left|\frac{c_n}{c_{n-1}}\right|^2 \frac{1}{h\omega}(\psi_{n-1}, (H+\frac{1}{2}h\omega)\psi_{n-1}) \\ &= \left|\frac{c_n}{c_{n-1}}\right|^2 n(\psi_{n-1},\psi_{n-1}) \; .\end{aligned}$$

Therefore

$$|c_n|^2 = \frac{1}{n}|c_{n-1}|^2 . \tag{9.27}$$

Using $c_o = 1$ we can solve this recursion relation to get

$$c_n = (n!)^{-1/2}$$

where we have made an arbitrary choice of phase. Thus

$$\psi_n(x) = (n!)^{-1/2}\left[\sqrt{\frac{m\omega}{2\hbar}}\ x - \sqrt{\frac{\hbar}{2m\omega}}\ \frac{d}{dx}\right]^n \psi_o(x)$$

or

$$\psi_n(x) = (-1)^n (n!)^{-1/2} \left(\frac{\hbar\omega}{2k}\right)^{n/2} \left(\frac{d}{dx} - \frac{k}{\hbar\omega}\ x\right)^n \left(\frac{k}{\pi\hbar\omega}\right)^{1/4} e^{-kx^2/2\hbar\omega} . \tag{9.28}$$

These are the hermite functions. They form a complete orthonormal set. Thus since $\psi_o(x)$ is unique, we have all the eigenfunctions of H.

It is a simple matter to check the orthogonality. Thus consider first

$$(\psi_o, \psi_1) = c_1(\psi_o, a^\dagger \psi_o)$$

$$= c_1(a\psi_o, \psi_o) = 0.$$

Similarly

$$(\psi_o, \psi_n) = c_n(\psi_o, a^{\dagger n}\psi_o)$$

$$= c_n(a\psi_o, a^{\dagger n-1}\psi_o) = 0.$$

If we now consider (ψ_n, ψ_m) with $n < m$ we are led to consider expressions of the form

$$(a^{\dagger n}\psi_o, a^{\dagger m}\psi_o) = (aa^{\dagger}a^{\dagger n-1}\psi_o, a^{\dagger m-1}\psi_o)$$

$$= \frac{1}{\hbar\omega}((H + \frac{1}{2}\hbar\omega)a^{\dagger n-1}\psi_o, a^{\dagger m-1}\psi_o)$$

$$= n(a^{\dagger n-1}\psi_o, a^{\dagger m-1}\psi_o).$$

Repeating this we get eventually

$$(a^{\dagger n}\psi_o, a^{\dagger m}\psi_o) = n!(\psi_o, a^{\dagger m-n}\psi_o)$$

$$= n!(a\psi_o, a^{\dagger m-n-1}\psi_o)$$

$$= 0 .$$

Notice, that throughout the crucial formula was

$$a\psi_o = 0.$$

Thus we have solved the simple harmonic oscillator using algebraic techniques except to find ψ_o explicitly. Notice that we did not need to find ψ_o explicitly. In fact all the results followed from the operator properties of H, p, and x. The usefulness of $a^{\dagger}$ and a do not end here. In fact all physical quantities with regard to the simple harmonic oscillator may be calculated using only a and $a^{\dagger}$. Later we shall have occasion to reexamine the time-dependence of the simple harmonic oscillator. We shall then find that a and $a^{\dagger}$ also have analogous quantities classically, and these classical quantities simplify the classical computation. Now, however, we illustrate some of the applications of a and $a^{\dagger}$.

9.1.1 Expectation Values

We have

$$P = \frac{i}{\sqrt{2}}(a^{\dagger} - a) \quad \text{and} \quad Q = \frac{1}{\sqrt{2}}(a^{\dagger} + a).$$

Thus since

$$p = \sqrt{m\hbar\omega}\, P \quad \text{and} \quad x = \sqrt{\hbar/m\omega}\, Q$$

we can evaluate expectation values of x and p in terms of a and $a^\dagger$. For example

$$(\phi_n, p\phi_n) = \sqrt{m\hbar\omega}\,(\phi_n, P\phi_n)$$

$$= i\sqrt{m\hbar\omega/2}\,\{(\phi_n, a^\dagger\phi_n) - (\phi_n, a\phi_n)\}$$

$$= 0.$$

Similarly, $(\phi_n, x\phi_n) = 0$. This is to be expected since the average momentum and position of a simple harmonic oscillator are indeed zero. Now consider the kinetic energy

$$T = \frac{p^2}{2m} = \frac{m\hbar\omega}{2m} P^2$$

$$= \frac{1}{2}\hbar\omega(-\frac{1}{2})(a^\dagger - a)(a^\dagger - a)$$

$$= \frac{1}{4}\hbar\omega(aa^\dagger + a^\dagger a - a^{\dagger 2} - a^2)$$

$$= \frac{1}{2}H - \frac{1}{4}\hbar\omega(a^{\dagger 2} + a^2) \; .$$

Thus

$$(\phi_n, T, \phi_n) = \frac{1}{2}(\phi_n, H, \phi_n) - \frac{1}{4}\hbar\omega(\phi_n, (a^{\dagger 2} + a^2)\phi_n)$$

$$= \frac{1}{2}(\phi_n, H, \phi_n) = \frac{1}{2}\,(n + \frac{1}{2})\hbar\omega \; .$$

Actually we have proven more since we showed that

$$\langle T\rangle = \frac{1}{2}\langle H\rangle \; . \qquad 9.29$$

This is the same as the classical virial theorem and states that the average kinetic energy equals one-half the average total energy. Also we have

$$\langle V\rangle = \langle H\rangle - \langle T\rangle = \frac{1}{2}\langle H\rangle \; . \qquad 9.30$$

Another set of useful formulae is obtained using 9.26 and 9.27. Thus we have

$$a^{\dagger}\psi_n = \frac{c_n}{c_{n+1}}\psi_{n+1} = \sqrt{n+1}\;\psi_{n+1} \; , \qquad 9.31$$

showing clearly why $a^{\dagger}$ is called a "step-up" operator. It also shows why $a^{\dagger}$ is known as a "creation" operator since it "creates" a higher eigenstate. Applying a to this we get

$$aa^{\dagger}\psi_n = \sqrt{n+1}\; a\psi_{n+1}$$

$$\therefore \qquad a\psi_n = \frac{1}{\sqrt{n}}\, aa^{\dagger}\psi_{n-1}$$

$$= \frac{1}{\sqrt{n}}\frac{1}{\hbar\omega}\,(H + \frac{1}{2}\hbar\omega)\psi_{n-1} \; .$$

Thus,

$$a\psi_n = \sqrt{n}\;\psi_{n-1} \; . \qquad 9.32$$

So a is a "lowering" or "step-down" operator. It is also sometimes known as an annihilation operator. To show how these are applied consider a matrix element of the form

$$(\phi_n, p\phi_\ell) = i\sqrt{\frac{m\hbar\omega}{2}}\,\{(\phi_n, a^{\dagger}\phi_\ell) - (\phi_n, a\phi_\ell)\}$$

$$= i\sqrt{\frac{m\hbar\omega}{2}}\,\{\sqrt{\ell+1}\;(\phi_n, \phi_{\ell+1}) - \sqrt{\ell}\;(\phi_n, \phi_{\ell-1})\} \; .$$

Thus,

$$(\phi_n, p\phi_\ell) = i\sqrt{\frac{m\hbar\omega}{2}}\,\{\sqrt{\ell+1}\,\delta_{n,\ell+1} - \sqrt{\ell}\,\delta_{n,\ell-1}\}\ . \qquad 9.33$$

In a similar manner we get

$$(\phi_n, x\phi_\ell) = \sqrt{\frac{\hbar}{2m\omega}}\,\{(\phi_n, a^\dagger\phi_\ell) + (\phi_n, a\phi_\ell)\}$$

or

$$(\phi_n, x\phi_\ell) = \sqrt{\frac{\hbar}{2m\omega}}\,\{\sqrt{\ell+1}\,\delta_{n,\ell+1} + \sqrt{\ell}\,\delta_{n,\ell-1}\}\ . \qquad 9.34$$

In this manner it is possible to evaluate simply all matrix elements between simple harmonic oscillator eigenstates. We now turn to another problem.

9.2 The Rigid Rotator

Consider a diatomic system which we approximate as follows. It is a "dumb-bell" with masses attached at the ends of a rigid rod and free to rotate in any direction (fig. 9.1). The moment of inertia about the line through the masses is assumed to be negligible (zero). If we assume that the center of mass of the system is fixed then the only motion possible is a rotation. In that case the classical hamiltonian is

$$H = \frac{\vec{L}^2}{2I} \qquad 9.35$$

where $\vec{L}$ is the total angular momentum and I is the moment of inertia about the axis through the center of mass and normal to the line

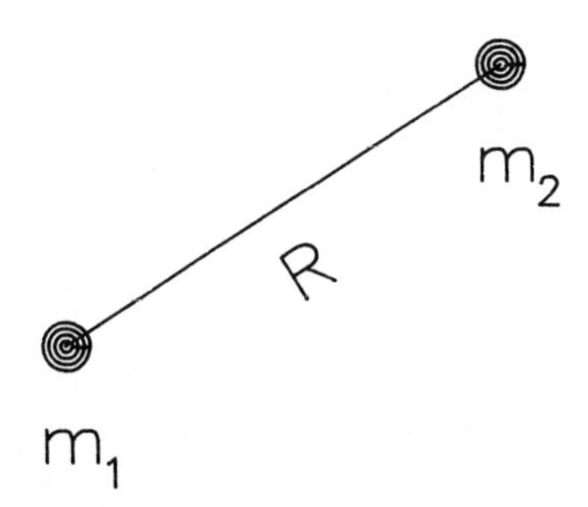

Fig. 9–1 Dumb–bell Molecule

connecting the masses. We wish to solve this problem quantum mechanically.

Now

$$\vec{L} = \vec{r}\times\vec{p} \ . \tag{9.36}$$

Thus,

$$L_x = yp_z - zp_y \ , \ L_y = zp_x - xp_z \ , \ L_z = xp_y - yp_x \ .$$

To quantize the system we simply replace $\vec{p}$ by $\frac{\hbar}{i}\vec{\nabla}$. So the quantum mechanical angular momentum operator is given by

$$\vec{L} = \frac{\hbar}{i}\vec{r}\times\vec{\nabla} \tag{9.37}$$

or

$$L_x = i\hbar\left(z\frac{\partial}{\partial y} - y\frac{\partial}{\partial z}\right) \ ,$$

$$L_y = i\hbar\left(x\frac{\partial}{\partial z} - z\frac{\partial}{\partial x}\right) \ ,$$

$$L_z = i\hbar\left(y\frac{\partial}{\partial x} - x\frac{\partial}{\partial y}\right) \ .$$

Now introducting spherical coordinates

$$x = r \sin\theta \cos\phi$$

$$y = r \sin\theta \sin\phi \qquad 9.38$$

$$z = r \cos\theta$$

we get

$$L_x = i\hbar\left(\sin\phi \frac{\partial}{\partial\theta} + \cos\phi \cot\theta \frac{\partial}{\partial\phi}\right)$$

$$L_y = i\hbar\left(\sin\phi \cot\theta \frac{\partial}{\partial\phi} - \cos\phi \frac{\partial}{\partial\theta}\right) \qquad 9.39$$

$$L_z = -i\hbar \frac{\partial}{\partial\phi} .$$

At this stage we can introduce a simplifying assumption in our problem. We assume that our "molecule" is constrained to rotation in one plane which we choose to be the x-y plane. Thus we are considering a <u>One-Dimensional Rotator</u>.

In this case our hamiltonian simplifies to

$$H = \frac{L_z^2}{2I} . \qquad 9.40$$

Writing out the eigenvalue problem we get

$$-\frac{\hbar^2}{2I}\frac{d^2\phi}{d\phi^2} = E\phi . \qquad 9.41$$

Before proceeding we must consider whether L_z is self- adjoint. Thus we look at the equation for the deficiency indices

$$-i\hbar \frac{df}{d\phi} = \pm\, i\hbar f . \qquad 9.42$$

The solutions are

$$f_{\pm} = e^{\mp\phi} \quad . \tag{9.43}$$

Since $0 \leq \phi \leq 2\pi$ both solutions are admissible and the deficiency indices are (1,1). Thus we have a one-parameter family of self-adjoint extensions and we must pick an appropriate one. We shall do this partly by symmetry considerations. Thus, suppose $f(\phi)$ is some physical state. In that case we want $f(\phi + 2\pi)$ to represent the same state. Thus

$$f(\phi + 2\pi) = e^{i\alpha} f(\phi) \ . \tag{9.44}$$

Here α is the parameter labelling the different self- adjoint extensions. What does this mean? It is easy to show in the manner we used for p that L_z is the generator of the translations of ϕ; that is, L_z is the generator of rotations about the z axis. Thus rotating by 2π about z brings us back to the original point and we must specify the "reflection" condition. The usual argument given is that ψ should be single-valued and hence $\alpha = 0$. This argument is not completely satisfactory since none of our assumptions about quantum mechanics require the wave function to be single-valued.* We shall now summarize the results of a careful investigation (ref. 9.1).

For a spherically symmetric system, it is physically necessary that not only should one component of the angular momentum (L_z in this case) be conserved, but in fact the total angular momentum

$$L^2 = L_x^2 + L_y^2 + L_z^2 \tag{9.45}$$

---

* In fact the Bohm-Aharanov effect (ref. 9.8) is based on a wave function that is <u>not</u> single-valued.

must be conserved. Thus we must be able to simultaneously diagonalize L^2 and some one component say

$$L_n = \hat{n}.\vec{L} \tag{9.46}$$

of the angular momentum. This requires

$$[L_n, L^2] = 0. \tag{9.47}$$

The above relation follows formally if one works out the commutation relations between L_x, L_y, L_z and L^2. We shall do this later. Furthermore since we can arbitrarily decide how we shall label a set of three orthogonal axes as x,y and z, the three operators L_x, L_y and L_z must be equivalent.

Consider now the self-adjoint extensions L_z^α corresponding to the boundary conditions 9.44. In this case there are corresponding self-adjoint extensions for L_x and L_y say L_x^α and L_y^α. The results for these extensions are as follows.

1) The eigenvalues of L_x^α, L_y^α are integral regardless of α. The eigenvalues of L_z^α are integral only for $\alpha = 0$.

2) For $\alpha = 0$

$$L^2\psi = L_x^2\psi + L_y^2\psi + L_z^2\psi$$

for every $\psi \in D_{L^2}$. For $\alpha \neq 0$ this relation is false.

3) If $\alpha = 0$

$$[L^2, L_x^\alpha] = [L^2, L_y^\alpha] = [L^2, L_z^\alpha] = 0.$$

If $\alpha \neq 0$ we still have

$$[L^2, L_z^\alpha] = 0$$

but L^2 does not commute with L_x^α and L_y^α. Thus neither L^2 and L_x^α

nor L^2 and L_y^α have simultaneous eigenfunctions for $\alpha \neq 0$.

These results show that for $\alpha \neq 0$ there is a special preferred z direction. However, this is not the case physically. Thus unless there is a physically preferred direction such as the direction of an external field the only physically acceptable self-adjoint extension is given by $\alpha = 0$. For this case we drop the superscript α. Notice, that mathematics cannot decide on the appropriate self-adjoint extension, the decision has to be made on purely physical grounds.

For more details on this question we list a series of references at the end of this chapter. We can now return to a consideration of our original problem of the one- dimensional rotator

$$H = \frac{L_z^2}{2I} = -\frac{\hbar^2}{2I}\frac{d^2}{d\phi^2} \qquad 9.40$$

where $f \in D_{L_z}$ requires

$$f(\phi + 2\pi) = f(\phi)$$

and f to be C^2. Clearly

$$[L_z,H] = 0 \ . \qquad 9.48$$

Thus we can simultaneously diagonalize L_z and H. The eigenvalue problem for L_z can be written

$$-i\hbar \frac{d\Phi}{d\phi} = m\hbar\Phi. \qquad 9.49$$

Thus

$$\Phi_m(\phi) = Ae^{im\phi} \ , \qquad 9.50$$

Applying the boundary condition with $\alpha = 0$ we get

$$e^{im(\phi+2\pi)} = e^{im\phi} .$$

Thus

$$e^{im2\pi} = 1.$$

Hence

$$m = 0, \pm 1, \pm 2, \ldots \tag{9.51}$$

If we use H we get

$$H\Phi_m = E\Phi_m = \frac{L_z^2}{2I}\Phi_m$$

and this reads

$$E\Phi_m = \frac{\hbar^2}{2I} m^2 \Phi_m$$

so the energy eigenvalues are

$$E_m = \frac{\hbar^2}{2I} m^2 . \tag{9.52}$$

Normalization:

$$\int_0^{2\pi} |\Phi_m(\phi)|^2 d\phi = 1$$

$$\therefore \qquad |A|^2 \int_0^{2\pi} d\phi = 1.$$

Thus choosing the phase we get

$$A = (2\pi)^{-1/2}$$

Hence,

$$\Phi = (2\pi)^{-1/2} e^{im\phi} . \tag{9.53}$$

9.3 Rigid Rotator in Three Dimensions - Angular Momentum

We are now ready for the full rigid rotator problem. Here

$$H = \frac{L^2}{2I} = \frac{1}{2I}(L_x^2 + L_y^2 + L_z^2) \ . \qquad 9.35$$

Clearly

$$[L^2, H] = 0 \qquad 9.54$$

and as we show next $[L_n, L^2] = [L_n, H] = 0$ where $n = x, y, z.$

We use the form 9.36 for $\vec{L}$ to compute several commutators. Consider

$$[L_x, L_y] = [yp_z - zp_y \ , \ zp_x - xp_z]$$

$$= [yp_z, zp_x] - [zp_y, zp_x] - [yp_z, xp_z] + [zp_y, xp_z].$$

Consider the first term

$$[yp_z, zp_x] = y[p_z, zp_x] + [y, zp_x]p_z$$

$$= y[p_z, z]p_x + yz[p_z, p_x] + [y, z]p_xp_z + z[y, p_x]p_z$$

$$= -i\hbar\, yp_x + 0 + 0 + 0 = -i\hbar y\, p_x \ .$$

The second and third terms both contribute nothing. The last term contributes $i\hbar\, xp_y$. Thus,

$$[L_x, L_y] = i\hbar(xp_y - yp_x) = i\hbar L_z \ . \qquad 9.55$$

Proceeding in this manner we get the commutation relations

$$[L_y, L_z] = i\hbar L_x \qquad 9.56$$

$$[L_z, L_x] = i\hbar L_y \quad . \qquad 9.57$$

Notice that these are just cyclic permutations of x,y,z as shown.

$$x \nearrow y \searrow z \rightarrow x$$

We further compute

$$[L_x, L^2] = [L_x, L_x^2] + [L_x, L_y^2] + [L_x, L_z^2]$$

$$= 0 + L_y[L_x, L_y] + [L_x, L_y]L_y + L_z[L_x, L_z] + [L_x, L_z]L_z$$

$$= i\hbar(L_yL_z + L_zL_y - L_zL_y - L_yL_z) = 0.$$

Thus $[L_x, L^2] = 0$.

Similarly $[L_y, L^2] = 0$ 9.58

and $[L_z, L^2]$

We have, therefore, shown that L^2 and one component of $\vec{L}$ can be simultaneously diagonalized. Since L_z has the simplest representation as a differential operator one chooses L_z.

Thus, calling the eigenvalues of L^2, $\ell(\ell+1)\hbar^2$ and of L_z, $m\hbar$ we have to solve the eigenvalue problems

$$L^2 Y_{\ell m}(\theta, \phi) = \ell(\ell+1)\hbar^2 Y_{\ell m}(\theta, \phi) \qquad 9.59$$

$$L_z Y_{\ell m}(\theta, \phi) = m\hbar Y_{\ell, m}(\theta, \phi). \qquad 9.60$$

The energy eigenvalues are then given by

$$E_\ell = \frac{\hbar^2}{2I}\,\ell(\ell+1)\ .$$

The factors $\ell(\ell+1)$ are chosen for later convenience. We could obviously have chosen anything else. If we now use 9.39 and the definition of L^2 we find

$$L^2 = -\hbar^2\left\{\frac{1}{\sin\theta}\frac{\partial}{\partial\theta}\left(\sin\theta\frac{\partial}{\partial\theta}\right) + \frac{1}{\sin^2\theta}\frac{\partial^2}{\partial\phi^2}\right\}\ . \qquad 9.61$$

This form becomes much more significant if we look at the Laplace operator in spherical coordinates

$$\nabla^2 = \frac{1}{r^2}\frac{\partial}{\partial r}\left(r^2\frac{\partial}{\partial r}\right) - \frac{L^2}{\hbar^2 r^2}\ . \qquad 9.62$$

Thus the angular portion of ∇^2 is given completely by L^2.

In order to solve 9.59, 9.60 we try to separate variables. Thus we set

$$Y_{\ell m}(\theta,\phi) = c_{\ell m}P_\ell^m(\cos\theta)\phi_m(\phi)$$

where $c_{\ell m}$ is a normalization constant. Then

$$Y_{\ell m}(\theta,\phi) = \frac{c_{\ell m}}{\sqrt{2\pi}}\,e^{im\phi}P_\ell^m(\cos\theta). \qquad 9.63$$

So that 9.60 is automatically satisfied and 9.59 becomes

$$\left\{\frac{1}{\sin\theta}\frac{d}{d\theta}\left(\sin\theta\frac{d}{d\theta}\right) - \frac{m^2}{\sin^2\theta}\right\}P_\ell^m(\cos\theta) = -\ell(\ell+1)P_\ell^m(\cos\theta). \qquad 9.64$$

If we set $\cos\theta = x$ then

$$\frac{d}{dx} = -\frac{1}{\sin\theta}\frac{d}{d\theta}\ .$$

and we get

$$\frac{d}{dx}\{(1-x^2)\frac{dP_\ell^m(x)}{dx}\} + \{\ell(\ell+1) - \frac{m^2}{1-x^2}\} P_\ell^m = 0 . \qquad 9.65$$

Writing this out it reads

$$(1-x^2)\frac{d^2P}{dx^2} - 2x\frac{dP}{dx} + \{\ell(\ell+1) - \frac{m^2}{1-x^2}\} P = 0. \qquad 9.66$$

This is the differential equation for the associated Legendre functions which are denoted by $P_\ell^m(x)$. Rather than solve this differential equation we shall proceed as for the simple harmonic oscillator and solve the problem algebraically. Since in the algebraic solution we are not restricted to having a representation of $\vec{L}$ as operators in ordinary space we shall not only find the solutions above but instead we shall also find some new solutions that do not correspond to orbital angular momentum but to something else called spin angular momentum.

9.4 Algebraic Approach to Angular Momentum

The assumptions involved are as follows.

1) Commutation Relations

$$[L_x, L_y] = i\hbar L_z$$

and cyclic permutations.

2) $\vec{L}$ is an observable. This means that every component of $\vec{L}$ is a self adjoint operator. It then follows that

$$\vec{L}^2 = L_x^2 + L_y^2 + L_z^2$$

is a positive, self-adjoint operator. Thus

$$L^2 \geq 0 \quad \text{and} \quad L^2 \geq L_z^2 \; . \tag{9.67}$$

Also as we saw before

$$[\vec{L}, L^2] = 0. \tag{9.68}$$

Thus we may diagonalize L^2 and one component of $\vec{L}$ simultaneously. It is conventional to choose L_z. We now introduce two non-hermitian operators $L_{\pm}$

$$L_{\pm} = L_x \pm iL_y \tag{9.69}$$

analogous to the raising or creation and lowering or annihilation operators $a^{\dagger}$ and a used with the simple harmonic oscillator. We also need the commutation relations for $L_{\pm}$. Thus,

$$[L_{\pm}, L^2] = 0 \tag{9.70}$$

as is obvious from 9.68. Also

$$[L_z, L_{\pm}] = [L_z, L_x] \pm i[L_z, L_y]$$

$$= i\hbar L_y \pm i(-i\hbar L_x) = \pm\hbar(L_x \pm iL_y).$$

Hence,

$$[L_z, L_{\pm}] = \pm \hbar L_{\pm} \; . \tag{9.71}$$

Finally we also get by straightforward multiplication that

$$L_{\pm}L_{\mp} = L_x^2 + L_y^2 \pm \hbar L_z$$

so that

$$L_\pm L_\mp = L^2 - L_z^2 \pm \hbar L_z \ . \tag{9.72}$$

We now consider the simultaneous eigenvalue problems

$$L^2\psi = a\hbar^2\psi \tag{9.73}$$

$$L_z\psi = b\hbar\psi \ . \tag{9.74}$$

Since

$$L^2 \geq L_z^2$$

we get that

$$a\hbar^2 \geq b^2\hbar^2$$

or $\quad a \geq b^2 \ .$ 9.75

Now apply L_+ to 9.74 and use 9.71. This gives

$$L_+L_z\psi = b\hbar L_+\psi = L_z(L_+\psi - \hbar L_+\psi).$$

Thus

$$L_z(L_+\psi) = (b+1)\hbar(L_+\psi) \ . \tag{9.76}$$

Thus $L_+\psi$ is a new eigenfunction of L_z with eigenvalue $(b+1)\hbar$. Since $L_\pm$ and L^2 commute (9.70), $L_+\psi$ is still an eigenfunction of L^2 with the eigenvalue $a\hbar^2$. However due to 9.75 we see that there must exist a maximum eigenvalue for L_z otherwise we could violate 9.75. Suppose ψ is this eigenfunction. In that case applying L_+ yields 9.76. Thus consistency can only be maintained if

$$L_+\psi = 0 \tag{9.77}$$

where ψ is the eigenfunction corresponding to the largest eigenvalue of L_z . Now apply L_- to 9.77 and use 9.72. Then we get

$$L_-L_+\psi = (L^2 - L_z^2 - \hbar L_z)\psi = 0 \qquad 9.78$$

or

$$(a\hbar^2 - b^2\hbar^2 - b\hbar^2)\psi = 0 \ .$$

Thus,

$$a = b^2 + b \ . \qquad 9.79$$

We now proceed downwards. Thus operate on 9.74 with L_- and use 9.71 to get

$$L_z(L_-\psi) = (b-1)\hbar(L_-\psi) \ . \qquad 9.80$$

Repeating this n times we get

$$L_z(L_-^n\psi) = (b-n)\hbar(L_-^n\psi) \ . \qquad 9.81$$

Now by making n large enough we can again violate 9.75. This means we must reach a point where

$$\psi' = L_-^n\psi \qquad 9.82$$

and

$$L_-\psi' = 0. \qquad 9.83$$

Applying L_+ and using 9.72 then gives:

$$L_+L_-\psi' = (L^2 - L_z^2 + \hbar L_z)(L_-^n\psi) = 0$$

or

$$[a\hbar^2 - (b-n)^2\hbar^2 + (b-n)\hbar^2](L_-^n\psi) = 0.$$

Thus,

$$a = (b-n)^2 - (b-n) \qquad 9.84$$

We combine this with 9.79 and get

$$2b(n+1) = n(n+1).$$

Thus,

$$b = \frac{1}{2} n = \ell , \qquad 9.85$$

where $\ell = 0 , 1/2, 1, 3/2, 2, 5/2, \ldots\ldots$.
Also

$$a = b(b+1) = \ell(\ell+1) . \qquad 9.86$$

Thus we can label the eigenstates with ℓ and m such that

$$L^2\psi_{\ell m} = \ell(\ell+1)\hbar^2\psi_{\ell m} \qquad 9.87$$

$$L_z\psi_{\ell m} = m\hbar\psi_{\ell m} \qquad 9.88$$

where ℓ is the largest value of $|m|$.

So $\ell \geq |m|$ or $m = \ell, \ell-1, \ell-2,\ldots,-\ell$. As we have already seen, for the orbital angular momentum ℓ is an integer, and the $\psi_{\ell m}$ are the spherical harmonics $Y_{\ell m}(\theta,\phi)$. We now reexamine them using the algebraic technique ; the case of half-odd-integer angular momentum is examined in section 9.6. Now consider

$$L_z = \frac{\hbar}{i}\frac{\partial}{\partial\phi}$$

$$L_\pm = \hbar e^{\pm i\phi}(\pm \frac{\partial}{\partial\theta} + i \cot\theta \frac{\partial}{\partial\phi})$$

and

$$L^2 = -\hbar^2\left[\frac{1}{\sin\theta}\frac{\partial}{\partial\theta}\left(\sin\theta\frac{\partial}{\partial\theta}\right) + \frac{1}{\sin^2\theta}\frac{\partial^2}{\partial\phi^2}\right] .$$

If we start with the state of lowest angular momentum, $Y_{\ell,-\ell}(\theta,\phi)$ we have

$$L^2 Y_{\ell,-\ell} = \hbar^2\ell(\ell+1)Y_{\ell,-\ell} \tag{9.89}$$

$$L_z Y_{\ell,-\ell} = -\hbar\ell Y_{\ell,-\ell} \tag{9.90}$$

and

$$L_- Y_{\ell,-\ell} = 0. \tag{9.91}$$

We can also work up from this point using L_+ . Thus

$$Y_{\ell,-\ell+1} = cL_+ Y_{\ell,-\ell} \tag{9.92}$$

and so on, eventually reaching the highest wavefunction $Y_{\ell,\ell}$. Then

$$L^2 Y_{\ell,\ell} = \hbar^2\ell(\ell+1)Y_{\ell,\ell}$$

$$L_z Y_{\ell,\ell} = \hbar\ell Y_{\ell\ell}$$

$$L_+ Y_{\ell\ell} = 0 .$$

To look for the eigenfunctions explicitly we proceed as before by writing

$$Y_{\ell m} = \frac{1}{\sqrt{2\pi}} e^{im\phi} P_\ell^m(\cos\theta).$$

We then get from

$$L_+ Y_{\ell\ell} = 0$$

that

$$e^{i\phi}\left(e^{i\ell\phi}\frac{dP_\ell^\ell}{d\theta} - \ell\cot\theta\, e^{i\ell\phi}P_\ell^\ell\right) = 0$$

or

$$\frac{dP_\ell^\ell}{d\theta} = \ell\cot\theta\, P_\ell^\ell\ . \tag{9.93}$$

Thus

$$P_\ell^\ell = A_\ell \sin^\ell\theta\ . \tag{9.94}$$

We get A_ℓ by normalization

$$|A_\ell|^2\int_0^\pi \sin^{2\ell}\theta\ .\ \sin\theta\, d\theta = 1. \tag{9.95}$$

Choosing the phase of A_ℓ to be $(-1)^\ell$ we get

$$P_\ell^\ell = (-1)^\ell\sqrt{\frac{(2\ell+1)!}{2}}\ \frac{1}{2^\ell \ell!}\sin^\ell\theta\ . \tag{9.96}$$

Thus,

$$Y_{\ell,\ell}(\theta,\phi) = \frac{(-1)^\ell}{\sqrt{2\pi}}\sqrt{\frac{(2\ell+1)!}{2}}\ \frac{\sin^\ell\theta}{2^\ell \ell!}\, e^{i\ell\phi}\ . \tag{9.97}$$

To get the other spherical harmonics we apply L_-. Thus we have

$$Y_{\ell,\ell-1} = c_{\ell,\ell-1}L_-Y_{\ell,\ell}$$

and finally

$$Y_{\ell,m} = c_{\ell,m}L_-^{\ell-m}Y_{\ell,\ell}\ .$$

Thus,

$$Y_{\ell,m} = \frac{c_{\ell,m}}{c_{\ell,m+1}} L_- Y_{\ell,m+1} .$$

To evaluate the normalization constants $c_{\ell,m}$ we simply use

$$(Y_{\ell m}, Y_{\ell m}) = \int_0^{2\pi} d\phi \int_0^{\pi} \sin\theta \, d\theta \, Y^*_{\ell m}(\theta,\phi) Y_{\ell m}(\theta,\phi) = 1 \qquad 9.98$$

to get

$$1 = \left| \frac{c_{\ell,m}}{c_{\ell,m+1}} \right|^2 (L_- Y_{\ell,m+1}, L_- Y_{\ell,m+1})$$

$$= \left| \frac{c_{\ell,m}}{c_{\ell,m+1}} \right|^2 (Y_{\ell,m+1}, L_+ L_- Y_{\ell,m+1})$$

$$1 = \left| \frac{c_{\ell,m}}{c_{\ell,m+1}} \right|^2 (Y_{\ell,m+1}, (L^2 - L_z^2 + \hbar L_z) Y_{\ell,m+1})$$

$$= \left| \frac{c_{\ell,m}}{c_{\ell,m+1}} \right|^2 \hbar^2 [\ell(\ell+1) - m(m+1)] .$$

Thus finally

$$L_- Y_{\ell,m+1} = \frac{c_{\ell,m+1}}{c_{\ell,m}} Y_{\ell,m} = \sqrt{\ell(\ell+1) - m(m+1)} \, \hbar Y_{\ell,m} . \qquad 9.99$$

Now applying L_+ to this equation and using 9.72 we get

$$[\ell(\ell+1) - (m+1)^2 + (m+1)] \hbar^2 Y_{\ell,m+1} = \sqrt{\ell(\ell+1) - m(m+1)} \, \hbar L_+ Y_{\ell,m}$$

or

$$L_+ Y_{\ell,m} = \sqrt{\ell(\ell+1) - m(m+1)} \, \hbar Y_{\ell,m+1} . \qquad 9.100$$

We can combine 9.99 and 9.100 into one very useful equation

$$L_{\pm}Y_{\ell,m} = \sqrt{\ell(\ell+1) - m(m\pm1)}\,\hbar Y_{\ell,m\pm1} \,. \qquad 9.101$$

Using L_- and iterating from $Y_{\ell,\ell}$ we finally get an explicit form for $Y_{\ell,m}$

$$Y_{\ell,m} = \sqrt{\frac{(\ell+m)!}{(2\ell)!(\ell-m)!}}\,\hbar^{-(\ell-m)}L_-^{\ell-m}Y_{\ell,\ell} \,. \qquad 9.102$$

Furthermore, using the explicit form for P_ℓ^m we get that

$$P_\ell^o = \sqrt{\frac{2\ell+1}{2}}\,\frac{1}{2^\ell \ell!}\,\left(\frac{d}{d\cos\theta}\right)^\ell (\cos^2\theta - 1)^\ell. \qquad 9.103$$

This is now recognized as a Rodrigues' formula for the Legendre polynomials $P_\ell(\cos\theta)$ which are normalized such that

$$P_\ell^o(\cos\theta) = \sqrt{\frac{2\ell+1}{2}}\,P_\ell(\cos\theta). \qquad 9.104$$

Writing out equation 9.102 explicitly in terms of the $P_\ell^m(\cos\theta)$, equation 9.63 becomes

$$Y_{\ell,m}(\theta,\phi) = \left(\frac{2\ell+1}{4\pi}\cdot\frac{(\ell-m)!}{(\ell+m)!}\right)^{1/2} P_\ell^m(\cos\theta)e^{im\phi}. \qquad 9.105$$

Since the $P_\ell^m(\cos\theta)$ are defined by

$$P_\ell^m\,(x) = \frac{(-1)^m}{2^\ell \ell!}\,(1-x^2)^{m/2}\,\frac{d^{\ell+m}}{dx^{\ell+m}}\,(x^2-1)^\ell \,, \qquad 9.106$$

it follows by some simple algebra (see problem 9.11) that

$$P_\ell^{-m}(x) = (-1)^m\,\frac{(\ell-m)!}{(\ell+m)!}\,P_\ell^m(x) \,, \qquad 9.107$$

From the normalization condition 9.98 we can now deduce the

Table 9.1 The Spherical Harmonics $Y_{\ell,m}(\theta,\phi)$.

$\ell = 0$	$Y_{0,0} = \frac{1}{\sqrt{4\pi}}$
$\ell = 1$	$Y_{1,1} = -\sqrt{\frac{3}{8\pi}} \sin\theta\ e^{i\phi}$
	$Y_{1,0} = \sqrt{\frac{3}{4\pi}} \cos\theta$
$\ell = 2$	$Y_{2,2} = \sqrt{\frac{15}{32\pi}} \sin^2\theta\ e^{2i\phi}$
	$Y_{2,1} = -\sqrt{\frac{15}{8\pi}} \sin\theta \cos\theta\ e^{i\phi}$
	$Y_{2,0} = \sqrt{\frac{6}{16\pi}} (2 \cos^2\theta - 1)$
$\ell = 3$	$Y_{3,3} = -\sqrt{\frac{35}{64\pi}} \sin^3\theta\ e^{3i\phi}$
	$Y_{3,2} = \sqrt{\frac{105}{32\pi}} \sin^2\theta \cos\theta\ e^{2i\phi}$
	$Y_{3,1} = -\sqrt{\frac{21}{64\pi}} \sin\theta(5 \cos^2\theta - 1) e^{i\phi}$
	$Y_{3,0} = \sqrt{\frac{7}{16\pi}} (5 \cos^3\theta - 3 \cos\theta)$

relationship:

$$\int_0^\pi P_\ell^m(\cos\theta) P_{\ell'}^m(\cos\theta) \sin\theta\ d\theta = \int_{-1}^1 P_\ell^m(x) P_\ell^m(x) dx = \frac{2}{2\ell+1} \frac{(\ell+m)!}{(\ell-m)!} \delta_{\ell,\ell'} \ . \qquad 9.108$$

Now using 9.105 and 9.107 it follows that

$$Y_{\ell,-m}(\theta,\phi) = (-1)^m Y_{\ell,m}^*(\theta,\phi) \ . \qquad 9.109$$

For later reference we list the spherical harmonics for $\ell = 0,1,2,3$ and $m \geq 0$. The spherical harmonics for negative m can be deduced from 9.109. The phases are as given by 9.97 and 9.102.

The utility of spherical harmonics extends to all systems with spherical symmetry. We shall have many occasions to use them in subsequent chapters. For the present we first examine several more aspects of angular momentum.

9.5 Rotations and Rotational Invariance

A rotation of a system about a point 0 is a displacement of all points P of the system such that the distance between any two points remains unchanged. Rotations of coordinate systems are implemented by orthogonal matrices. Thus if the point $\vec{x}'$ is obtained from $\vec{x}$ by a rotation R about the origin of the coordinate system we write

$$\vec{x}' = \overrightarrow{Rx} \quad ,$$

or in component form

$$x_i' = \sum_j R_{ij} x_j . \qquad 9.110$$

Since we need invariance of the distance form the origin we get as a condition

$$\sum_i x_i' x_i' = \sum_{i,j,k} R_{ij} x_j R_{ik} x_k = \sum_k x_k x_k \ . \qquad 9.111$$

This requires that

$$\sum_i R_{ij} R_{ik} = \delta_{jk} \qquad 9.112$$

and states that R, considered as a matrix, is orthogonal, namely

$$RR^t = R^t R = 1. \qquad 9.113$$

Here R^t is the "transpose" of the matrix R. This equation immediately shows that

$$\det R \det R^t = (\det R)^2 = 1.$$

Thus

$$\det R = \pm 1. \tag{9.114}$$

However if we use the parity operator (section 4.13)

$$P\vec{x} = -\vec{x} \tag{9.115}$$

we see that

$$\det P = -1 . \tag{9.116}$$

And if $\det R = -1$, then $\det PR = \det P \det R = 1$. Conversely any R with determinant -1 can be written as

$$R = PR_p \tag{9.117}$$

where

$$\det R_p = 1.$$

We call rotations with determinant +1, proper rotations. Those with determinant -1 are called improper and involve a parity transformation. Henceforth we consider only proper rotations and drop the subscript p.

Suppose we are in a coordinate system S with a quantum mechanical single particle state specified by a wavefunction ψ. Then if we rotate the coordinate system to a new system S' we get a new state ψ'. Since none of the physics has changed we require that all probabilities remain unchanged. So if ϕ is another state and ϕ' its image after rotation, we require that

$$|(\phi',\psi')|^2 = |(\phi,\psi)|^2 . \tag{9.118}$$

E. Wigner (ref. 9.2) has shown that 9.118 implies that ϕ',ψ' can always be obtained from ϕ,ψ by either unitary or antiunitary transformations. The antiunitary transformations correspond to having time-reversal (section 4.6). We therefore consider unitary transformations. Thus

$$\psi' = U_R\psi \tag{9.119}$$

where

$$U_R^\dagger U_R = U_R U_R^\dagger = 1. \tag{9.120}$$

Clearly if R_1 and R_2 are two rotations such that

$$R_1R_2 = R_3 \tag{9.121}$$

then we need

$$U_{R_1R_2} = U_{R_3} \quad . \tag{9.122}$$

This statement together with 9.120 says that the operators U_R yield a <u>unitary representation of the rotation group</u>.

To proceed further we need to decide on how to specify a rotation. There are very many different methods available. The procedure we use is to specify an axis of rotation (unit vector) $\vec{n}$ and angle of rotation θ about this axis. Consider a rotation about the z-axis by an angle θ, i.e., $R_{z,\theta}$. Then

$$\begin{pmatrix} x' \\ y' \\ z' \end{pmatrix} = \begin{pmatrix} \cos\theta & \sin\theta & 0 \\ -\sin\theta & \cos\theta & 0 \\ 0 & 0 & 1 \end{pmatrix} \begin{pmatrix} x \\ y \\ z \end{pmatrix} \quad . \tag{9.123}$$

And if we apply this transformation to the coordinates of a wavefunction $\psi(x,y,z)$ we get

$$U_{z,\theta}\psi(x,y,z) = \psi(x\cos\theta + y\sin\theta, -x\sin\theta + y\cos\theta, z). \quad 9.124$$

Now let $\theta \to 0$, i.e., be infinitesimal. Then to lowest order in θ

$$U_{z,\theta}\psi(x,y,z) = \psi(x + \theta y, -x\theta + y, z) .$$

Taylor expanding (again to first order in θ) we obtain

$$\begin{aligned} U_{z,\theta}\psi(x,y,z) &= \psi(x,y,z) + \theta[y\frac{\partial}{\partial x} - x\frac{\partial}{\partial y}]\psi(x,y,z) \\ &= \psi(x,y,z) - \frac{i\theta}{\hbar} L_z\,\psi(x,y,z) . \end{aligned} \quad 9.125$$

Thus,

$$U_{z,\theta} = 1 - \frac{i\theta}{\hbar} L_z \quad 9.126$$

for infinitesimal θ.

Similarly we find for $n = x$ or y and infinitesimal θ that:

$$U_{x,\theta} = 1 - \frac{i\theta}{\hbar} L_x , \quad U_{y,\theta} = 1 - \frac{i\theta}{\hbar} L_y . \quad 9.127$$

Thus more generally:

$$U_{n,\theta} = 1 - \frac{i\theta}{\hbar}\vec{n}.\vec{L} \quad 9.128$$

for infinitesimal θ.

Now we use equation 9.122 and 9.128 for a finite angle θ together with an infinitesimal angle ε

$$U_{n,\theta+\varepsilon} = U_{n,\varepsilon}U_{n,\theta} \quad 9.129$$

$$= (1 - \frac{i\varepsilon}{\hbar}\vec{n}.\vec{L})U_{n,\theta} . \quad 9.130$$

Thus

$$\lim_{\varepsilon\to o} \frac{1}{\varepsilon} [U_{n,\theta+\varepsilon} - U_{n,\theta}] = \lim_{\varepsilon\to o} [-\frac{i}{\hbar} \vec{n}.\vec{L}\, U_{n,\theta}] \ . \tag{9.131}$$

Or since the limit defines a derivative we get

$$\frac{d}{d\theta} U_{n,\theta} = -\frac{i}{\hbar} \vec{n}.\vec{L}\, U_{n,\theta} \tag{9.132}$$

where

$$U_{n,o} = 1 \ . \tag{9.133}$$

Thus integrating we find

$$U_{n,\theta} = \exp(-\frac{i}{\hbar} \vec{n}.\vec{L}\theta). \tag{9.134}$$

From 9.132 and 9.134 we see that the angular momentum operators $\frac{1}{\hbar}\vec{L}$ are the "generators of rotations" in the sense of Stone's Theorem (section 6.6). As always, the function exp A of an operator A is defined by its power series expansion.

Now consider a classical Hamiltonian H that remains invariant under rotations of the coordinate system. Typically such an H will be of the form

$$H = \frac{\vec{p}^{\,2}}{2m} + V(|\vec{r}|) \ . \tag{9.135}$$

If the invariance is to remain in the transition to quantum mechanics, we require that the Schrödinger equation also remain invariant under rotations. Thus we need that

$$H\psi = E\psi \tag{9.136}$$

should imply that $U_R\psi$ is a solution whenever ψ is a solution. But

$$U_R H\psi = U_R H U_R^{-1} U_R\psi = E U_R\psi. \tag{9.137}$$

So for $U_R\psi$ to be a solution of 9.136 requires that

$$U_R H U_R^{-1} = H. \tag{9.138}$$

This is the condition that the quantum mechanical Hamiltonian be invariant under rotations. Multiplying equation 9.138 on the right by U_R and using infinitesimal rotations about the x,y, and z axis respectively we obtain (problem 9.12) the equation

$$[H,\vec{L}] = 0. \tag{9.139}$$

Equation 9.139 is equivalent to 9.138 and states necessary and sufficient conditions for rotational invariance of the hamiltonian H.

9.6 Spin Angular Momentum

In finding the eigenvalues of L^2 and L_z by algebraic techniques we obtained (equation 9.85) that ℓ could be $\frac{1}{2}$ integral as well as integral. We now examine the special case of $\ell = 1/2$. In this case ℓ can not represent orbital angular momentum since by solving the differential equation for the eigenvalues of L_z (equation 9.49) we found that the eigenvalues m (equation 9.51) had to be integers. The case of angular momentum $1/2\hbar$ represents a new intrinsic or internal quantum number called spin. It is as much a fundamental property of a particle as its charge or mass.

If the total angular momentum which we now call spin s, is $s = 1/2$ then the z component of spin s_z can have eigenvalues

$$m_s = s,\ s-1, \ldots - |s| = 1/2\ ,\ -1/2. \tag{9.140}$$

Thus there are only these two eigenvalues of m_s possible. We call the corresponding eigenfunctions u_+ and u_- "spinors".

$$s^2 u_\pm = (s_x^2 + s_y^2 + s_z^2) u_\pm = \frac{1}{2}(\frac{1}{2} + 1)\hbar^2 u_\pm \qquad 9.141$$

and

$$s_z u_\pm = \pm \frac{1}{2} \hbar \, u_\pm \, . \qquad 9.142$$

If we further write

$$s_\pm = s_x \pm i \, s_y \qquad 9.143$$

and use equations 9.101 for $\ell = \frac{1}{2}$ we get

$$s_+ u_+ = 0 \; , \; s_- u_- = 0 \qquad 9.144$$

$$s_+ u_- = \sqrt{\frac{1}{2}\,(\frac{1}{2} + 1) + \frac{1}{2}\,(-\,\frac{1}{2} + 1)}\; \hbar u_+ = \hbar u_+ \qquad 9.145a$$

$$s_- u_+ = \sqrt{\frac{1}{2}\,(\frac{1}{2} + 1) - \frac{1}{2}\,(\frac{1}{2} - 1)}\; \hbar u_- = \hbar u_- \qquad 9.145b$$

Evaluating s_z , $s_\pm$ on the two-dimensional sub-space spanned by $u_\pm$ they become simple matrix operators

$$s_z = \frac{\hbar}{2} \begin{pmatrix} 1 & 0 \\ 0 & -1 \end{pmatrix} \; , \; s_+ = \hbar \begin{pmatrix} 0 & 1 \\ 0 & 0 \end{pmatrix} \; , \; s_- = \hbar \begin{pmatrix} 0 & 0 \\ 1 & 0 \end{pmatrix} \qquad 9.146$$

where the spinors $u_\pm$ are explicitly given by

$$u_+ = \begin{pmatrix} 1 \\ 0 \end{pmatrix} \; , \; u_- = \begin{pmatrix} 0 \\ 1 \end{pmatrix} \, . \qquad 9.147$$

But

$$s_x = \frac{1}{2}\,(s_+ + s_-) = \frac{\hbar}{2} \begin{pmatrix} 0 & 1 \\ 1 & 0 \end{pmatrix} \qquad 9.148$$

$$s_y = \frac{i}{2}(s_- - s_+) = \frac{\hbar}{2}\begin{pmatrix} 1 & -i \\ i & 0 \end{pmatrix} \qquad 9.149$$

and multiplying out and adding we find

$$s^2 = s_x^2 + s_y^2 + s_z^2 = \frac{3}{4}\hbar^2\begin{pmatrix} 1 & 0 \\ 0 & 1 \end{pmatrix} . \qquad 9.150$$

Thus we have a simple matrix representation of the spin 1/2 operators and eigenfunctions. It is conventional to introduce the following three matrices

$$\sigma_x = \begin{pmatrix} 0 & 1 \\ 1 & 0 \end{pmatrix} , \quad \sigma_y = \begin{pmatrix} 0 & -i \\ i & 0 \end{pmatrix} , \quad \sigma_z = \begin{pmatrix} 1 & 0 \\ 0 & -1 \end{pmatrix} \qquad 9.151$$

called the Pauli spin matrices. In terms of these we can express the spin matrices $\vec{s}$ by

$$\vec{s} = \frac{\hbar}{2}\vec{\sigma} . \qquad 9.152$$

By explicit multiplication it is easy to verify that the Pauli matrices anti-commute. This means

$$\sigma_x\sigma_y + \sigma_y\sigma_x = \sigma_x\sigma_z + \sigma_z\sigma_x = \sigma_y\sigma_z + \sigma_z\sigma_y = 0. \qquad 9.153$$

The commutation relations among the Pauli matrices can also be obtained either by explicit multiplication or by using 9.151. and the commutation relations for s_x, s_y, s_z. In either case we get

$$\begin{aligned} \sigma_x\sigma_y - \sigma_y\sigma_x &= 2i\sigma_z \\ \sigma_y\sigma_z - \sigma_z\sigma_y &= 2i\sigma_x \\ \sigma_z\sigma_x - \sigma_x\sigma_z &= 2i\sigma_y . \end{aligned} \qquad 9.154$$

Combining these results with 9.151 we get

$$\sigma_x\sigma_y = i\sigma_z \ , \quad \sigma_y\sigma_z = i\sigma_x \ , \ \sigma_z\sigma_x = i\sigma_y \ . \tag{9.155}$$

To understand the effect of spin, consider a free electron. Since an electron has spin 1/2, the energy is now 4-fold not just 2-fold degenerate. The eigenfunctions of the Hamiltonian

$$H = \frac{\vec{p}^{\,2}}{2m}$$

are

$$\psi_{\vec{k}\uparrow} = \frac{1}{(2\pi)^{3/2}} e^{i\vec{k}\cdot\vec{x}}\begin{pmatrix}1\\0\end{pmatrix} \qquad \vec{p} = \hbar\vec{k} \ , \ s_z = \hbar/2 \tag{9.156}$$

$$\psi_{\vec{k}\downarrow} = \frac{1}{(2\pi)^{3/2}} e^{i\vec{k}\cdot\vec{x}}\begin{pmatrix}0\\1\end{pmatrix} \qquad \vec{p} = \hbar\vec{k} \ , \ s_z = -\hbar/2 \tag{9.157}$$

$$\psi_{-\vec{k}\uparrow} = \frac{1}{(2\pi)^{3/2}} e^{-i\vec{k}\cdot\vec{x}}\begin{pmatrix}1\\0\end{pmatrix} \qquad \vec{p} = -\hbar\vec{k} \ , \ s_z = \hbar/2 \tag{9.158}$$

$$\psi_{-\vec{k}\downarrow} = \frac{1}{(2\pi)^{3/2}} e^{-i\vec{k}\cdot\vec{x}}\begin{pmatrix}0\\1\end{pmatrix} \qquad \vec{p} = -\hbar\vec{k} \ , \ s_z = -\hbar/2 \tag{9.159}$$

They are just the product of the momentum and spin eigenfunctions since all three spin operators $\vec{s}$ commute with H. In general, the Hamiltonian may have a spin-dependent part and then $\vec{s}$ does not commute with H. In that case we have to solve a pair of coupled differential equations to obtain the eigenfunctions and eigenvalues of H. This occurs whenever an external magnetic field is applied because a particle with spin carries a magnetic moment $\vec{\mu}$ proportional to the spin, i.e.,

$$\vec{\mu} = g\left(\frac{\mu_B}{\hbar}\right)\vec{s} \ . \tag{9.160}$$

Here g is the so-called gyromagnetic ratio (a pure number usually of the order of 1 in magnitude) and μ_B is the "magneton" of the

Table 9.2 Typical Magnetons and g-Factors

Particle	Magneton (erg/gauss)	g-Factor
electron	0.927×10^{-20}	-2.00
proton	0.505×10^{-23}	2×2.79
neutron	0.504×10^{-23}	-2×1.91

particle. Thus

$$\mu_B = \frac{|q|\hbar}{2mc} \tag{9.161}$$

where q is the charge and m is the mass of the particle involved. Typical numbers are given in Table 9.2 below. In the case of the electron, the quantity $\frac{e\hbar}{2mc} = 0.927\times10^{-20}$ erg/gauss is known as one "Bohr magneton". The value of $\frac{e\hbar}{2M_p c}$ where M_p = mass of the proton is 0.505×10^{-23} erg/gauss and is known as one "nuclear magneton".

We shall examine the effect of the electronic magnetic moment in Chapter 16. For the present we leave further considerations of spin.

Notes, References and Bibliography

9.1 C. van Winter - Orbital Angular Momentum and Group Representations - Annals of Phys. N.Y. 47 (1968) 232-274.

Several other papers which also consider the question of the single-valuedness of the wavefunction are listed below

E. Merzbacher, Am. J. Phys. 30 (1962) 237.

M.L. Whippmann, Am. J. Phys. 34 (1966) 656.

W. Pauli, Helv. Phys. Acta 12 (1939) 147.

P. Carruthers and M.M. Nieto - Phase and Angle Variables in Quantum Mechanics - Reviews of Modern Physics 40 (1968) 411.

E.T. Whittacker and G.N. Watson - A Course of Modern Analysis, Cambridge at the University Press, 4th edition (1963).

P.M. Morse and H. Feshbach - Methods of Theoretical Physics - McGraw-Hill Book Co., Inc. (1953).

9.2 Y. Aharanov and D. Bohm - Phys. Rev. 115 (1959) 485.

Chapter 9 Problems

9.1 Use algebraic techniques to evaluate the following expectation values as a function of time for a state which at $t = 0$ is given by

$$\psi(x,0) = au_0(x) + bu_3(x)$$

where $u_o(x)$ is the ground state and $u_3(x)$ is the third excited state of a S.H.O.: $\langle H \rangle$, $\langle p^2/2m \rangle$, $\langle 1/2kx^2 \rangle$, $\langle p \rangle$, $\langle x \rangle$, $(\Delta p)^2$, $(\Delta x)^2$.

9.2 a) Compute the 3×3 matrices

$$(L_j)_{mm'} = (Y_{1m}, L_j Y_{1m'})$$

where $j = x, y, z$.

b) Show that they satisfy the cyclic commutation relation

$$[L_x, L_y] = i\hbar L_z \text{ etc.}$$

c) Furthermore show that each matrix L_j satisfies the characteristic equation

$$[L_j^2 - \hbar^2 1)L_j = 0.$$

9.2 (Cont'd)

d) Evaluate in closed form the expression for the matrices

$$e^{iL_x\alpha}\ ,\ e^{iL_y\beta}\ ,\ e^{iL_z\gamma}\ .$$

Hint: Write out their series and resum them using the characteristic equation. Compare these results to rotation matrices i.e. matrices corresponding to a rotation through angles (α,β,γ) about the x, y, z axes respectively.

9.3 Diagonalize the following hamiltonian, that is, find a unitary operator U that brings the following hamiltonian to diagonal form.

$$H = Ea^{\dagger}a + V(a + a^{\dagger})\ ;\quad E,\ V\ \text{constants},$$

where

$$[a,a^{\dagger}] = \beta^2 \quad \text{a positive constant.}$$

Hint: Try operators $b = ua + v$

$$b^{\dagger} = \bar{u}a^{\dagger} + \bar{v}$$

where u and v are complex numbers that you choose, and recall the simple harmonic oscillator.

9.4 Evaluate matrix elements of the form $(Y_{\ell m}, zY_{\ell' m'})$. These occur in the evaluation of dipole radiation rates.

9.5 Let A be an operator such that

$$[A,L_x] = [A,L_y] = 0.$$

Calculate $[A,L^2]$.

Hint: First compute $[A,L_z]$.

9.6 Let a wave function be

$$\psi = A\left(\frac{x^2-y^2+2ixy}{x^2+y^2+z^2} + \frac{3z}{\sqrt{x^2+y^2+z^2}} + 5\right)e^{-\alpha\sqrt{x^2+y^2+z^2}}$$

Find the probability of obtaining any (ℓ,m) value.

9.7 A measurement of the x-component of angular momentum is made on a particle in a state of total angular momentum 1 and z-component 1. What are the probabilities for obtaining the values 1, 0, -1?

In the same situation a measurement is made of the component of angular momentum along an axis lying in the x-z plane and making an angle θ with the z-axis. What is the probability of getting the values 1, 0, -1?

9.8 Compute $[\vec{x},\vec{L}]$ and $[\vec{p},\vec{L}]$ and compare the results with $[\vec{L},\vec{L}]$. What does this suggest about the commutator $[\vec{A},\vec{L}]$ where $\vec{A}$ is an arbitrary vector operator?

9.9 Formally one can derive the relation

$$[L_z,\phi] = -i\hbar$$

and deduce from it that

$$\Delta L_z \Delta\phi \geq \hbar/2.$$

Now $\Delta\phi$ is of necessity $\leq 2\pi$, and in an eigenstate of L_z, $\Delta L_z = 0$. This violates $\Delta L_z \Delta\phi \geq \hbar/2$. Explain this apparent paradox.
Hint: Examine the domain of L_z on which it is self-adjoint. A similar argument also holds for $[p,x] = -i\hbar$ and a particle confined to a finite interval on the line. See also ref. 9.5

9.10 A system is in a state of angular momentum given by

$$\Psi = aY_{1,1} + bY_{1,0} + cY_{1,-1}$$

where

$$|a|^2 + |b|^2 + |c|^2 = 1.$$

a) Compute the expectation value of L_x (Hint: the following formulae may be useful.

$$L_{\pm} Y_{\ell,m} = \sqrt{\ell(\ell+1 - m(m\pm1)}\, \hbar\, Y_{\ell,m\pm1}$$

b) Compute the expectation value of L^2.

c) What are possible values of the coefficients a,b,c in order that

$$L_x\Psi = \hbar\Psi \quad ?$$

It may again be useful to recall that

$$L_x = \frac{1}{2}(L_+ + L_-)$$

9.11 Consider $P_\ell^m(x)$ given by equation 9.106 and expand $(x^2-1)^\ell$ i.e. write the general term in this binomial expression. Hence obtain an explicit expression for $P_\ell^m(x)$. Similarly obtain an explicit form for $P_\ell^{-m}(x)$. By comparing these two results deduce equation 9.107:

$$P_\ell^{-m}(x) = (-1)^m \frac{(\ell-m)!}{(\ell+m)!} P_\ell^m(x).$$

9.12 Starting from equation 9.138,

$$U_R H U_R^{-1} = H$$

and using $U_R = \exp \frac{i\varepsilon}{\hbar} \vec{L}.\vec{n}$ with $|\varepsilon| \ll 1$ obtain to first order in ε that

$$[H\ ,\ \vec{n}\cdot\vec{L}] = 0\ .$$

Hence conclude that

$$[H,\vec{L}] = 0$$

Chapter 10

Central Force Problems

There are many systems such that the potential is a function of only the distance r from the center of force. In these cases the hamiltonian is invariant under rotations and thus commutes with all components of the angular momentum operator. Such problems are called central force problems.

We begin by extracting the angular variables (separation of variables) by using the eigenfunctions $Y_{\ell m}$ of L^2 and L_z. The Schrödinger equation is thus effectively reduced to an equivalent Schrödinger equation in one variable. We then proceed to solve this radial Schrödinger equation for three different potentials: the infinite square well, the isotropic harmonic oscillator and the Coulomb potential.

The isotropic simple harmonic oscillator is solved in Cartesian coordinates as well as in spherical coordinates. This latter solution allows us to introduce the parity operator in spherical coordinates. The associated Laguerre polynomials also arise naturally in these solutions.

The hydrogenic atom (Coulomb potential) is discussed next. There we also give a further discussion of the associated Laguerre polynomials.

Finally we show how to reduce a two-body problem to an equivalent one-body problem by extracting the center of mass motion.

10.1 The Radial Equation

In this chapter we shall only consider hamiltonians of the form

$$H = \frac{\vec{p}^{\,2}}{2m} + V(r) \ . \qquad 10.1$$

Since this hamiltonian shows spherical symmetry we already saw (problem 9.12) that

$$[H,\vec{L}] = 0. \qquad 10.2$$

So H commutes with every component of $\vec{L}$ and hence

$$[H,L^2] = 0 \ . \qquad 10.3$$

Thus it is possible to diagonalize H, L^2 and L_z simultaneously. As always we could have chosen L_x or L_y instead of L_z, but the choice of L_z is conventional. If we now consider the eigenvalue problem

$$H\phi_{E\ell m_\ell} = E\phi_{E\ell m_\ell} \qquad 10.4$$

then we can proceed by separation of variables or use the results of the previous chapter. We take the latter approach and set

$$\phi_{E\ell m_\ell} = R_{E\ell m_\ell}(r)Y_{\ell m_\ell}(\theta,\phi) \ . \qquad 10.5$$

This is possible because

$$\vec{p}^2\psi = -\hbar^2\nabla^2\psi = -\hbar^2 \frac{1}{r}\frac{\partial^2}{\partial r^2} r\psi + \frac{L^2}{r^2}\psi. \tag{10.6}$$

Thus 10.4 reads

$$-\frac{\hbar^2}{2m}\left(\frac{1}{r}\frac{\partial^2}{\partial r^2} r\psi\right) + \frac{L^2}{2mr^2}\psi + V(r)\psi = E\psi$$

and using 10.5 this becomes

$$-\frac{\hbar^2}{2m}\frac{1}{r}\frac{d^2}{dr^2} rR_{E,\ell}(r) + \left[\frac{\ell(\ell+1)\hbar^2}{2mr^2} + V(r)\right]R_{E\ell}(r) = ER_{E\ell}(r). \tag{10.7}$$

We have dropped the m_ℓ dependence of R since 10.7 shows that R is independent of m_ℓ.

If we now call

$$U_{E\ell}(r) = rR_{E\ell}(r) \tag{10.8}$$

we get:

$$-\frac{\hbar^2}{2m}\frac{d^2U_{E\ell}(r)}{dr^2} + \left[\frac{\ell(\ell+1)\hbar^2}{2mr^2} + V(r)\right]U_{E\ell}(r) = EU_{E\ell}(r). \tag{10.9}$$

This is identical in form to the one-dimensional Schrödinger equation. The effective potential is given here by

$$V_{eff}(r) = \frac{\ell(\ell+1)\hbar^2}{2mr^2} + V(r) . \tag{10.10}$$

There is, however, one very major and significant difference here, namely $0 \leq r \leq \infty$ and $U(0) = 0$ as can be seen from 10.8. Thus to make this problem completely equivalent to a one-dimensional problem we need an effective potential

$$V'_{eff}(r) = V_{eff}(r) \qquad r > 0$$

$$V'_{eff}(r) = \infty \qquad r \leq 0 \tag{10.11}$$

On the other hand $\frac{\hbar}{i}\frac{d}{dr}$ as we saw before (section 6.4) is not a self-adjoint operator and has no self-adjoint extensions and thus cannot be considered as an observable, such as for instance the momentum operator in the radial direction. A possible candidate for a radial momentum operator is

$$p_r = \frac{\hbar}{i}\frac{1}{r}\frac{\partial}{\partial r} r . \tag{10.12}$$

If we consider its deficiency indices for $L^2(0,\infty)$ with the inner product

$$(f,g) = \int_0^\infty f(r)g(r)r^2 dr$$

we get

$$p_r f = \pm i\hbar f . \tag{10.13}$$

Thus,

$$\frac{d}{dr}(rf) = \mp rf . \tag{10.14}$$

The solutions are

$$f_{\mp} = \frac{A}{r} e^{\pm r} .$$

Thus the deficiency indices are (1,0) and this is not a possible observable. In fact there does not seem to be an observable corresponding to the radial momentum. Nevertheless p_r is a useful operator in that we can write the kinetic energy

$$T = \frac{\vec{p}^2}{2m} = \frac{p_r^2}{2m} + \frac{L^2}{2mr^2} . \qquad 10.15$$

This is an easy way to remember the Laplace operator in spherical coordinates.

Let us now reconsider the equation 10.9 with the effective potential V_{eff} given at 10.11. We are primarily interested in bound state problems so that $V(r)$ has to be attractive (negative). In fig. 10.1 we have sketched $V'_{eff}(r)$ for $r > 0$ for several different ℓ values.

$$V_{eff}(r) = \frac{\ell(\ell+1)\hbar^2}{2mr^2} + V(r) \text{ for } r > 0$$

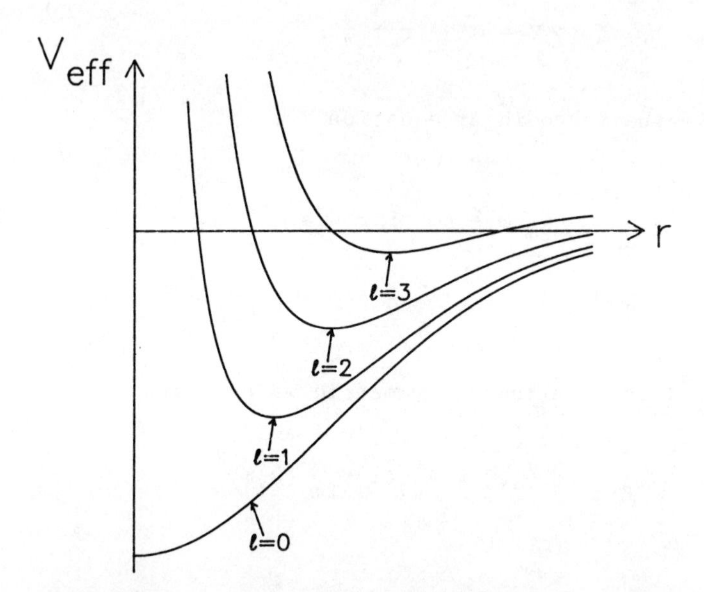

Fig. 10–1 Typical Effective Potentials for Central Force Problem

Thus as ℓ increases, the depth of the well decreases and the well's minimum shifts to the right. Hence a particle tends to be less tightly bound. Also, since the wave function tends to concentrate above the minimum of the potential the most probable point for finding a particle moves further out. This is, of course, the same as in classical mechanics. A particle with high angular momentum is less tightly bound and tends to be in an orbit with a larger radius. We now consider some specific problems.

10.2 Infinite Square Well

The simplest problem we considered in one dimension was the infinite square well. We now reconsider it in three dimensions. The potential we choose is

$$\begin{aligned} V &= 0 \qquad r < a \\ V &= \infty \qquad r \geq a \, . \end{aligned} \tag{10.16}$$

Thus we must solve the Schrödinger equation

$$\frac{p^2}{2m}\psi = E\psi \qquad r < a \tag{10.17}$$

with

$\psi = 0$ at $r = a$.

Since the potential is spherically symmetric we can set

$$\psi_{E\ell m_\ell}(r,\theta,\phi) = R_{E\ell}(r)Y_{\ell m_\ell}(\theta,\phi) \, . \tag{10.18}$$

Then,

$$L^2\psi_{E\ell m_\ell} = \ell(\ell+1)\hbar^2 \, \psi_{E\ell m_\ell} \tag{10.19}$$

$$L_z \psi_{E\ell m_\ell} = m_\ell \hbar \, \psi_{E\ell m_\ell} \tag{10.20}$$

and

$$H\psi_{E\ell m_\ell} = E\psi_{E\ell m_\ell} \; . \tag{10.21}$$

This last equation reduces to an equation for $R_{E\ell}(r)$, namely:

$$-\frac{\hbar^2}{2m}\frac{1}{r}\frac{d^2}{dr^2}(rR_{E,\ell}) + \left[\frac{\ell(\ell+1)\hbar^2}{2mr^2} - E\right]R_{E\ell} = 0 \; . \tag{10.22}$$

Setting

$$k^2 = \frac{2mE}{\hbar^2} \tag{10.23}$$

and

$$x = kr \;, \quad R_{E\ell}(r) = y_{E\ell}(x) \tag{10.24}$$

we get

$$\frac{1}{x}\frac{d^2}{dx^2}(xy_{E\ell}) + \left[1 - \frac{\ell(\ell+1)}{x^2}\right] y_{E\ell} = 0$$

or

$$\left[\frac{d^2}{dx^2} + \frac{2}{x}\frac{d}{dx} + 1 - \frac{\ell(\ell+1)}{x^2}\right] y_{E\ell}(x) = 0. \tag{10.25}$$

The solutions are the spherical Bessel functions which we examine in detail in section 18.5. They are written

$$j_\ell(x) \quad \text{and} \quad n_\ell(x)$$

where

$$j_\ell(x) = \left(\frac{\pi}{2x}\right)^{1/2} J_{\ell+1/2}(x) \tag{10.26}$$

$$n_\ell(x) = (-1)^\ell \left(\frac{\pi}{2x}\right)^{1/2} J_{-\ell-1/2}(x) \; . \tag{10.27}$$

Here $J_n(x)$ is the ordinary Bessel function which satisfies the differential equation

$$\frac{d^2 J_n}{dx^2} + \frac{1}{x}\frac{dJ_n}{dx} + \left(1 - \frac{n^2}{x^2}\right) J_n = 0. \qquad 10.28$$

To bring 10.25 to this form set

$$y = x^{-1/2} u .$$

Then u satisfies 10.28.

Now the behaviours of $j_\ell(x)$ and $n_\ell(x)$ for small x are given by:*

$$j_\ell(x) \underset{x \to 0}{\longrightarrow} \frac{x^\ell}{(2\ell+1)!!} \qquad 10.29$$

$$n_\ell(x) \underset{x \to 0}{\longrightarrow} -(2\ell-1)!!x^{-\ell-1} . \qquad 10.30$$

One of our boundary conditions is that $y_{E\ell}(x)$ be finite at the origin. Thus we have

$$y_{E\ell}(x) = A_\ell j_\ell(x) . \qquad 10.31$$

The equation for the energy eigenvalues is now given by the boundary condition

$$R_{E\ell}(a) = 0$$

or

$$j_\ell(ka) = 0. \qquad 10.32$$

---

* All results pertaining to spherical Bessel functions are derived in Section 18.5.

Thus if $\lambda_{n,\ell}$ is the n'th zero of $j_\ell(x)$ then we have the energy $E_{n,\ell}$ given by

$$E_{n,\ell} = \frac{\hbar^2}{2ma^2}\lambda_{n,\ell}^2 \quad . \tag{10.33}$$

We shall not carry this problem any further. Instead we now consider the eigenvalue problem for the three dimensional simple harmonic oscillator.

10.3 Simple Harmonic Oscillator: Separation in Cartesian Coordinates

The hamiltonian in this case is

$$H = \frac{\vec{p}^{\,2}}{2m} + \frac{1}{2}kr^2 \quad . \tag{10.34}$$

We first solve this problem in Cartesian coordinates.

Clearly we can write H as

$$H = H_x + H_y + H_z \tag{10.35}$$

where

$$H_x = \frac{p_x^2}{2m} + \frac{1}{2}kx^2 \tag{10.36}$$

and similarly for H_y and H_z. Each of these is just a one dimensional simple harmonic oscillator of the kind we solved before (section 9.1). Furthermore, all of these hamiltonians are mutually commuting.

$$[H_x,H_y] = [H_x,H_z] = [H_y,H_z] = 0 \quad . \tag{10.37}$$

Thus we can diagonalize them simultaneously. Hence the eigenfunction ϕ of

$$H\psi_n = E_n\psi_n$$

can be written

$$\psi = \psi(x)\psi(y)\psi(z)$$

where

$$H_x\psi_{n_1}(x) = E_{n_1}\psi_{n_1}(x) \qquad 10.38$$

and so forth. As we saw before, the eigenvalues are

$$E_{n_1} = (n_1 + 1/2)\hbar\omega$$

$$E_{n_2} = (n_2 + 1/2)\hbar\omega \qquad 10.39$$

$$E_{n_3} = (n_3 + 1/2)\hbar\omega.$$

Thus,

$$E_n \equiv E_{n_1 n_2 n_3} = (n_1 + n_2 + n_3 + \tfrac{3}{2})\hbar\omega \qquad 10.40$$

or

$$E_n = (n + \tfrac{3}{2})\,\hbar\omega \qquad 10.41$$

where

$$n = n_1 + n_2 + n_3. \qquad 10.42$$

10.3.1 Degeneracy

We mentioned this concept before and simply recall the definition. An eigenvalue is degenerate of order d if there are d linearly independent eigenfunctions yielding this eigenvalue. In this case as a glance at 10.41 and 10.42 shows all eigenvalues except the ground-state (lowest state) are degenerate. Generally a degeneracy indicates that the hamiltonian has some special symmetry property. In this case the hamiltonian is spherically symmetric, hence the degeneracy. Another general property is that the ground-state is non-degenerate. We now examine the degeneracy of our oscillator more closely.

$n = 0$. In this case $n_1 = n_2 = n_3 = 0$ and this state is non-degenerate.

$n = 1$. In this case we have 3 possibilities;
$n_1 = 1,\ n_2 = n_3 = 0;\quad n_2 = 1,\ n_1 = n_3 = 0;\quad n_3 = 1,\ n_1 = n_2 = 0$.

Thus we have a three-fold degeneracy and the following eigenvalue solutions:

$$H\psi_{100} = \frac{5}{2}\hbar\omega\,\psi_{100}$$

$$H\psi_{010} = \frac{5}{2}\hbar\omega\,\psi_{010}$$

$$H\psi_{001} = \frac{5}{2}\hbar\omega\,\psi_{001}\ .$$

$n = 2$. In this case the degeneracy is 6-fold. The general formula for d, the order of the degeneracy, is

$$d = \frac{(n+2)(n+1)}{2}\ . \qquad 10.43$$

10.4 Simple Harmonic Oscillator: Separation in Spherical Coordinates

We now solve this problem again making explicit use of the spherical harmonics and orbital angular momentum. Since

$$H = \frac{p^2}{2m} + \frac{1}{2} kr^2 \qquad 10.34$$

we can write

$$\psi = \psi_{n\ell m} = R_{n\ell}(r)Y_{\ell m}(\theta,\phi). \qquad 10.44$$

This follows because

$$[H,L^2] = [H,L_z] = 0. \qquad 10.45$$

In addition we find that the hamiltonian 10.34 is invariant under parity transformations. Now under parity transformations both the position operator $\vec{r}$ and the momentum operator $\vec{p}$ change sign. Therefore under a parity transformation the angular momentum operator $\vec{L} = \vec{r}\times\vec{p}$ remains invariant. Thus,

$$P\vec{L} = \vec{L}P \qquad 10.46$$

This allows us to choose simultaneous eigenstates of H, L^2, L_z and P. Writing out the parity transformation $P\vec{r}P^{-1} = \vec{r}\,' = -\vec{r}$ in spherical coordinates, we find (figure. 10.2)

$$\begin{aligned} r' &= r \\ \theta' &= \pi-\theta \\ \phi' &= \phi+\pi \end{aligned} \qquad 10.47$$

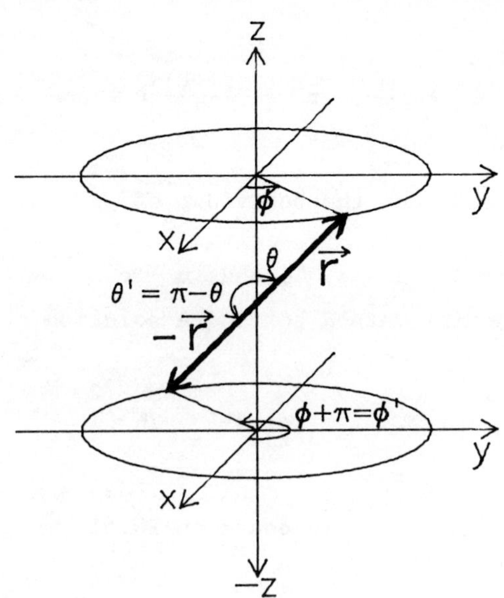

Fig. 10–2 Parity Transformation in Spherical Coordinates

On the other hand, using the explicit form (equations 9.105, 9.106) for the spherical harmonics, we find that

$$Y_{\ell,m}(\pi-\theta,\phi+\pi) = (-1)^{\ell} Y_{\ell,m}(\theta,\phi) \qquad 10.48$$

Thus,

$$PY_{\ell,m} = (-1)^{\ell} Y_{\ell,m} \qquad 10.49$$

showing that the spherical harmonics are eigenstates of the parity operator with the eigenvalue +1 if ℓ is even and −1 if ℓ is odd.

Now we are ready to consider the radial equation for the S.H.O.. Setting, as usual,

$$rR_{n,\ell} = u_{n,\ell} \qquad 10.50$$

we find

$$\frac{d^2u}{dr^2} - [\frac{m^2\omega^2}{\hbar^2} r^2 + \frac{\ell(\ell+1)}{r^2}]u = - \frac{2mE}{\hbar^2} u \qquad 10.51$$

To solve we first consider the behaviour of u for $r \to 0$ and $r \to \infty$. These are respectively $u \sim r^{\ell+1}$ and $u \sim e^{-\frac{m\omega}{2\hbar} r^2}$. We again use Sommerfeld's polynomial method to find a solution of the form

$$u = A(r) r^{\ell+1} e^{-\frac{m\omega}{2\hbar} r^2} \qquad 10.52$$

Substituting this expression in equation 10.51, we get the following equation for $A(r)$

$$\frac{d^2A}{dr^2} + 2(\frac{\ell+1}{r} - \frac{m\omega}{\hbar} r)\frac{dA}{dr} + [\frac{2mE}{\hbar^2} - \frac{m\omega}{\hbar} (2\ell+3)]A = 0 \qquad 10.53$$

Since we already have the asymptotic behaviour for $u(r)$ the solution for $A(r)$ must be a polynomial. Thus we try

$$A = \sum_{n=0}^{M'} a_n r^n \qquad 10.54$$

Substituting this expression into equation 10.53 and collecting terms with the same power of r we get:

$$\sum_{n=0}^{M'} \{n(n+\ell+2)a_{n+2} - [\frac{2m\omega}{\hbar} (n+\ell+3/2) - \frac{2mE}{\hbar^2}]a_n\}r^n = 0 \qquad 10.55$$

Thus a solution exists if

$$a_{n+2} = \frac{\frac{2m\omega}{\hbar} (n+\ell+3/2) - \frac{2mE}{\hbar^2}}{n(n+\ell+2)} a_n \qquad 10.56$$

If $a_{M'+2} = 0$ so that $a_{M'}$ is the highest term in the expansion 10.54 then the numerator on the right side of 10.56 must vanish for $n = M'$. This gives us the energy quantization

$$E_{M',\ell} = \hbar\omega(M'+\ell+3/2) \qquad M' = 0,1,\ldots \tag{10.57}$$

or

$$E_N = \hbar\omega(N+3/2) \tag{10.58}$$

where

$$N = M'+\ell \tag{10.59}$$

Thus the energy is highly degenerate since we can have for a given M' all ℓ values from 0 to $N-M'$. However, if we insist on parity as a good quantum number then according to equation 10.49 we can either have all even ℓ or all odd ℓ between 0 an $N-M'$ for a given M'. This means that M' must be an even integer, $M' = 2M$. It is now a simple matter to count the degeneracy and see that we get the same result as equation 10.43.

There is another way to proceed for the radial equation that allows us to make contact with functions (polynomials) that occur in the solution of the hydrogenic system considered next in section 10.5. The trick is to introduce a new independent variable

$$x = \frac{\hbar r^2}{m\omega} \quad \text{and} \quad y(x) = A(r). \tag{10.60}$$

With these changes equation 10.53 becomes

$$x\,\frac{d^2y}{dx^2} + (\ell+3/2-x)\frac{dy}{dx} + \frac{1}{4}\left(\frac{2E}{h\omega} - 2\ell-3\right)y = 0. \tag{10.61}$$

This is just a special case of the equation for the associated Laguerre polynomials $L^b_{a-b}(x)$, namely

$$x \frac{d^2}{dx^2} L^b_{a-b}(x) + (b+1-x)\frac{d}{dx} L^b_{a-b}(x) + (a-b)L^b_{a-b}(x) = 0 \qquad 10.62$$

Thus to express $y(x)$ in terms of $L^b_{a-b}(x)$ we need the identification

$$b = \ell + 1/2 \qquad 10.63$$

$$a - b = \frac{1}{4}\left(\frac{2E}{\hbar\omega} - 2\ell - 3\right) \qquad 10.64$$

It is a general property of equation 10.62 (see section 10.5) that for the solutions to be polynomials we need that $a-b = M$, an integer. Thus

$$a = \ell + 1/2 + M \quad , \; M = 0,1,2,\ldots \qquad 10.65$$

and

$$\frac{1}{4}\left(\frac{2E}{\hbar\omega} - 2\ell - 3\right) = M$$

So

$$E_{M,\ell} = \hbar\omega(\ell + 2M + 3/2) = \hbar\omega(N + 3/2) \qquad 10.66$$

where

$$N = \ell + 2M \; . \qquad 10.67$$

This is the same result we discovered above by using the parity operator, namely

$$E_N = \hbar\omega(N + 3/2) \; , \quad N = \ell + 2M \; , \; M = 0,1,2,\ldots \qquad 10.68$$

We introduced the associated Laguerre polynomials here in a rather ad hoc fashion. In the next section we shall encounter them again and discuss them in more detail.

10.5 The Hydrogenic Atom

We consider a particle of charge $-e$ in the electromagnetic field of a charge Ze. Then

$$V(r) = -\frac{Ze^2}{r} \tag{10.69}$$

and

$$H = \frac{\vec{p}^{\,2}}{2m} - \frac{Ze^2}{r}\ . \tag{10.70}$$

Since the potential is central, angular momentum is conserved and we have

$$[L^2,H] = [L_z,H] = [L^2,L_z] = 0\ . \tag{10.71}$$

Therefore the Schrödinger equation

$$H\psi = E\psi$$

separates and we can write

$$\psi_{n,\ell,m} = R_{n,\ell}(r)Y_{\ell,m}(\theta,\phi) \tag{10.72}$$

or putting $U_{n,\ell}(r) = rR_{n,\ell}(r)$ as in equation 10.8 we get the radial equation

$$\left[-\frac{\hbar^2}{2m}\frac{d^2}{dr^2} + \frac{\ell(\ell+1)\hbar^2}{2mr^2} - \frac{Ze^2}{r}\right]U = EU\ . \tag{10.73}$$

It is convenient to introduce dimensionless quantities. To this end we define

$$\kappa = \sqrt{-\frac{2mE}{\hbar^2}}\ . \tag{10.74}$$

The minus sign is due to the fact that we are looking for bound states so that E is negative. Now let

$$x = 2\kappa r \text{ and } U(r) = y(x)\ . \tag{10.75}$$

Then with

$$\nu = \frac{mZe^2}{\kappa\hbar^2} , \tag{10.76}$$

equation 10.73 becomes

$$y'' - [\frac{\ell(\ell+1)}{x^2} - \frac{\nu}{x} + \frac{1}{4}]y = 0 . \tag{10.77}$$

We again first consider the asymptotic forms of y and then use Sommerfeld's polynomial method. For large x 10.77 becomes

$$y'' - \frac{1}{4} y \simeq 0 , \qquad x \to \infty$$

so

$$y \sim e^{\pm x/2} .$$

Since y must be square-integrable only the minus sign is acceptable. For small x the equation becomes

$$y'' - \frac{\ell(\ell+1)}{x^2} y \simeq 0 , \quad x \to 0 .$$

The solutions are $y \simeq x^{\ell+1}$ and $y \simeq x^{-\ell}$. Since we must have $y \to 0$ as $x \to 0$ only $y \simeq x^{\ell+1}$ is acceptable. We therefore set

$$y = e^{-x/2} x^{\ell+1} v(x) \tag{10.78}$$

and substitute this into 10.77 to obtain an equation for $v(x)$. The resultant equation is

$$xv'' + (2\ell+2-x)v' + (\nu-\ell-1)v = 0. \tag{10.79}$$

This is precisely of the same form as equation 10.62 if we set

$$b = 2\ell + 1 , \ a = \ell + \nu . \tag{10.80}$$

To get polynomial solutions requires that $a-b = \nu-\ell-1$ be a non-negative integer. Thus

$$\nu = \ell+N = n \ , \quad N = 1,2,\ldots \ . \tag{10.81}$$

This is precisely the quantization condition for the energy since

$$\nu = \frac{mZe^2}{\kappa\hbar^2} = \frac{mZe^2}{\hbar^2}\sqrt{-\frac{\hbar^2}{2mE}} = n.$$

Thus

$$E_n = -\frac{1}{2}\, m\, \frac{Z^2e^4}{\hbar^2}\, \frac{1}{n^2} \tag{10.82}$$

or introducing the fine structure constant

$$\alpha = \frac{e^2}{\hbar c} \simeq \frac{1}{137} \tag{10.83}$$

we get

$$E_n = -\frac{1}{2}\, mc^2 (Z\alpha)^2 \cdot \frac{1}{n^2} \quad , \ n = 1,2,\ldots \ . \tag{10.84}$$

This is the same result we obtained using Bohr-Sommerfeld quantization (section 2.4).

We now return to a systematic study of equation 10.62, or equivalently 10.79. If we try a series solution

$$L^b_{a-b}(x) = 1 + a_1 x + a_2 x^2 + \ldots + a_{N-1} x^{N-1} \tag{10.85}$$

and substitute this into equation 10.62 and equate the coefficient of equal powers of x we get:

$$(n+1)(n+1+b)a_{n+1} = (n+b-a)a_n \tag{10.86}$$

If we want $a_{N-1} \neq 0$ but $a_N = 0$ then we need that

$$N-1+b-a = 0$$

or

$$N = a-b+1 \tag{10.87}$$

This is the condition we used before to ensure that our solutions are polynomials. If our solutions were not polynomials we would have to examine the convergence of the series $\sum_{n=0}^{\infty} a_n x^n$, where equation 10.86 yields a recursion relation for the coefficients a_n. From 10.86 we see that

$$\frac{a_n}{a_{n-1}} = \frac{n+b-a-1}{n(n+b)} \underset{n\to\infty}{\longrightarrow} \frac{1}{n} \tag{10.88}$$

Thus the asymptotic behaviour of the series would be such that $\frac{a_n}{a_{n-1}} = \frac{1}{n}$ or $a_n = \frac{1}{n!}$. This would lead to behaviour like e^x and in view of the asymptotic behaviour displayed in equation 10.78 is unacceptable.

Now consider equation 10.62 with $y = L^b_{a-b}(x)$.

$$x \frac{d^2y}{dx^2} + (b+1-x) \frac{dy}{dx} + (a-b)y = 0 . \tag{10.62}$$

We look for a convenient integral representation for the solution of this equation. Thus we try

$$y(x) = \int_C e^{-xt} f(t)dt \tag{10.89}$$

where we shall later choose the integration path for our convenience. Substituting into 10.62 and differentiating under the integral sign yields:

$$\int_C f(t)[xt^2 - (b+1-x)t + a-b]e^{-xt}dt = 0. \tag{10.90}$$

This can be rewritten as

$$\int_C f(t)[-t(t+1)\frac{d}{dt} + a-b-(b+1)t]e^{-xt}dt = 0. \tag{10.91}$$

We now rewrite the first term as a total differential by subtracting and adding $-e^{-xt}\frac{d}{dt}[f(t)t(t+1)]$. Then we get

$$\int_C \frac{d}{dt}[-e^{-xt}t(t+1)f(t)]dt + \int_C e^{-xt}\{t(t+1)\frac{df}{dt} + [a-b+1-(b-1)t]f(t)\}dt = 0. \tag{10.92}$$

If the contour C forms a closed path then the first integral vanishes. To make the second integral vanish we now simply choose f to satisfy

$$t(t+1)\frac{df}{dt} = -[a-b+1-(b-1)t]f(t) \ . \tag{10.93}$$

The solution of this equation is:

$$f(t) = A\,\frac{(t+1)^a}{t^{a-b+1}} \tag{10.94}$$

Now in order to be able to close the contour we must not cross any branch points. Thus we need a = integer, a-b+1 = integer. This allows us to choose the contour to be a circle enclosing the origin. We then find

$$y(x) = A\oint e^{-xt}\,\frac{(t+1)^a}{t^{a-b+1}}\,dt \tag{10.95}$$

Now we have only the contribution from the simple pole at t = 0. This residue is obtained by writing

$$e^{-xt} = \sum_{n=0}^{\infty} \frac{(-x)^n}{n!} t^n \qquad 10.96$$

and

$$(t+1)^a = \sum_{r=0}^{a} \binom{a}{r} t^r \qquad 10.97$$

so that the integrand is

$$\sum_{n=0}^{\infty} \sum_{r=0}^{a} \frac{(-x)^n}{n!} \binom{a}{r} t^{n+r+b-a-1} \qquad 10.98$$

Thus we have a simple pole whenever

$$n+r+b-a = 0$$

or

$$n = a-b-r \qquad 10.99$$

The contribution of this pole is

$$\int \frac{(-x)^{a-b-r}}{(a-b-r)!} \binom{a}{r} t^{-1} dt = 2\pi i \frac{(-x)^{a-b-r}}{(a-b-r)!} \binom{a}{r} \qquad 10.100$$

Thus up to normalization we have

$$L_{a-b}^{b}(x) = A \sum_{n=0}^{a-b} \frac{(-x)^{a-b-r}}{(a-b-r)!} \binom{a}{r} \cdot 2\pi i \qquad 10.101$$

The normalization is by convention such that

$$L_a^o(0) \equiv L_a(0) = a!. \qquad 10.102$$

The functions $L_a(x)$ are known as the Laguerre polynomials. Subsituting this into 10.101 we get

$$A = \frac{a!}{2\pi i} \qquad 10.103$$

So we have

$$L_a(x) \equiv L_a^o(x) = \frac{a!}{2\pi i} \oint e^{-xt} \frac{(t+1)^a}{t^{a+1}} dt \qquad 10.104$$

We can rewrite this as

$$L_a(x) = \frac{a!}{2\pi i} e^x (-1)^a \frac{d^a}{dx^a} \oint \frac{e^{-x(t+1)}}{t^{a+1}} dt$$

$$= (-1)^a \frac{a!}{2\pi i} e^x \frac{d^a}{dx^a} \oint \frac{e^{-xt}}{t^{a+1}} dt \, e^{-x}$$

or

$$L_a(x) = e^x \frac{d^a}{dx^a} (x^a e^{-x}) \qquad 10.105$$

This is a Rodrigues' formula for the Laguerre polynomials $L_a(x)$. To get the associated Laguerre polynomials $L_{a-b}^b(x)$ we first note that the Laguerre polynomials $L_a(x)$ satisfy the equation

$$x \frac{d^2 L_a}{dx^2} + (1-x) \frac{dL_a}{dx} + aL_a = 0 \, . \qquad 10.106$$

If we differentiate this equation b times we find that

$$y_a^b = \frac{d^b L_a}{dx^b} \qquad 10.107$$

satifies

$$x \frac{d^2 y_a^b}{dx^2} + (b+1-x) \frac{dy_a^b}{dx} + (a-b)y_a^b = 0 \qquad 10.108$$

which is the same as equation 10.62. Thus we have

$$L_{a-b}^b(x) = c_a^b \frac{d^b L_a(x)}{dx^b} \qquad 10.109$$

where c_a^b is a constant which by convention is chosen as $(-1)^b$. Thus,

$$L_{a-b}^b(x) = (-1)^b \frac{d^b L_a(x)}{dx^b} \qquad 10.110$$

or using 10.95 and 10.103

$$L_{a-b}^b(x) = \frac{a!}{2\pi i} \oint e^{-xt} \frac{(t+1)^a}{t^{a-b+1}} dt. \qquad 10.111$$

In terms of these functions we are now ready to write down the bound-state wave functions for the hydrogenic atom. A note of caution is in order: the functions $L_{a-b}^b(x)$ are also written sometimes as $L_a^b(x)$. Since both notations are fairly common, care must be exercised when mixing formulas from different books.

Up to normalization the hydrogen atom wave functions are given by

$$\psi_{n,\ell,m}(r,\theta,\phi) = A_{n,\ell} e^{-x/2} x^{\ell} L_{n-\ell-1}^{2\ell+1}(x) Y_{\ell,m}(\theta,\phi). \qquad 10.112$$

The constant $A_{n,\ell}$ is the normalization constant and may be evaluated in a number of ways (most often a generating function is used). The result is

$$\psi_{n,\ell,m}(\vec{r}) = \left(\left(\frac{2Z}{na_o}\right)^3 \frac{(n-\ell-1)!}{2n[(n+\ell)!]^3} \right)^{1/2} \left(\frac{2Zr}{na_o}\right)^\ell \exp - \frac{Zr}{na_o}$$

$$\cdot\; L^{2\ell+1}_{n-\ell-1}\left(\frac{2Zr}{na_o}\right) Y_{\ell,m}(\theta,\phi) \qquad 10.113$$

where

$$a_o = \frac{\hbar^2}{me^2} = \text{Bohr radius.} \qquad 10.114$$

Since the hydrogen atom has a high degree of symmetry (it is invariant under rotations) we expect at least the usual $(2\ell+1)$-fold degeneracy associated with a central potential. In fact the degeneracy is even higher. This so-called "accidental" degeneracy is due to the fact that the hydrogen hamiltonian does not change under an even larger group of transformations than just the three-dimensional rotations. We shall now calculate the degree d_n of the degeneracy for each level n. Since the value of E_n depends only on n we have degeneracy with respect to both m (rotational) and ℓ (accidental). For each fixed valued of n, ℓ can vary from 0 to $n-1$ and for each of these ℓ values m can vary over $2\ell+1$ values from $-\ell$ to ℓ. Thus the degree of degeneracy (ignoring spin) is given by

$$d_n = \sum_{\ell=0}^{n-1} (2\ell+1) = 2\,\frac{n(n-1)}{2} + n = n^2 . \qquad 10.115$$

The problem we have solved yields only the bound state (square-integrable) wave-functions for the hydrogenic atom. These do <u>not</u> form a complete set. In addition there are solutions that are only δ-function normalizable. These solutions correspond to continuous positive values of the energy and represent solutions for particles scattered by a Coulomb potential. Due to the fact that the Coulomb potential decreases only very slowly (as r^{-1}), these solutions are rather complicated and will not be considered by us. They are expressible in terms of hypergeometric functions but are usually avoided.

In practice the Coulomb potential is usually "screened" by other charges and thus yields an effective potential that decreases more rapidly. This yields a simpler scattering problem and is considered in section 18.4 where we obtain the quantum mechanical analogue of Rutherford scattering by this means.

In our treatment of the hydrogen atom we have not yet justified the use of a fixed center of force. We shall now do so and show how this is accomplished for a general two-body problem. The procedure is identical to that used in classical mechanics for removing the center of mass motion.

10.6 Reduction of the Two-Body Problem

Consider a general hamiltonian for two particles interacting only with each other. Then using the fact that space is homogeneous (no preferred origin) restricts the possible interaction potential to a function of $\vec{r}_1-\vec{r}_2$ where $\vec{r}_1$ and $\vec{r}_2$ are the position vectors of particles 1 and 2 respectively. The hamiltonian is therefore of the form

$$H = \frac{\vec{p}_1^{\,2}}{2m_1} + \frac{\vec{p}_2^{\,2}}{2m_2} + V(\vec{r}_1 - \vec{r}_2) \ . \qquad 10.116$$

We now introduce the center of mass and relative coordinates

$$\vec{R} = \frac{m_1\vec{r}_1 + m_2\vec{r}_2}{m_1 + m_2} \qquad 10.117$$

$$r = \vec{r}_1 - \vec{r}_2 \ . \qquad 10.118$$

It is also convenient to introduce the total and reduced masses

$$M = m_1 + m_2 \tag{10.119}$$

$$m = \frac{m_1 m_2}{m_1 + m_2} , \tag{10.120}$$

as well as the total momentum $\vec{P} = -i\hbar\vec{\nabla}_R$, and the relative momentum $\vec{p}$ $-i\hbar\vec{\nabla}_r$. The hamiltonian then can be rewritten (problem 10.4) to read

$$H = \frac{\vec{P}^2}{2M} + \frac{\vec{p}^2}{2m} + V(\vec{r}) . \tag{10.121}$$

Furthermore

$$[\vec{P},\vec{p}] = 0 . \tag{10.122}$$

Also

$$[\vec{P},H] = 0. \tag{10.123}$$

Thus we can diagonalize $\vec{P}$ and H simultaneously. This amounts of course to extracting the center of mass motion.

In fact if we call

$$H_o = \frac{\vec{p}^2}{2m} + V(\vec{r}) . \tag{10.124}$$

Then H_o , H and $\vec{P}$ can all be simultaneously diagonalized to give

$$\vec{P}\phi(\vec{R}) = \hbar\vec{K}\phi(\vec{R}) \tag{10.125}$$

$$H_o\psi(\vec{r}) = E_o\psi(\vec{r}) \tag{10.126}$$

and

$$H\Psi(\vec{R},\vec{r}) = \left(\frac{\hbar^2\vec{K}^2}{2M} + E_o\right)\Psi(\vec{R},\vec{r}) \ . \qquad 10.127$$

This is, in fact, accomplished simply by separation of variables, i.e.

$$\Psi(\vec{R},\vec{r}) = \phi(\vec{R})\psi(\vec{r}) \ . \qquad 10.128$$

Then

$$\phi(R) = \left(\frac{1}{2\pi}\right)^{3/2} e^{i\vec{K}\cdot\vec{R}} \ . \qquad 10.129$$

Equation 10.126 is now nothing more than the Schrödinger equation for a particle in a fixed center of force. Thus, we get the equivalent one-body problem from a given two-body problem by simply using the reduced mass in the relative coordinate system. For the hydrogen atom this amounts to replacing the electron mass m_e by the reduced mass

$$m = \frac{m_e m_p}{m_e + m_p} \qquad 10.130$$

where m_p, the proton mass is about 1840 m_e. Thus this produces a correction of about .05% and is well within the limits of observable effects.

This concludes our treatment of exactly solvable problems. There are several more classes of potentials for which closed-form solutions are known. The principal merit of these exact solutions is that they produce a point of departure for approximate solutions. This will be the subject matter of a later chapter. In preparation for this we next develop some more formalism and in the process improve our notation as well.

Notes, References and Bibliography

The original paper treating the hydrogen atom is still very readable. It is:

E. Schrödinger - Annalen der Physik 79 (1926).

Many detailed solutions of central force problems are to be found in the "Pauli Lectures on Physics" Vol. 5 Wave Mechanics edited by C.P. Enz - The MIT Press, Cambridge, Mass. 1973.

See also ref. 4.3.

For some of the difficulties and interpretation of the "radial momentum operator" see:

F.S. Crawford Jr. - Amer. J. Phys. 32 (1964) 611.

The classical limit of the hydrogen atom is discussed in:

L.S. Brown - Amer. J. Phys. 41 (1973) 525.

Chapter 10 Problems

10.1 Solve the isotropic simple harmonic oscillator problem in two dimensions in both Cartesian and cylindrical coordinates. Hint: L_z commutes with the hamiltonian as does the parity operator.

10.2 Consider the attractive potential $V(r) = -V_o e^{-2\alpha r}$. This is one of the few solvable potentials.
Hint: Change variables to $u = e^{-\alpha r}$. The resultant equation is Bessel's equation. Discuss carefully the boundary conditions to be obeyed by $\phi(u) = R(r)$.

10.3 Normalize the hydrogenic wave functions.

10.4 Show that the hamiltonian given in 10.116 reduces to the hamiltonian in 10.121 under the transformations 10.117–10.120.

10.5 A particle is in a spherical potential well

$$V(r) = \begin{cases} -V_o & r < a \\ 0 & r > a \end{cases} .$$

Find the transcendental equation which yields the energy eigenvalue for the state with angular momentum ℓ. What is the minimum degeneracy of this state? If a proton and neutron are bound in an $\ell = 0$ state with an energy of 2.2 MeV, determine V_o for $a \simeq 2 \times 10^{-13}$ cm.

Chapter 11

Transformation Theory

The choice of a set of coordinate axes or basis is completely arbitrary in Euclidean space. The same thing is true in hilbert space. It is therefore very useful to know how to change from one basis to another. This is known as transformation theory. Since the basis set is competely arbitrary it is also useful to work as much as possible in a basis independent manner. This is what we do in Euclidean space when we work with equations for vectors rather than with equations for the components of vectors. There is a completely analogous procedure available in hilbert space involving a special vector notation called Dirac notation. This powerful notation is discussed in this chapter.

Transformations from one basis set to another are carried out by means of unitary transformations. If we also permit time-dependent unitary transformations we can throw all, or part, of the time dependence onto the operators. We refer to these time transformed representations as "pictures" and discuss three of them, the Schrödinger, Heisenberg and Dirac in detail.

11.1 Rotations in a Vector Space

Since by definition, a basis set is a complete orthonormal set, say $\{u_n(x_1,x_2,x_3)\}$, we can expand any given wave function $\psi(x_1,x_2,x_3)$ in terms of this basis set. The index n may be a multi-index consisting of a set of discrete indices as in the three-dimensional oscillator where $n \equiv (n_1,n_2,n_3)$ or it may be a set of continuous indices or a mixture of discrete and continuous indices.

We shall use a summation sign as a generic symbol for summation over discrete and integration over continuous indices. Thus in the case stated,

$$\psi(x_1,x_2,x_3) = \sum_n a_n u_n(x_1,x_2,x_3) \tag{11.1}$$

where

$$a_n = (u_n,\psi) \ . \tag{11.2}$$

It is clear that once we have picked the basis set $\{u_n\}$, the wave function is completely specified by the set of ordered numbers $(a_1,a_2,\ldots)$. This is completely analogous to the representation of vectors by ordered n-tuples in ordinary analytic geometry.

For example in E_3 we may pick an orthonormal triad $\hat{e}_1$, $\hat{e}_2$, $\hat{e}_3$. Then we can write a vector $\vec{v}$ as

$$\vec{v} = \sum_{n=1}^{3} a_n \hat{e}_n$$

where

$$a_n = (\hat{e}_n,\vec{v})$$

It is now quite common to suppress the basis vectors $\{\hat{e}_n\}$ and write

$$\vec{v} = (a_1, a_2, a_3) \quad .$$

The change from one basis set to another is accomplished in E_3 by means of rotations or orthogonal transformations. To see this consider a second basis set $\{\hat{f}_1, \hat{f}_2, \hat{f}_3\}$ obtained from the set $\{\hat{e}_1, \hat{e}_2, \hat{e}_3\}$ by a rotation R . Thus we have:

$$\hat{f}_\ell = \sum_n R_{\ell n} \hat{e}_n \quad . \tag{11.3}$$

The statement, that the set $\{\hat{f}_\ell\}$ is still orthonormal, when written out, reads

$$(\hat{f}_\ell, \hat{f}_k) = \delta_{\ell,k} = (\sum_n R_{\ell n} \hat{e}_n \, , \, \sum_m R_{\ell m} \hat{e}_m)$$

$$= \sum_{n,m} R_{\ell n} R_{km} (\hat{e}_n, \hat{e}_m)$$

$$= \sum_{n,m} R_{\ell n} R_{km} \delta_{nm} = \sum_n R_{\ell n} R_{k,n} \tag{11.4}$$

In terms of the matrix R, equation 11.4 reads

$$RR^t = R^t R = 1 \quad . \tag{11.5}$$

Here the superscript t means "transpose". Equation 11.5 states that the matrices R are orthogonal. Conversely if the matrices R are orthogonal so that equation 11.5 holds then the vectors $\hat{f}_\ell$ defined by equation 11.3 also form an orthonormal basis set if the $\hat{e}_n$ do.

A completely analogous condition holds for vectors in hilbert space H. In this case since the vectors are complex, we have complex orthogonal or <u>unitary</u> transformations. To see how this works consider two different orthonormal basis sets $\{u_n\}$ and $\{v_n\}$. Then, since the sets are by definition complete we can expand one set in terms of the other. Thus we have

$$v_n = \sum_m V_{nm} u_m \tag{11.6}$$

and

$$u_n = \sum_m V^{-1}_{nm} v_m \tag{11.7}$$

That V^{-1} exists is obvious from the fact that the u_n may be expanded in terms of the v_n.

Now using the orthonormality of the sets we get

$$(v_n, v_\ell) = \delta_{n,\ell} = \sum_{m,k} V^*_{nm} V_{\ell k} (u_m, u_k)$$

$$= \sum_{m,k} V^*_{nm} V_{\ell k} \delta_{m,k} = \sum_m V^*_{nm} V_{\ell m}$$

$$= \sum_m V_{\ell m} (V^{*t})_{mn}$$

or in operator form

$$VV^{*t} \equiv VV^\dagger = 1 . \tag{11.8}$$

In a similar fashion using 11.7 we obtain:

$$\delta_{n,\ell} = \sum_m V^{-1}_{\ell m} \, (V^{*t-1})_{mn}$$

or in operator form

$$V^{-1} V^{-1\dagger} = 1 . \tag{11.9}$$

Taking the inverse of this last equation we get

$$V^\dagger V = 1. \tag{11.10}$$

Thus

$$V^{\dagger}V = VV^{\dagger} = 1. \tag{11.11}$$

Definition

An operator V is unitary if and only if it satisfies equations 11.11.

Thus V is a unitary operator and we can change bases in H by means of unitary transformations.

11.2 Example 1. Fourier Transform of Hermite Functions

A complete orthonormal basis in H for one-dimensional problems is provided by the hermite functions $u_n(x)$ given by equation 9.28.

If we choose units such that $k/\hbar\omega = 1$ then we have

$$u_n(x) = (-1)^n (n!)^{-1/2}\, 2^{-n/2} \left(\frac{d}{dx} - x\right)^n \pi^{1/4}\, e^{-x^2/2}.$$

These functions as stated form a complete orthonormal set. If we also admit continuous eigenvalues (rigged hilbert space) then the functions

$$v_k(x) = (2\pi)^{-1/2} e^{-ikx}$$

also form a complete orthonormal set. We shall now find the unitary operator V with matrix elements $V_{k,n}$ connecting the 2 sets. Thus

$$v_k(x) = \sum_n V_{k,n} u_n(x) .$$

Using the orthonormality of the u_n we get:

$$(v_k, u_m) = \sum_n V^*_{k,n} (u_n, u_m) = V^*_{k,m} . \tag{11.12}$$

We therefore need only evaluate the inner product (v_k, u_m). We leave it as an exercise (problem 11.1) to show that the result is

$$V_{k,m} = i^m u_m(k) \ . \tag{11.13}$$

This shows furthermore that the Fourier transform of a hermite function is again a hermite function.

11.3 Dirac Notation

So far we have always, in a sense, used a representation of our states in only one definite basis, the position basis. The basis has, however, been suppressed in all cases. This characterization is really true only in a rigged hilbert space but we continue to use the language of ordinary hilbert space as explained at the end of Chapter 8. To make clear what is meant by the statement above, recall that the eigenfunctions of position are δ functions. Thus

$$x\delta(x-a) = a\delta(x-a). \tag{11.14}$$

These form a complete orthonormal set and any function $\psi(x)$ can be considered as an expansion in terms of this set.

$$\psi(x) = \int_{-\infty}^{\infty} \delta(x-a)\psi(a)\, da \ . \tag{11.15}$$

This is not as purely formal as it seems and Dirac (ref 11.2) very early devised an ingenious notation to take advantage of this. We now explain this notation.

To begin with we consider an abstract vector space of states. Thus linear superposition of states is defined but not an inner product. The elements of the space are called <u>kets</u> and are denoted by $|\ \rangle$. If we want to specify a specific ket we insert a label $|a\rangle$. This specification of kets is completely basis independent. This means that we do not explicitly write any wavefunction $\psi_a(x)$ but only write the symbol $|a\rangle$. Now just as in the case of $\psi_a(x)$, the label "a" usually refers to the eigenvalue of some operator A. This statement we write not as

$$A\psi_a(x) = a\psi_a(x) \tag{11.16}$$

but as

$$A|a\rangle = a|a\rangle \; . \tag{11.17}$$

Both equation 11.16 and 11.17 say the same thing, except that in equation 11.17 we have not committed ourselves to a definite function or its Fourier transform or what have you.

Corresponding to the space of kets we introduce the dual space (see section 8.1) of continuous linear functionals defined on the kets. This is called the space of bras and they are written $\langle|$. Specific bras are labelled in the same manner as kets. Furthermore if A is an operator on the space of kets then the corresponding operator on the space of bras is $A^\dagger$. Thus the equation corresponding to 11.17 is given on the space of bras by

$$\langle a|A^\dagger = a^*\langle a| \; . \tag{11.18}$$

It corresponds to taking the dagger of equation 11.16, namely

$$A^\dagger\psi_a^* = a^*\psi_a^* \; . \tag{11.19}$$

Since the bras are linear functionals over the kets they give a mapping from the kets into the complex numbers. We write this as $\langle|\rangle$ or for two specific ones as $\langle a|b\rangle$. Again in terms of functions the corresponding expression is (ψ_a,ψ_b). The completeness relation for these states, as we verify later, reads

$$1 = \sum_n |n\rangle\langle n| \tag{11.20}$$

if the label n is discrete or

$$1 = \int |k\rangle dk \langle k| \tag{11.21}$$

if the index k is continuous.

These equations correspond to the equations

$$\delta(x-y) = \sum_n \psi_n(x)\psi_n^*(y) \tag{11.22}$$

if the index n is discrete or

$$\delta(x-y) = \int dk\ \psi_k(x)\psi_k^*(y) \tag{11.23}$$

if the index k is continuous.

The expectation value of an operator is now written as $\langle a|A|b\rangle$. It is important to remember that <u>A acts to right</u> in this formula. The operator acting to the left in this expression is $A^\dagger$. Again in terms of wave functions this is written as $(\psi_a, A\psi_b)$ and clearly if A acts to the left we have $(A^\dagger\psi_a, \psi_b)$. In the compressed notation of Dirac we have to <u>remember</u> that $A^\dagger$ acts to the left.

To establish precisely the connection between Dirac's bra, ket notation and the usual wave-function formalism consider a specific eigenket $|n\rangle$ of the hamiltonian H. Thus,

$$H|n\rangle = E_n|n\rangle\ . \tag{11.24}$$

Now let the ket $|x\rangle$ be an eigenket of the position operator x_{op} so that

$$x_{op}|x\rangle = x|x\rangle \tag{11.25}$$

We have written x_{op} but clearly this operator is just "multiplication by x". We now state that the eigenfunctions $\phi_n(x)$ of the hamiltonian H in configuration space are given by

$$\phi_n(x) = \langle x|n\rangle \qquad 11.26$$

$$\phi_n^*(x) = \langle n|x\rangle$$

The orthogonality relation of two eigenfunctions ϕ_n and ϕ_m now follows from the orthogonality of the ket $|n\rangle$ and the bra $|m\rangle$, namely

$$\langle m|n\rangle = \delta_{n,m} \qquad 11.27$$

and the closure condition

$$\int |x\rangle dx\langle x| = 1. \qquad 11.28$$

This last equation is the same as equation 11.21. Thus consider

$$(\phi_m,\phi_n) = \int \phi_m^*(x)\phi_n(x)dx$$

$$= \int \langle m|x\rangle dx\langle x|n\rangle$$

$$= \langle m|n\rangle = \delta_{m,n} \qquad 11.29$$

To make contact between equation 11.24 and the usual form of the time-independent Schrödinger equation we use 11.25 as well as the fact that if p is the momentum operator then

$$p|x\rangle = \frac{\hbar}{i}\frac{d}{dx}|x\rangle \quad . \qquad 11.30$$

Then if $H = \frac{p^2}{2m} + V(x)$ we take the inner product of equation 11.24 with the bra $\langle x|$ to get

$$\langle x|H|n\rangle = E_n\langle x|n\rangle = E_n\phi_n(x) \quad . \tag{11.31}$$

The left side of this equation can be rewritten as follows:

$$\langle x|H|n\rangle = \int \langle x|H|y\rangle dy\langle y|n\rangle = \int \langle x|H|y\rangle\phi_n(y)dy$$

But using the explicit form for the hamiltonian we have

$$\langle x|V(x)|y\rangle = V(y)\langle x|y\rangle = V(y)\delta(x-y) \tag{11.32}$$

where we have used the fact that $|y\rangle$ is an eigenket of the position operator x and hence also of $V(x)$.

Furthermore,

$$\langle x|\frac{p^2}{2m}|y\rangle = -\frac{\hbar^2}{2m}\langle x|\frac{d^2}{dy^2}|y\rangle = -\frac{\hbar^2}{2m}\frac{d^2}{dy^2}\langle x|y\rangle$$

$$= -\frac{\hbar^2}{2m}\frac{d^2}{dy^2}\delta(x-y) \quad . \tag{11.33}$$

Substituting equations 11.32 and 11.33 into equation 11.31 we get:

$$\langle x|H|n\rangle = \int \{[-\frac{\hbar^2}{2m}\frac{d^2}{dy^2} + V(y)]\delta(x-y)\}\phi_n(y)dy$$

$$= -\frac{\hbar^2}{2m}\frac{d^2}{dx^2}\phi_n(x) + V(x)\phi_n(x) \tag{11.34}$$

so that equation 11.30 now reads explicitly

$$-\frac{\hbar^2}{2m}\frac{d^2\phi_n}{dx^2} + V(x)\phi_n(x) = E_n\phi_n(x) \tag{11.35}$$

which is the usual form of the Schrödinger equation.

We have carried out these computations with excessive detail to illustrate all the steps that are involved. In practice, one frequently treats the bras and kets as if they were, in fact, wavefunctions with a little triangular bracket stuck on to them. This is all right in most circumstances; however, on occasion it is important to remember how they are really related to the wavefunction. This latter circumstance occurs when products of operators are operating on a ket.

For example if we have the expression $AB|\eta\rangle$ and we wish to write it in terms of wavefunctions then we can, for example, use the basis set $\{|x\rangle\}$ consisting of the eigenkets of the position operator. The expression is then rewritten as follows

$$\begin{aligned}\langle x|AB|\eta\rangle &= \int \langle x|A|y\rangle dy\langle y|B|z\rangle dz\langle z|\eta\rangle \\ &= \int dydz\langle x|A|y\rangle\langle y|B|z\rangle\phi_\eta(z)\end{aligned} \qquad 11.36$$

In a few paragraphs we shall see that this expression has a quite simple interpretation.

We next verify the completeness relations 11.20; the expression 11.21 is obtained in exactly the same way. Let $\{u_n\}$ be a complete basis and let $\{|n\rangle\}$ and $\{\langle n|\}$ denote the corresponding complete sets of kets and bras respectively so that

$$\begin{aligned}u_n(x) &= \langle x|n\rangle \\ u_n^*(x) &= \langle n|x\rangle\end{aligned} \qquad 11.37$$

Now any wave-function $\psi(x)$ and corresponding ket $|\psi\rangle$ can be expanded as

$$\psi = \sum a_n u_n \qquad 11.38$$

or

$$|\psi\rangle = \sum a_n |n\rangle \ . \qquad 11.39$$

In both cases we have

$$a_n = (u_n,\psi) = \langle n|\psi\rangle \ . \qquad 11.40$$

Thus,

$$|\psi\rangle = \sum_n |n\rangle\langle n|\psi\rangle \qquad 11.41$$

which implies

$$\sum |n\rangle\langle n| = 1. \qquad 11.42$$

This is the completeness relation expressed in bra, ket notation.

If we now consider matrix elements of any operator A between wave-functions ϕ and ψ we can write the whole expression in bra and ket notation. Using 11.41 and 11.42 we see that the matrix element $(\phi,A\psi)$ can be written as

$$\langle\phi|A|\psi\rangle = \sum_{m,n} \langle\phi|m\rangle\langle m|A|n\rangle\langle n|\psi\rangle \ . \qquad 11.43$$

Then $\langle m|A|n\rangle$ is a "matrix" for the operator A in the standard basis we have used for labelling our bras and kets. In fact the right side of 11.43 is symbolically a matrix product where $\langle n|\psi\rangle$ are column and $\langle\phi|m\rangle$ row matrix elements. Similarly $\langle m|A|n\rangle$ are the elements of a square matrix. This also shows that the right hand side of equation 11.36 may be interpreted as multiplying from left to right the square "matrices" $\langle x|A|y\rangle$ and $\langle y|B|z\rangle$ with the "column matrix" $\phi_\eta(z)$.

Formulated in this way, quantum mechanics is historically referred to as matrix mechanics in contrast to the Schrödinger formulation which is called wave mechanics. Clearly they are just two different versions of the same thing. It is very useful to have many different formulations of the same theory. This allows us to choose the most convenient formulation for a particular computation and also gives us deeper insight into the structure of the theory. From now on we freely employ the Dirac notation and switch to different formulations as the mood or convenience strikes us. In fact, we shall soon lose track of exactly which formulation we are using because in Dirac notation no commitment to a particular formulation is required.

To begin the process of familiarizing ourselves with this notation we reconsider angular momentum and obtain matrix representations for the various operators.

11.4 Example 2 - Angular Momentum

Our whole discussion of angular momenta (if we do not restrict ourselves to orbital angular momenta) could have been carried out using only the algebraic properties of the operators. In this case it is conventional to denote the total angular momentum by $\vec{J}$ and the eigenvalues by j and m. Thus the equations corresponding to 9.55, 9.56 and 9.57 are

$$[J_x, J_y] = i\hbar J_z \qquad 11.44$$

$$[J_y, J_z] = i\hbar J_x \qquad 11.45$$

and

$$[J_z, J_x] = i\hbar J_y \quad . \qquad 11.46$$

Also

$$[J^2, \vec{J}] = 0 \qquad 11.47$$

We again define

$$J_\pm = J_x \pm iJ_y \quad . \qquad 11.48$$

The simultaneous eigenkets of J^2, J_z are now denoted by $|j,m\rangle$.

Thus we have

$$J^2|j,m\rangle = j(j+1)\hbar^2|j,m\rangle \qquad 11.49$$

and

$$J_z|j,m\rangle = m\hbar|j,m\rangle \qquad 11.50$$

and corresponding to 9.101 we have

$$J_\pm|j,m\rangle = \sqrt{j(j+1)-m(m\pm 1)}\,\hbar|j,m\pm 1\rangle \ . \qquad 11.51$$

It is now a straightforward matter to evaluate the matrix elements of J^2, J_x, J_y and J_z. Thus

$$\langle j',m'|J^2|j,m\rangle = j(j+1)\hbar^2\langle j',m'|j,m\rangle$$

$$= j(j+1)\hbar^2\delta_{jj'}\delta_{mm'} \ . \qquad 11.52$$

So in this basis J^2 is given by a diagonal matrix. If the values of j, j' are fixed so that j = j' then the matrix is 2j+1 by 2j+1. Thus for j = j' = 1

$$\langle 1,m|J^2|1,m'\rangle = 2\hbar^2\begin{pmatrix}1 & 0 & 0\\ 0 & 1 & 0\\ 0 & 0 & 1\end{pmatrix}.$$

Similarly J_z is given by a diagonal matrix.

$$\langle j',m'|J_z|j,m\rangle = m\hbar\ \delta_{jj'}\ \delta_{mm'} \qquad 11.53$$

and for j = j' = 1 this becomes (recall problem 9.2)

$$\langle 1m'|J_z|1m\rangle = \hbar\begin{pmatrix}1 & 0 & 0\\ 0 & 0 & 0\\ 0 & 0 & -1\end{pmatrix} \qquad 11.54$$

To evaluate the matrix elements for J_x and J_y it is more convenient to first evaluate $J_\pm$. Thus,

$$\langle j',m'|J_\pm|j,m\rangle = \sqrt{j(j+1)-m(m\pm1)}\,\hbar\langle j',m'|j,m\pm1\rangle$$

$$= \sqrt{j(j+1)-m(m\pm1)}\,\hbar\,\delta_{j',j}\,\delta_{m'm\pm1}\ . \qquad 11.55$$

Again for $j = j' = 1$ these take the following form.

$$\langle 1m'|J_+|1m\rangle = \hbar\begin{pmatrix} 0 & \sqrt{2} & 0 \\ 0 & 0 & \sqrt{2} \\ 0 & 0 & 0 \end{pmatrix} \qquad 11.56$$

and

$$\langle 1m'|J_-|1m\rangle = \hbar\begin{pmatrix} 0 & 0 & 0 \\ \sqrt{2} & 0 & 0 \\ 0 & \sqrt{2} & 0 \end{pmatrix} \qquad 11.57$$

They are obviously the hermitian conjugates of each other as is clear from

$$J_\pm = J_x \pm iJ_y\ .$$

We can now solve for $J_x = \frac{1}{2}(J_+ + J_-)$ and $J_y = -\frac{i}{2}(J_+ - J_-)$

to get

$$J_x = \frac{\hbar}{\sqrt{2}}\begin{pmatrix} 0 & 1 & 0 \\ 1 & 0 & 1 \\ 0 & 1 & 0 \end{pmatrix} \qquad 11.58$$

$$J_y = \frac{\hbar}{\sqrt{2}}\begin{pmatrix} 0 & -i & 0 \\ i & 0 & -i \\ 0 & i & 0 \end{pmatrix} \qquad 11.59$$

Both of these correspond to observables and are thus represented by hermitian matrices. The general scheme is given by 11.52, 11.53 and 11.55 and looks as follows:

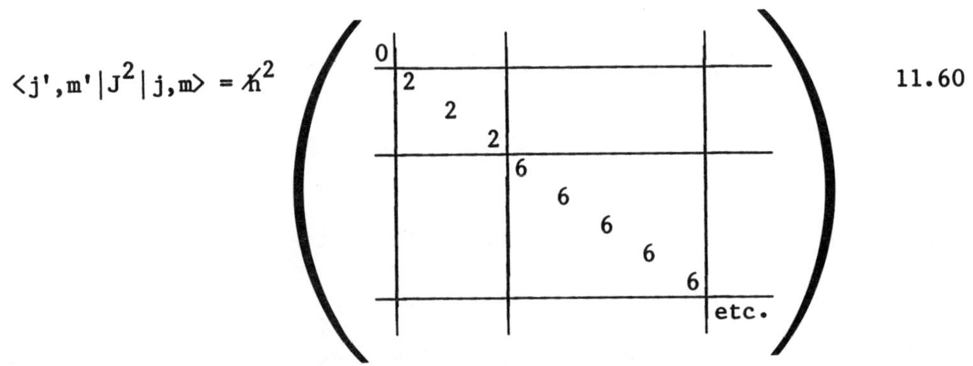

$$\langle j',m'|J^2|j,m\rangle = \hbar^2 \begin{pmatrix} 0 & & & & & & & & \\ & 2 & & & & & & & \\ & & 2 & & & & & & \\ & & & 2 & & & & & \\ & & & & 6 & & & & \\ & & & & & 6 & & & \\ & & & & & & 6 & & \\ & & & & & & & 6 & \\ & & & & & & & & 6 \\ & & & & & & & & & \text{etc.} \end{pmatrix} \qquad 11.60$$

for j, j' integral. The blocks are labelled by j, j' and the elements within the blocks by m, m'. All elements that are omitted are zero. We also have

$$\langle j',m'|J_z|j,m\rangle = \hbar \begin{pmatrix} 0 & & & & & & & & \\ & 1 & & & & & & & \\ & & 0 & & & & & & \\ & & & -1 & & & & & \\ & & & & 2 & & & & \\ & & & & & 1 & & & \\ & & & & & & 0 & & \\ & & & & & & & -1 & \\ & & & & & & & & -2 \\ & & & & & & & & & \text{etc.} \end{pmatrix} \qquad 11.61$$

and

$$\langle j',m'|J_-|j,m\rangle = \hbar \begin{pmatrix} 0 & & & & & & & & \\ & 0 & 0 & 0 & & & & & \\ & \sqrt{2} & 0 & 0 & & & & & \\ & 0 & \sqrt{2} & 0 & & & & & \\ & & & & 0 & 0 & 0 & 0 & 0 \\ & & & & 2 & 0 & 0 & 0 & 0 \\ & & & & 0 & \sqrt{6} & 0 & 0 & 0 \\ & & & & 0 & 0 & \sqrt{6} & 0 & 0 \\ & & & & 0 & 0 & 0 & 2 & 0 \\ & & & & & & & & & \text{etc.} \end{pmatrix} \qquad 11.62$$

A specific example of the above general case (namely $J = 1/2$) has already been examined in detail in section 9.6. A brief re-reading of that section might be useful now. The case for $J = 1$ given by equations 11.54, 11.58 and 11.59 is completely equivalent to using the differential operators for L_x, L_y, L_z on the three-dimensional basis set $Y_{1,1}$, $Y_{1,0}$, $Y_{1,-1}$. In many instances it may be more convenient to use these matrix representations of the angular momentum operators rather than the differential operators.

11.5 Schrödinger Picture

There is another use of the word "representation " in the literature. We shall call these pictures throughout this text. They occur when time-dependent unitary transformations are used. So far we have always worked in the so-called Schrödinger picture. In this picture the operators are generally time-independent and all the time dependence is carried by the states as given by the Schrödinger equation.

The matrix representations of operators in this picture, as stated, are time-independent unless they are <u>explicitly</u> time-dependent as in the case of an interaction that is switched on and off. We leave the discussion of such explicitly time-dependent operators for later (chapter 15).

Consider a basis $\{u_n\}$ or $\{|n\rangle\}$. Then,

$$|\Psi(t)\rangle = \sum a_n(t)|n\rangle \qquad 11.63$$

and

$$i\hbar \frac{\partial}{\partial t} |\Psi(t)\rangle = H|\Psi(t)\rangle \qquad 11.64$$

can be rewritten to read

$$i\hbar \sum_n \frac{da_n}{dt} |n\rangle = \sum_n a_n H|n\rangle \quad . \qquad 11.65$$

Or taking matrix elements with $\langle m|$ and using the fact that $\langle m|n\rangle = \delta_{mn}$ we get

$$i\hbar \frac{da_m}{dt} = \sum_n H_{mn} a_n \tag{11.66}$$

where

$$H_{mn} = \langle m|H|n\rangle \tag{11.67}$$

is clearly time-independent if H does not explicitly depend on time. Similarly any other operator A has time independent matrix elements $\langle m|A|n\rangle$. Of course the expectation value of A is time dependent through the time dependence of $|\Psi(t)\rangle$.

11.6 Heisenberg Picture

We have previously seen (eqns. 7.24 and 7.27) that the evolution of a state in the Schrödinger picture may be described by a unitary operator, the so-called evolution operator

$$U(t,t_o) = \exp - \frac{iH}{\hbar}(t-t_o) \ . \tag{11.68}$$

Suppose we apply the inverse of this unitary operator (with $t_o = 0$) to every state of our Hilbert space. For concreteness let $|\Psi_s(t)\rangle$ be a state in the Schrödinger picture which has evolved according to $U(t) \equiv U(t,0)$ from $t = 0$. Then define a state in the Heisenberg picture $|\Psi_H\rangle$ by

$$|\Psi_H\rangle = \exp \frac{iH}{\hbar} t|\Psi_s(t)\rangle = U^\dagger(t)|\Psi_s(t)\rangle \tag{11.69}$$

$$= U^\dagger(t)U(t)|\Psi_s(0)\rangle$$

$$= |\Psi_s(0)\rangle \tag{11.70}$$

since $U^\dagger U = UU^\dagger = 1$ by virtue of the unitarity of the operator U. This shows that the states $|\Psi_H\rangle$ in the Heisenberg picture are time-independent and coincide with the states in the Schrödinger picture at time $t = 0$.

Since the physically interesting objects are expectation values or matrix elements of the type $\langle\Psi_s(t)|A_s|\Phi_s(t)\rangle$, these must remain invariant under our unitary transformation. Thus it is necessary that the operators in the Schrödinger picture A_s transform to operators A_H in the Heisenberg picture. This gives

$$\langle\Psi_s(t)|A_s|\Phi_s(t)\rangle = \langle\Psi_H|A_H|\Phi_H\rangle \; .$$

Thus,

$$\langle\Psi_s(0)|U^\dagger(t)A_sU(t)|\Phi_s(0)\rangle = \langle\Psi_H|U^\dagger(t)A_sU(t)|\Phi_H\rangle$$

$$= \langle\Psi_H|A_H|\Phi_H\rangle \; .$$

Hence,

$$A_H(t) = U^\dagger(t)A_sU(t). \qquad 11.71$$

This shows that in the Heisenberg picture the time dependence is carried by the operators. To find the equation of motion they satisfy we simply use the fact that

$$i\hbar \frac{dU(t)}{dt} = HU(t)$$

and (remembering that $\frac{dH}{dt} = 0$) differentiate 11.71 to get

$$i\hbar \frac{dA_H}{dt} = i\hbar \frac{dU^\dagger}{dt} A_s U + i\hbar U^\dagger A_s \frac{dU}{dt}$$

$$= -HU^\dagger A_s U + U^\dagger A_s UH$$

$$= -HA_H + A_H H$$

or

$$i\hbar \frac{dA_H}{dt} = [A_H, H] \ . \qquad 11.72$$

These are the famous Heisenberg equations of motion. They are almost identical with the classical Hamilton's equations if we replace $-\frac{i}{\hbar}$ $[A_H,H]$ by the classical Poisson bracket $\{A,H\}$. We do not pursue this formal similarity any further. (See however ref. 11.2, Chapter 4).

To illustrate this Heisenberg approach we once more consider the simple harmonic oscillator. Then

$$H = \frac{p^2}{2m} + \frac{1}{2} m\omega^2 x^2 \ .$$

In the Heisenberg picture the equations of motion are

$$i\hbar\dot{x} = [x,H] = [x, \frac{p^2}{2m}] = i\hbar \frac{p}{m}$$

or

$$\dot{x} = \frac{p}{m} \qquad 11.73$$

and

$$i\hbar\dot{p} = [p,H] = \frac{1}{2} m\omega^2 [p,x^2] = -i\hbar m\omega^2 x$$

or

$$\dot{p} = -m\omega^2 x \ . \qquad 11.74$$

These are clearly identical with the classical equations of motion. Hence,

$$\ddot{x} = \frac{\dot{p}}{m} = -\omega^2 x \tag{11.75}$$

$$\ddot{p} = -m\omega^2 \dot{x} = -\omega^2 p \ . \tag{11.76}$$

Thus,

$$x = x_o \cos \omega t + \dot{x}_o \sin \omega t \tag{11.77}$$

$$p = p_o \cos \omega t + \dot{p}_o \text{ sinc } \omega t \tag{11.78}$$

These solutions also appear identical with the classical ones. Now,

$$m\dot{x} = p = -m\omega x_o \sin \omega t + m\omega \dot{x}_o \cos \omega t \tag{11.79}$$

Therefore

$$\dot{p}_o = -m\omega x_o \tag{11.80}$$

$$p_o = m\omega \dot{x}_o \tag{11.81}$$

so that

$$x = x_o \cos \omega t + \frac{p_o}{m\omega} \sin \omega t \tag{11.82}$$

$$p = p_o \cos \omega t - m\omega x_o \sin \omega t. \tag{11.83}$$

Note that, in spite of the similarity with the classical solutions, x_o and p_o are operators. In fact they are the same as the Schrödinger picture operators. It is interesting to note that in this case also the creation and annihilation operators are useful. We have

$$\dot{x} = \frac{p}{m} \qquad 11.84$$

$$\dot{p} = -m\omega^2 x \quad . \qquad 11.85$$

Defining (see 9.4, 9.5 and 9.8, 9.9)

$$a = \frac{1}{\sqrt{2m\omega\hbar}} [m\omega x + ip] \qquad 11.86$$

$$a^\dagger = \frac{1}{\sqrt{2m\omega\hbar}} [m\omega x - ip] \qquad 11.87$$

we get

$$i\hbar\dot{a} = [a,H] = \hbar\omega a \qquad 11.88$$

$$i\hbar\dot{a}^\dagger = [a^\dagger,H] = -\hbar\omega a^\dagger \quad . \qquad 11.89$$

The solutions are

$$a = a_o e^{-i\omega t} \qquad 11.90$$

$$a^\dagger = a_o^\dagger e^{i\omega t} \quad . \qquad 11.91$$

So in this case we have only first order equations to solve. The commutators in 11.88, 11.89 were obtained from 9.15, 9.16. We could of course just have used 11.86 and 11.77 directly to get

$$i\hbar\dot{a} = i\hbar \frac{1}{\sqrt{2m\omega\hbar}} [m\omega\dot{x} + i\dot{p}]$$

$$= i\hbar \frac{1}{\sqrt{2m\omega\hbar}} [\omega p - im\omega^2 x]$$

or

$$i\hbar\dot{a} = h\omega \frac{1}{\sqrt{2m\omega\hbar}} [m\omega x + ip] = \hbar\omega a$$

which coincides with 11.88. This shows that introducing a, $a^\dagger$ would also simplify the classical problem for the simple harmonic oscillator. The states in this case are of course time independent and coincide with the states of the Schrödinger picture at time zero.

There is one more picture which is of great utility in applications and we discuss it next. It is clear, however, that it is possible to define as many different pictures as there are time-dependent unitary transformations. They vary, of course, in their utility. The so-called Interaction or Dirac picture is one of the most useful and has played a very important role in the development of Quantum Electrodynamics.

11.7 Dirac or Interaction Pictures

As the name implies, this picture is useful in the case of interactions and can be thought of as lying half- way between the Schrödinger and Heisenberg pictures.

Suppose we have an, of necessity self-adjoint, hamiltonian H which is itself the sum of two self adjoint operators H_o and H' such that they have a common dense domain. That is,

$$H = H_o + H'. \tag{11.92}$$

In practice H_o will usually be an exactly diagonalizable hamiltonian and H' a complicated interaction part. Again let $|\Psi_s(t)\rangle$ be a state in the Schrödinger picture and define the unitary operator

$$U(t) = \exp - \frac{iH_o}{\hbar} t. \tag{11.93}$$

The evolution of $|\Psi_s(t)\rangle$ is according to the Schrödinger equation with the full hamiltonian

$$i\hbar \frac{\partial}{\partial t} |\Psi_s(t)\rangle = (H_o + H')|\Psi_s(t)\rangle. \qquad 11.94$$

We then define a state in the Dirac picture $|\Psi_D(t)\rangle$ by

$$|\Psi_D(t)\rangle = U^\dagger(t)|\Psi_s(t)\rangle . \qquad 11.95$$

Since $U^\dagger(t)$ satisfies the equation

$$i\hbar \frac{\partial}{\partial t} U^\dagger(t) = -H_o U^\dagger(t) . \qquad 11.96$$

The equation of motion for $|\Psi_D(t)\rangle$ is found from

$$\begin{aligned} i\hbar \frac{\partial}{\partial t} |\Psi_D(t)\rangle &= i\hbar \frac{\partial}{\partial t} U^\dagger(t)|\Psi_s(t)\rangle + U^\dagger(t) i\hbar \frac{\partial}{\partial t} |\Psi_s(t)\rangle \\ &= -H_o U^\dagger|\Psi_s(t)\rangle + U^\dagger(H_o + H')|\Psi_s(t)\rangle \\ &= U^\dagger H'|\Psi_s(t)\rangle = U^\dagger H' U U^\dagger|\Psi_s(t)\rangle \end{aligned}$$

or

$$i\hbar \frac{\partial}{\partial t} |\Psi_D(t)\rangle = H'_D|\Psi_D(t)\rangle \qquad 11.97$$

where we have used $[U^\dagger, H_o] = 0$ and defined

$$H'_D = U^\dagger H' U. \qquad 11.98$$

Thus in the interaction picture the state evolves only according to the interaction part H'_D of the hamiltonian. The price we have paid for this is that all operators including H'_D are now time dependent. In fact they satisfy a Heisenberg type of equation of motion. Thus if A_D is any operator in the interaction picture, that is

$$A_D = U^\dagger A_s U . \qquad 11.99$$

Then,

$$i\hbar \frac{\partial}{\partial t} A_D = i\hbar \frac{\partial U^\dagger}{\partial t} A_s U + U^\dagger A_s i\hbar \frac{\partial U}{\partial t}$$

$$= -H_o U^\dagger A_s U + U^\dagger A_s U H_o$$

or

$$i\hbar \frac{\partial}{\partial t} A_D = [A_D, H_o] \quad . \qquad 11.100$$

This is of course also the equation of motion for H'_D. Thus,

$$i\hbar \frac{\partial}{\partial t} H'_D = [H'_D, H_o] \quad . \qquad 11.101$$

To see how this looks in matrix form consider solving first the equation

$$H_o|k\rangle = E_k|k\rangle \quad . \qquad 11.102$$

Then,

$$|k,t\rangle = |k\rangle e^{-i \frac{E_k}{\hbar} t} \quad . \qquad 11.103$$

Thus we can write

$$|\Psi(t)\rangle = \sum_k a_k(t)|k\rangle e^{-i \frac{E_k}{\hbar} t} \qquad 11.104$$

where

$$i\hbar \frac{\partial}{\partial t} |\Psi(t)\rangle = H|\Psi(t)\rangle = (H_o + H')|\Psi(t)\rangle. \qquad 11.105$$

Hence we get:

$$i\hbar \sum_k (\dot{a}_k |k\rangle e^{-i\frac{E_k}{\hbar}t} - \frac{iE_k}{\hbar} a_k |k\rangle e^{-i\frac{E_k}{\hbar}t})$$

$$= \sum_k a_k H_o |k\rangle e^{-i\frac{E_k}{\hbar}t} + \sum_k a_k H' |k\rangle e^{-i\frac{E_k}{\hbar}t} .$$

This simplifies using 11.102 to

$$i\hbar \sum_k \dot{a}_k |k\rangle e^{-i\frac{E_k}{\hbar}t} = \sum_k a_k H' |k\rangle e^{-i\frac{E_k}{\hbar}t} .$$

Multiplying by $\langle \ell | e^{-i\frac{E_\ell}{\hbar}t}$ we get

$$i\hbar \frac{d}{dt} a_\ell = \sum_k H'_{\ell k} e^{i\frac{(E_\ell - E_k)}{\hbar}t} a_k \tag{11.106}$$

where

$$H'_{\ell k} = \langle \ell | H' | k \rangle \qquad \text{and} \tag{11.107}$$

$e^{i\frac{E_\ell}{\hbar}t} H'_{\ell k} e^{-i\frac{E_k}{\hbar}t}$ is the ℓ,k matrix element of H' in the interaction picture and could be written as

$$H'_{D\ell k} = e^{i\frac{E_\ell}{\hbar}t} H'_{\ell k} e^{-i\frac{E_k}{\hbar}t} . \tag{11.108}$$

Thus,

$$i\hbar \frac{d}{dt} a_\ell = \sum_k H'_{D\ell k}\, a_k \; . \qquad 11.109$$

This is the matrix form of the Schrödinger equation in the interaction picture.

We shall return to these equations again when we discuss time-dependent perturbation theory.

Notes, References and Bibliography

The transformation theory is due in large measure to

11.1 J. von Neumann - Mathematical Foundations of Quantum Mechanics - Princeton University Press, Princeton (1955).

It is of very great value to also read

11.2 P.A.M. Dirac - Principles of Quantum Mechanics, Oxford University Press, Oxford, 3rd edition (1947).

Chapter 11 Problems

11.1 Verify equation 11.13, that is, compute the Fourier transform of the hermite functions.

11.2 Repeat problem 11.7 using the matrix representation for the angular momentum operators.

11.3 Find and solve the Heisenberg equation of motion for a particle with a hamiltonian

$$H = \frac{p^2}{2m} .$$

If at $t = 0$ the particle is in the state $|0,0,0\rangle$ of a harmonic oscillator basis, find $\langle r^2 \rangle$ as a function of t.

11.4 Transform the displaced one-dimensional S.H.O.

$$H = \frac{p^2}{2M} + \frac{1}{2} kx^2 + \beta x = H_o + \beta x$$

to the Dirac picture.
Solve for both $|\psi_D(t)\rangle$, the state of the system and $x_D(t)$, the position operator of the system. Assume $|\psi_D(0)\rangle = |0\rangle$.

11.5 Solve problem 11.4 in the Heisenberg picture.

Chapter 12

Time-Independent Non-Degenerate Perturbation Theory

So far we have only discussed problems that permitted exact analytic solutions. This is, in fact, seldom the case and one is forced to use approximation procedures. The exact solutions obtained so far then help to give one an intuitive feeling for what to expect in complicated situations. Thus, faced with what may be a more or less exact, but complicated, hamiltonian H describing a given physical situation one can adopt several different approaches.

In many situations it is convenient to view the hamiltonian as a mathematical model of the situation. It then becomes reasonable to modify the model until one arrives at a simpler hamiltonian that still retains the essential physical features. To check the model dependence of the results one can then construct many different models all of which, in some sense, approximate the situation of interest and study how the results vary from model to model. This approach requires a good deal of physical insight and has been much employed in nuclear and solid state physics.

On the other hand, even using model hamiltonians one is frequently still left with a problem that cannot be solved exactly and such that further approximations on the model would destroy the physics of interest. In this case one requires techniques for handling such hamiltonians in an approximate fashion. Here also the problems divide into essentially two classes. In one of these one is interested in the modification of the stationary states of the system under the perturbation. In the second case one is interested not in the shifts in the stationary states but rather in the transition between stationary states due to the perturbation. This latter case usually arises with time-dependent perturbations. We shall not consider the second case for several chapters but concentrate instead on the first case for now.

A particularly happy situation occurs if in a given physical problem, say a hamiltonian H, it is possible to split the problem into a relatively simple (solvable) part H_o and a "small" part $\lambda H'$ so that

$$H = H_o + \lambda H' \; . \qquad 12.1$$

Small here means that in come sense H' and H_o are comparable and $\lambda \ll 1$ or else $\lambda \approx 1$ and H' is small compared to H_o. Both these statements are vague since both H_o and H' are unbounded. To make them concrete, we shall assume for the time being that for any eigenket $|n\rangle$ of H_o

$$|\langle n|H_o|n\rangle| \geq |\langle n|H'|n\rangle| \; . \qquad 12.2$$

In such a case it is possible to treat $\lambda H'$ as a perturbation on H_o. A number of different and useful techniques have been developed for this purpose and it is the purpose of this chapter to study some of these techniques.

12.1 Rayleigh-Schrödinger Pertubation Theory for Stationary States

Consider a hamiltonian of the form

$$H = H_o + \lambda H' \qquad 12.1$$

where $\lambda H'$ is small with respect to H_o. We further suppose that H_o is sufficiently simple that we are able to solve its eigenvalue problem exactly. Thus we have

$$H_o|n\rangle^{(o)} = E_n^{(o)}|n\rangle^{(o)} \tag{12.3}$$

where $|n\rangle^{(o)}$, $E_n^{(o)}$ are the exact eigenkets and eigenvalues of H_o. We assume throughout this section that the eigenvalue of interest say $E_n^{(o)}$ is <u>non-degenerate</u>. The reason for this assumption will become obvious shortly. We further assume that $E_n^{(o)}$ is a discrete eigenvalue and that as $\lambda \to 0$ the exact eigenvalue E_n of H approaches $E_n^{(o)}$. In this case it is reasonable that for small λ we can expand the eigenfunction $|n\rangle$ and eigenvalue E_n of H in a fairly rapidly converging power series in λ. The eigenfunction $|n\rangle$ is specified only up to an arbitrary constant which we fix by requiring that

$$^{(o)}\langle n|n\rangle = {}^{(o)}\langle n|n\rangle^{(o)} = 1\ . \tag{12.4}$$

Also our assumption of a power series leads us to write

$$E_n = E_n^{(o)} + \lambda E_n^{(1)} + \lambda^2 E_n^{(2)} + \ldots \tag{12.5}$$

$$|n\rangle = |n\rangle^{(o)} + \lambda|n\rangle^{(1)} + \lambda^2|n\rangle^{(2)} + \ldots \tag{12.6}$$

Substituting this in the Schrödinger equation

$$H|n\rangle = E_n|n\rangle \tag{12.7}$$

and equating the coefficients of equal powers of λ we get:

$$(H_o - E_n^{(o)})|n\rangle^{(o)} = 0 \tag{12.8$^{(o)}$}$$

$$(H_o - E_n^{(o)})|n\rangle^{(1)} + (H' - E_n^{(1)})|n\rangle^{(o)} = 0 \tag{12.8$^{(1)}$}$$

$$(H_o - E_n^{(o)})|n\rangle^{(2)} + (H' - E_n^{(1)})|n\rangle^{(1)} - E_n^{(2)}|n\rangle^{(o)} = 0 \qquad 12.8^{(2)}$$

$$(H_o - E_n^{(o)})|n\rangle^{(r)} + (H' - E_n^{(1)})|n\rangle^{(r-1)}$$
$$- E_n^{(2)}|n\rangle^{(r-2)} - \dots - E_n^{(r)}|n\rangle^{(o)} = 0 \qquad 12.8^{(r)}$$

Clearly $12.8^{(o)}$ is identical with 12.3 and simply states that $E_n^{(o)}$ and $|n\rangle^{(o)}$ are eigenvalues and eigenfunctions respectively of H_o.

The normalization condition 12.4 when substituted into 12.6 now gives

$${}^{(o)}\langle n|n\rangle^{(o)} = {}^{(o)}\langle n|n\rangle^{(o)} + \lambda\,{}^{(o)}\langle n|n\rangle^{(1)} + \lambda^2\,{}^{(o)}\langle n|n\rangle^{(2)} + \dots .$$

Thus we get

$${}^{(o)}\langle n|n\rangle^{(1)} = {}^{(o)}\langle n|n\rangle^{(2)} = \dots\ {}^{(o)}\langle n|n\rangle^{(r)} = \dots = 0. \qquad 12.9$$

Now $12.8^{(o)}$ determines the zeroth order approximation for E_n and $|n\rangle$, $12.8^{(1)}$ the first order approximation in terms of the zeroth and finally $12.8^{(r)}$ the r'th order approximation in terms of all the lower order approximations. To see this we simply form the inner product with ${}^{(o)}\langle n|$ of equation $12.8^{(r)}$ to get

$${}^{(o)}\langle n|H'|n\rangle^{(r-1)} - E_n^{(r)} = 0 \qquad 12.10$$

where we have used 12.9 and the fact that H_o is self- adjoint.

If we now further take the inner product of equation $(12.8^{(r)})$ with ${}^{(o)}\langle \ell|$ for $\ell \neq n$ we get the components of $|n\rangle^{(r)}$ along $|\ell\rangle^{(o)}$, and together with 12.9 this specifies $|n\rangle^{(r)}$ completely. Thus we get

$$(E_\ell^{(o)} - E_n^{(o)})\,{}^{(o)}\langle \ell|n\rangle^{(r)} + {}^{(o)}\langle \ell|H' - E_n^{(1)}|n\rangle^{(r-1)}$$
$$- E_n^{(2)}\,{}^{(o)}\langle \ell|n\rangle^{(r-2)} - \dots - E_n^{(r-1)}\,{}^{(o)}\langle \ell|n\rangle^{(1)} = 0 \qquad \ell \neq n$$

or

$$^{(o)}\langle\ell|n\rangle^{(r)} = \frac{1}{E_n^{(o)}-E_\ell^{(o)}}\Big[{}^{(o)}\langle\ell|H'-E_n^{(1)}|n\rangle^{(r-1)} - \sum_{s=2}^{r-1} E_n^{(s)}\ {}^{(o)}\langle\ell|n\rangle^{(r-s)}\Big] \qquad \ell \neq n . \tag{12.11}$$

We shall now look at the lowest orders of perturbation in detail.

12.2 First Order Perturbations

If we set $r = 1$ in 12.10 and 12.11 we get that

$$E_n^{(1)} = {}^{(o)}\langle n|H'|n\rangle^{(o)} \tag{12.12}$$

and

$$\begin{aligned} ^{(o)}\langle\ell|n\rangle^{(1)} &= \frac{1}{E_n^{(o)}-E_\ell^{(o)}}\ {}^{(o)}\langle\ell|H'-E_n^{(1)}|n\rangle^{(0)} \qquad \ell \neq n \\ &= \frac{1}{E_n^{(o)}-E_\ell^{(o)}}\ {}^{(o)}\langle\ell|H'|n\rangle^{(o)} \qquad \ell \neq n . \end{aligned} \tag{12.13}$$

Hence to first order in λ

$$E_n = E_n^{(o)} + \lambda E_n^{(1)} = {}^{(o)}\langle n|H_o+\lambda H'|n\rangle^{(o)} . \tag{12.14}$$

This is just the expectation value of the total hamiltonian H in the unperturbed state $|n\rangle^{(o)}$.

Also from 12.13 and since $^{(o)}\langle n|n\rangle^{(1)} = 0$ we see that to first order in λ the eigenket $|n\rangle = |n\rangle^{(o)} + \lambda|n\rangle^{(1)}$ is

$$|n\rangle = |n\rangle^{(o)} + \lambda \sum_{\ell \neq n} \frac{|\ell\rangle^{(o)}\ {}^{(o)}\langle \ell|H'|n\rangle^{(o)}}{E_n^{(o)} - E_\ell^{(o)}} \quad . \qquad 12.15$$

We can also write this as

$$|n\rangle = |n\rangle^{o} + \lambda \sum_{\ell \neq n} (E_n^{(o)} - H_o)^{-1} |\ell\rangle^{(o)(o)}\langle \ell|H'|n\rangle^{(o)}. \qquad 12.16$$

Since $|n\rangle^{(o)}$ is already normalized it might appear from 12.15 or 12.16 that $|n\rangle$ is not. However to order λ, $|n\rangle$ is indeed normalized since

$$\langle n|n\rangle = {}^{(o)}\langle n|n\rangle^{o} + \lambda^2 \sum_{\ell \neq n} \frac{{}^{(o)}\langle n|H'|\ell\rangle^{(o)(o)}\langle \ell|H'|n\rangle^{(o)}}{(E_n^{(o)} - E_\ell^{(o)})^2}$$

$$= {}^{(o)}\langle n|n\rangle^{(o)} + \lambda^2 \sum_{\ell \neq n} \left| \frac{H'_{n,\ell}}{E_n^{(o)} - E_\ell^{(o)}} \right|^2$$

$$= {}^{(o)}\langle n|n\rangle^{(o)} + \text{terms of order } \lambda^2 \ .$$

The cross-terms cancel in view of the self-adjointness of H_o and H'. Clearly the degree of convergence of E_n is determined by the ratio

$$\left| \frac{{}^{(o)}\langle n|H'|\ell\rangle^{(o)}}{E_n^{(o)} - E_\ell^{(o)}} \right| \quad .$$

It is clear now why the assumption of non-degeneracy was made for otherwise $E_n^{(o)} - E_\ell^{(o)}$ could vanish giving a divergent rather than a small finite result. In fact if H_o has close-lying levels so that $E_n^{(o)} - E_\ell^{(o)}$ is small, these levels will have to be treated as if they were degenerate.

We next illustrate the above procedure by means of various examples.

12.3 Example 1 - Anharmonic Oscillator

The hamiltonian is

$$H = \frac{p^2}{2m} + \frac{1}{2} m\omega^2 x^2 + \lambda x^4$$

$$= H_o + \lambda x^4 . \qquad 12.17$$

We shall consider the first order perturbation of the ground state $|0\rangle$ of H_o . Thus

$$H_o|0\rangle = \frac{1}{2}\hbar\omega|0\rangle$$

and in terms of the annihilation operator a

$$a|0\rangle = 0,$$

$$x = \sqrt{\frac{\hbar}{2m\omega}}\ (a+a^\dagger) \qquad 12.18$$

where we have used 9.5 and 9.11. Using 12.14 we get the ground state energy to first order in λ as

$$\langle 0|H_o + \lambda x^4|0\rangle = \frac{1}{2}\hbar\omega + \lambda\langle 0|x^4|0\rangle . \qquad 12.19$$

But

$$\langle 0|x^4|0\rangle = \left(\frac{\hbar}{2m\omega}\right)^2 \langle 0|a+a^\dagger)^4|0\rangle$$

$$= \left(\frac{\hbar}{2m\omega}\right)^2 \langle 0|a^4 + a^3a^\dagger + a^2a^\dagger a + a^2a^{\dagger 2} + aa^\dagger a^2 + aa^\dagger aa^\dagger +$$

$$+ aa^{\dagger 2}a + aa^{\dagger 3} + a^\dagger a^3 + a^\dagger a^2 a^\dagger + a^\dagger a a^\dagger a + a^\dagger aa^{\dagger 2}$$

$$+ a^{\dagger 2}a^2 + a^{\dagger 2}aa^\dagger + a^{\dagger 3}a + a^{\dagger 4}|0\rangle$$

$$= \left(\frac{\hbar}{2m\omega}\right)^2 \langle 0|a^2a^{\dagger 2} + aa^\dagger aa^\dagger|0\rangle = 3\left(\frac{\hbar}{2m\omega}\right)^2$$

Hence to first order in λ the ground state energy is given by

$$E_o \simeq \frac{1}{2}\hbar\omega\left(1+\frac{3\lambda}{2}\frac{\hbar}{m^2\omega^3}\right). \tag{12.20}$$

12.4 Example 2 - Ground State of Helium-like Ions

The hamiltonian in this case is

$$H = \frac{p_1^2}{2m} + \frac{p_2^2}{2m} - \frac{Ze^2}{r_1} - \frac{Ze^2}{r_2} + \frac{e^2}{r_{12}} \tag{12.21}$$

where (fig. 12.1) $\vec{r}_1$ and $\vec{r}_2$ are the position vectors of the first and second electron respectively and $r_{12} = |\vec{r}_1-\vec{r}_2|$ is the separation between the two electrons. Also $Z = 2$ for helium but we leave it arbitrary for the time being since we also consider helium-like ions. If we at first neglect the repulsion of the 2 electrons we get

$$H_o = H_{o1} + H_{o2} \tag{12.22}$$

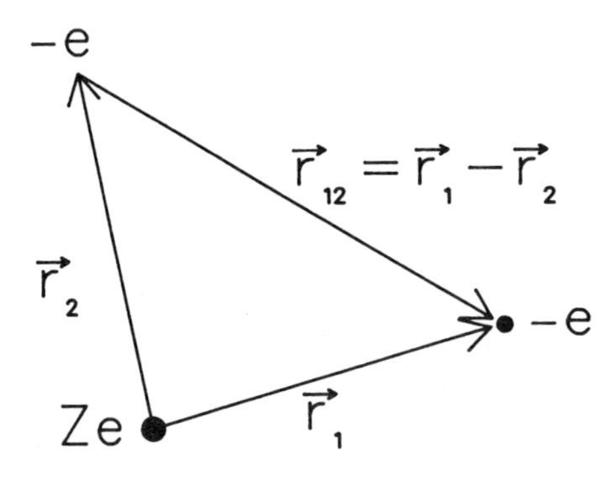

Fig. 12-1 Geometry for Helium-Like Atoms

where

$$H_{oi} = \frac{p_i^2}{2m} - \frac{Ze^2}{r_i} \qquad 12.23$$

is just the hydrogenic hamiltonian. Furthermore

$$[H_{o1}, H_{o2}] = 0\ . \qquad 12.24$$

Therefore we can diagonalize H_{o1} , H_{o2} and hence $H_o = H_{o1} + H_{o2}$ simultaneously. The solution for the ground state energy is then

$$E_o^{(o)} = -2\left(\frac{1}{2}\, mc^2\alpha^2 Z^2\right) \qquad 12.25$$

and the ground state eigenfunction of H_o is

$$\psi_o^{(o)}(\vec{r}_1, \vec{r}_2) = \frac{1}{\pi a^3}\, e^{-(r_1+r_2)/a} \qquad 12.26$$

where

$$a = \frac{a_o}{Z} = \frac{\hbar^2}{Zme^2} \qquad 12.27$$

The correction $E_o^{(1)}$ due to the electrostatic repulsion of the two electrons is then, to first order, given by

$$E_o^{(1)} = \int \psi_o^{(o)*}\, \frac{e^2}{r_{12}}\, \psi_o^{(o)}\, d^3r_1 d^3r_2$$

$$= \frac{e^2}{\pi^2 a^6} \int \frac{e^{-2(r_1+r_2)/a}}{|\vec{r}_1 - \vec{r}_2|}\, d^3r_1 d^3r_2\ . \qquad 12.28$$

This is just the electrostatic energy due to two spherical charge distributions with charge densities

$$\rho(\vec{r}) = -\frac{e}{\pi a^3} e^{-2r/a} . \qquad 12.29$$

To evaluate the integrals in 12.28 we first expand $1/|\vec{r}_1-\vec{r}_2|$ in terms of spherical harmonics (problem 12.1)

$$\frac{1}{|\vec{r}_1-\vec{r}_2|} = \sum_{\ell=0}^{\infty} \frac{r_<^{\ell}}{r_>^{\ell+1}} \frac{4\pi}{2\ell+1} Y^*_{\ell m}(\hat{r}_1)Y_{\ell m}(\hat{r}_2) . \qquad 12.30$$

where

$$r_< = \begin{cases} r_1 & \text{if } r_1 < r_2 \\ r_2 & \text{if } r_2 < r_1 \end{cases} \quad \text{and} \quad r_> = \begin{cases} r_2 & \text{if } r_1 < r_2 \\ r_1 & \text{if } r_2 < r_1 \end{cases}$$

Then,

$$E_o^{(1)} = 16\pi^2 \int_0^{\infty} \rho(r_1)r_1 dr_1 \left(\int_0^{r_1} \rho(r_2)r_2^2 dr_2 + r_1 \int_{r_1}^{\infty} \rho(r_2)r_2 dr_2\right)$$

$$= \frac{5}{4} Z\cdot \frac{1}{2} mc^2\alpha^2$$

where we have used the explicit form of $\rho(r)$ and 12.30. Hence the ground state energy to first order in the electrostatic repulsion of the two electrons is given by

$$E_o \simeq -mc^2\alpha^2 Z^2(1- \frac{5}{8Z}) . \qquad 12.31$$

Since the perturbation decreases with increasing Z we expect the accuracy of our result to improve as we go from He to Be^{++} . The results are as presented in table 12.1.

Table 12.1. Ground State Energies of Helium and Helium-Like Ions.

Atom	Z	E_o (calculated) eV	E_o (experimental) eV	% Error
He	2	- 74	- 79	6
Li^+	3	-193	-197	2
Be^{++}	4	-366	-370	1

12.5 Second Order Perturbations

We shall now take a closer look at perturbations to second order. From 12.10 we get that

$$E_n^{(2)} = {}^{(o)}\langle n|H'|n\rangle^{(1)} \quad . \tag{12.32}$$

But we also have that

$$|n\rangle^{(1)} = \sum_{\ell\neq n} \frac{|\ell\rangle^{(o)}\ {}^{(o)}\langle\ell|H'|n\rangle^{(o)}}{E_n^{(o)} - E_\ell^{(o)}} \quad . \tag{12.15}$$

Hence we get

$$E_n^{(2)} = \sum_{\ell\neq n} \frac{{}^{(o)}\langle n|H'|\ell\rangle^{(o)}\,{}^{(o)}\langle\ell|H'|n\rangle^{(o)}}{E_n^{(o)} - E_\ell^{(o)}} \quad . \tag{12.33}$$

Or calling, as before

$$H'_{n,\ell} = {}^{(o)}\langle n|H'|\ell\rangle^{(o)} \tag{12.34}$$

we have to second order in λ

$$E_n = E_n^{(o)} + \lambda\,{}^{(o)}\langle n|H'|n\rangle^{(o)} + \lambda^2 \sum_{\ell\neq n} \frac{|H'_{n,\ell}|^2}{E_n^{(o)} - E_\ell^{(o)}} . \tag{12.35}$$

We leave it as an exercise (problem 12.2) to show that by setting $r = 2$

in equation 12.11 one gets the second order correction to the wave function

$$|n\rangle^{(2)} = \sum_{m,\ell\neq n} \frac{|\ell\rangle^{(o)}\ {}^{(o)}\langle\ell|H'|m\rangle^{(o)}\ {}^{(o)}\langle m|H'|n\rangle^{(o)}}{(E_n^{(o)}-E_\ell^{(o)})(E_n^{(o)}-E_m^{(o)})}$$

$$- \sum_{\ell\neq n} \frac{|\ell\rangle^{(o)}\ {}^{(o)}\langle\ell|H'|n\rangle^{(o)}\ {}^{(o)}\langle n|H'|n\rangle^{(o)}}{(E_n^{(o)}-E_\ell^{(o)})^2} \qquad 12.36$$

The use of the second order perturbation formulae is considerably more complicated than the first order perturbations due to the infinite sum over intermediate states. In case there are only a finite number of intermediate states that contribute, the infinite sum reduces to a finite sum. We shall now illustrate this with an example and give a physical interpretation of the perturbation sum.

12.6 Example 3 - Displaced Simple Harmonic Oscillator

The hamiltonian we want to consider is

$$H = \frac{p^2}{2m} + \frac{1}{2} m\omega^2 x^2 + \lambda x \qquad 12.37$$

$$= H_o + \lambda x \ .$$

We want to treat λx as a perturbation. Since in terms of creation and annihilation operators

$$x = \sqrt{\frac{\hbar}{2m\omega}}\,(a+a^\dagger)$$

We have that to first order in λ

$$E_n = \langle n|H_o + \lambda x|n\rangle = (n+\frac{1}{2})\hbar\omega + \lambda\langle n|x|n\rangle$$

$$= (n+\frac{1}{2})\hbar\omega \ .$$

Then to second order in λ we have

$$E_n = (n+\frac{1}{2})\hbar\omega + \lambda^2 \sum_{\ell\neq n} \frac{\langle n|x|\ell\rangle\langle\ell|x|n\rangle}{(n-\ell)\hbar\omega}$$

$$= (n+\frac{1}{2})\hbar\omega + \frac{\lambda^2}{2m\omega^2} \sum_{\ell\neq n} \frac{\langle n|a+a^\dagger|\ell\rangle\langle\ell|a+a^\dagger|n\rangle}{n-\ell} \ .$$

Using 9.31 and 9.32 namely

$$a^\dagger|n\rangle = \sqrt{n+1}\ |n+1\rangle$$

and

$$a|n\rangle = \sqrt{n}|n-1\rangle$$

we get that to second order in λ

$$E_n = (n+\frac{1}{2})\hbar\omega + \frac{\lambda^2}{2m\omega^2} \sum_{\ell} \frac{n\delta_{n,\ell+1} + (n+1)\delta_{n,\ell-1}}{n-\ell}$$

or

$$E_n = (n+\frac{1}{2})\hbar\omega - \frac{\lambda^2}{2m\omega^2} \ . \qquad 12.38$$

If we compare this with the exact solution, which is easily obtained (problem 9.3 with $\beta = 1$ and $V = \lambda\sqrt{\hbar/2m\omega}$, or by completing the square in the hamiltonian 12.37) we get

$$E_n = (n+\frac{1}{2})\hbar\omega - \frac{\lambda^2}{2m\omega^2}$$

which agrees exactly with the second order perturbation theory. The reason for this is that the exact eigenvalue has only quadratic

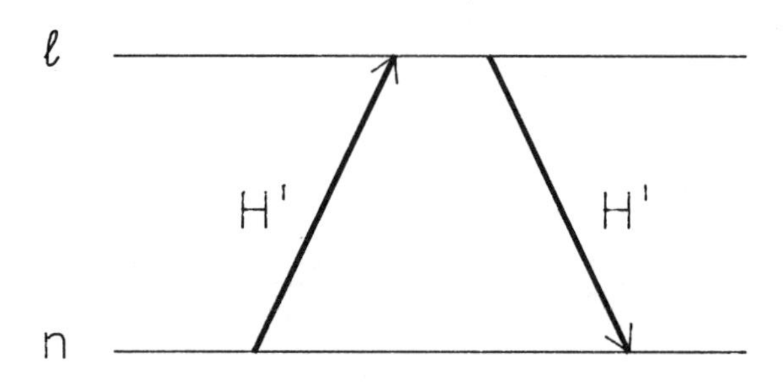

Fig. 12–2 Second Order Perturbation

dependence on λ. Thus all higher order perturbation terms must vanish. We shall now try to make plausible without computation the absence of third order terms.

In first order perturbation theory the perturbation H' must connect the given unperturbed state $|n\rangle$ with itself if it is to contribute. In second order H' must connect $|n\rangle$ with some other state and then back. Pictorially this is as shown in (fig. 12.2). This corresponds to the term

$$\frac{\langle n|H'|\ell\rangle\langle\ell|H'|n\rangle}{E_n - E_\ell} .$$

We then sum over all such intermediate states. In the example above we have $\ell = n\pm 1$ as the only possible intermediate states.

In third order the picture is as in Figure 12.3. This corresponds to a term

$$\frac{\langle n|H'|\ell\rangle\langle\ell|H'|\ell'\rangle\langle\ell'|H'|n\rangle}{(E_n - E_\ell)(E_\ell - E_{\ell'})} , \quad \text{with} \quad \ell \neq n , \ \ell \neq \ell' .$$

This also explains why there are not third order terms in our previous example. There we could only have either

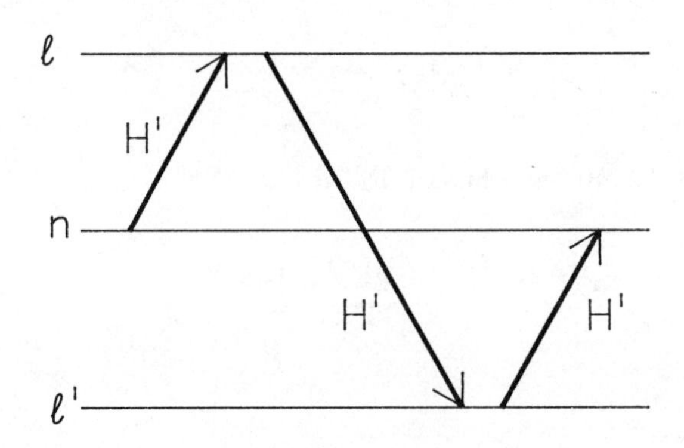

Fig. 12–3 Third Order Perturbation

$$n \longrightarrow n+1 \begin{matrix} \nearrow n+2 \\ \searrow n \end{matrix} \quad \text{(second order)}$$

$$n \longrightarrow n-1 \begin{matrix} \nearrow n \\ \searrow n-2 \end{matrix} \quad \text{(second order)} \quad .$$

In either case there is no non-vanishing matrix element connecting $n\pm2$ with n. Hence there are no third order contributions.

We next develop non-degenerate perturbation theory up to arbitrary order. The technique we employ will allow us to recover the Rayleigh-Schrödinger method as well as a new method known as the Brillouin-Wigner.

12.7 Non-degenerate Perturbations to all Orders

We again start with the hamiltonian

$$H = H_o + \lambda V$$

and assume that the eigenvalues $E_n^{(o)}$ of H_o are non- degenerate with eigenkets $|n\rangle^{(o)}$. Suppose we wish to calculate the eigenvalue E_n and corresponding eigenket $|n\rangle$ of H. Thus

$$(H_o + \lambda V - E_n)|n\rangle = 0. \tag{12.39}$$

We retain the normalization

$$^{(o)}\langle n|n\rangle = 1. \qquad 12.40$$

Then using 12.39 and 12.40 we obtain by using

$$\langle n|H_o|n\rangle^{(o)} = E_n^{(o)}\langle n|n\rangle^{(o)} \qquad 12.41$$

that

$$E_n^{(o)} + \lambda^{(o)}\langle n|V|n\rangle = E_n \qquad 12.42$$

This gives us the level shift due to the perturbation.

$$E_n - E_n^{(o)} = \lambda^{(o)}\langle n|V|n\rangle\ . \qquad 12.43$$

Unfortunately 12.43 involves the unknown exact eigenket $|n\rangle$.

To avoid having to exclude the term $\ell = n$ in the sum over ℓ in the expansion for $|n\rangle$ we define the projection operator (see problem 6.4)

$$P_n = |n\rangle^{(o)(o)}\langle n| \qquad 12.44$$

which has the property of projecting any ket along $|n\rangle^{(o)}$. The operator

$$Q_n = 1 - P_n \qquad 12.45$$

is also a projection operator and has the property that $Q_n|\phi\rangle$ is always orthogonal to $|n\rangle^{(o)}$. Any ket $|\ \rangle$ can now be written as a sum of a term parallel to $|n\rangle^{(o)}$ and a part orthogonal to $|n\rangle^{(o)}$

$$|\ \rangle = P_n|\ \rangle + Q_n|\ \rangle\ . \qquad 12.46$$

Hence in particular for the ket $|n\rangle$ we have

$$|n\rangle = P_n|n\rangle + Q_n|n\rangle$$

$$= |n\rangle^{(o)} + Q_n|n\rangle \qquad 12.47$$

where we have used 12.40. Now let z be an arbitrary complex number then we may write

$$(H_o-z)|n\rangle = (E_n - \lambda V - z)|n\rangle \qquad 12.48$$

and hence

$$|n\rangle = \frac{1}{H_o-z}(E_n-\lambda V-z)|n\rangle \; . \qquad 12.49$$

Now multiply this equation by $1 = P_n + Q_n$ to get

$$|n\rangle = P_n \frac{1}{H_o-z}(E_n-\lambda V-z)|n\rangle + Q_n \frac{1}{H_o-z}(E_n-\lambda V-z)|n\rangle$$

$$= P_n|n\rangle + Q_n \frac{1}{H_o-z}(E_n-\lambda V-z)|n\rangle \qquad 12.50$$

Thus

$$|n\rangle = |n\rangle^{(o)} + Q_n \frac{1}{H_o-z}(E_n-\lambda V-z)|n\rangle \qquad 12.51$$

where in the last step we have used equation 12.47.

To get the perturbation series we simply iterate this equation by replacing $|n\rangle$ on the right side in terms of $|n\rangle^{(o)}$.

After the first step we get

$$|n\rangle = |n\rangle^{(o)} + Q_n \frac{1}{H_o-z}(E_n-\lambda V-z)|n\rangle^{(o)}$$

$$+ Q_n \frac{1}{H_o-z}(E_n-\lambda V-z)Q_n \frac{1}{H_o-z}(E_n-\lambda V-z)|n\rangle \; . \qquad 12.52$$

Repeating this procedure indefinitely we finally obtain

$$|n\rangle = \sum_{\ell=0}^{\infty} [Q_n \frac{1}{H_o-z} (E_n-\lambda V-z)]^{\ell}|n\rangle^{(o)} \qquad 12.53$$

Substituting this in the expression for the energy shift we then get

$$E_n-E_n^{(o)} = \lambda \sum_{\ell=0}^{\infty} {}^{(o)}\langle n|V[Q_n \frac{1}{H_o-z} (E_n-\lambda V-z)]^{\ell}|n\rangle^{(o)}. \qquad 12.54$$

By choosing appropriate values for z we now either recover the Rayleigh-Schrödinger theory or else obtain the Brillouin-Wigner Theory.

Thus choosing $z = E_n^{(o)}$ we recover Rayleigh- Schrödinger perturbation theory.

$$E_n = E_n^{(o)} + \lambda \sum_{\ell=0}^{\infty} {}^{(o)}\langle n|V [Q_n \frac{1}{H_o-E_n^{(o)}} (E_n-E_n^{(o)}-\lambda V)]^{\ell}|n\rangle^{(o)}. \qquad 12.55$$

and

$$|n\rangle = \sum_{\ell=0}^{\infty} [Q_n \frac{1}{H_o-E_n^{(o)}} (E_n-E_n^{(o)}-\lambda V)]^{\ell}|n\rangle^{(o)} . \qquad 12.56$$

If instead we choose $z = E_n$ we obtain the Brillouin- Wigner perturbation theory. Here

$$E_n = E_n^{(o)} + \lambda \sum_{\ell=0} {}^{(o)}\langle n|V[Q_n \frac{-1}{H_o-E_n} \lambda V]^{\ell}|n\rangle^{(o)} \qquad 12.57$$

and

$$|n\rangle = \sum_{\ell=0}^{\infty} [Q_n \frac{-1}{H_o-E_n} \lambda V]^{\ell}|n\rangle^{(o)} . \qquad 12.58$$

We can use either set to obtain E_n and $|n\rangle$ to arbitrary order in λ. Thus to first order 12.57 yields

$$E_n = E_n^{(o)} + \lambda \, {}^{(o)}\langle n|V|n\rangle^{(o)}$$

and to second order it yields

$$E_n = E_n^{(o)} + \lambda \, {}^{(o)}\langle n|V|n\rangle^{(o)} + \lambda^2 \sum_{\ell \neq n} \frac{|{}^{(o)}\langle n|V|\ell\rangle^{(o)}|^2}{E_n - E_\ell^{(o)}} .$$

This equation must still be solved for E_n. That is why Rayleigh-Schrödinger perturbation theory is used more often; it gives E_n directly. On the other hand, the Brillouin-Wigner theory converges more rapidly in certain situations and may therefore be more convenient.

These formal expansions to all orders are not any more useful than our previous result in Section 12.1 for obtaining numerical values. They are mainly of utility in trying to prove general results.

12.8 Sum Rule for Second Order Perturbation Theory

Second order perturbation theory inevitably gives us sums of the form

$$E_n^{(2)} = \sum_{\ell \neq n} \frac{{}^{(o)}\langle n|H'|\ell\rangle^{(o)(o)}\langle \ell|H'|n\rangle^{(o)}}{E_n^{(o)} - E_\ell^{(o)}} . \qquad 12.33$$

These sums are infinite in general and can not be done in closed form. An approximation procedure that has been used on occasion (ref. 12.1) is to set $E_\ell^{(o)} = E$, a constant. Then

$$E_n^{(2)} \simeq \frac{1}{E_n^{(o)} - E} \sum_{\ell \neq n} {}^{(o)}\langle n|H'|\ell\rangle^{(o)(o)}\langle \ell|H'|n\rangle^{(o)}$$

$$= \frac{1}{E_n^{(o)} - E} [\sum_{\ell} {}^{(o)}\langle n|H'|\ell\rangle^{(o)(o)}\langle \ell|H'|n\rangle^{(o)} - {}^{(o)}\langle n|H'|n\rangle^{(o)(o)}\langle n|H'|n\rangle^{(o)}].$$

Using the completeness relation $\sum_\ell |\ell\rangle^{(o)(o)}\langle\ell| = 1$, the above expression reduces to

$$E_n^{(2)} = \frac{1}{E_n^{(o)}-E}\,[{}^{(o)}\langle n|(H')^2|n\rangle^{(o)} - |{}^{(o)}\langle n|H'|n\rangle^{(o)}|^2] \qquad 12.59$$

Thus we have a neat closed form expression. Unfortunately it still contains the parameter E which must somehow be chosen in an arbitrary fashion. This presents the major drawback for this method.

Fortunately there are quite a few cases in which the sum 12.33 can be evaluated exactly if one can solve a certain differential equation. Since one can then do the sum we obtain a sum rule. In practice there are many different kinds of sum rules (ref. 12.2); we concentrate here only on the particular sum expressed by equation 12.33.

To evaluate this perturbation sum we look for an operator F_n such that

$$H'|n\rangle^{(o)} = [F_n,H_o]|n\rangle^{(o)}. \qquad 12.60$$

Here as before $H = H_o + \lambda H'$. Since the right hand side of 12.60 vanishes if we take an inner product with ${}^{(o)}\langle n|$ we modify this equation slightly to read

$${}^{(o)}\langle\ell|H'|n\rangle^{(o)} = E_n^{(1)}\delta_{\ell,n} + {}^{(o)}\langle\ell|[F_n,H_o]|n\rangle^{(o)} \qquad 12.61$$

This expression clearly simplifies since

$$\begin{aligned} {}^{(o)}\langle\ell|[F_n,H_o]|n\rangle^{(o)} &= {}^{(o)}\langle\ell|F_nH_o-H_oF_n|n\rangle^{(o)} \\ &= {}^{(o)}\langle\ell|F_nE_n^{(o)}-E_\ell^{(o)}F_n|n\rangle^{(o)} \\ &= (E_n^{(o)}-E_\ell^{(o)})\,{}^{(o)}\langle\ell|F_n|n\rangle^{(o)} \end{aligned} \qquad 12.62$$

Substituting this result on the right side of 12.33 we find

$$E_n^{(2)} = \sum_{\ell \neq n} {}^{(o)}\langle n|H'|\ell\rangle^{(o)(o)}\langle \ell|F_n|n\rangle^{(o)}$$

$$= \sum_{\ell} {}^{(o)}\langle n|H'|\ell\rangle^{(o)(o)}\langle \ell|F_n|n\rangle^{(o)}$$

$$- {}^{(o)}\langle n|H'|n\rangle^{(o)(o)}\langle n|F_n|n\rangle^{(o)} \qquad 12.63$$

or

$$E_n^{(2)} = {}^{(o)}\langle n|H'F_n|n\rangle^{(o)} - E_n^{(1)(o)}\langle n|F_n|n\rangle^{(o)}. \qquad 12.64$$

For the last step we have again used the completeness relation $\sum_{\ell} |\ell\rangle^{(o)(o)}\langle \ell| = 1$ as well as the fact that

$$E_n^{(1)} = {}^{(o)}\langle n|H'|n\rangle^{(o)} .$$

Thus if we can find such an F_n then the sum for the energy in second order perturbation theory can be reduced to an integral.

To see how this F_n may be evaluated in practice consider a hamiltonian for a system in one dimension (ref. 12.3)

$$H = \frac{p^2}{2m} + V(x) + \lambda V_1(x) \qquad 12.65$$

where the unperturbed hamiltonian

$$H_o = \frac{p^2}{2m} + V(x) \qquad 12.66$$

has the eigenfunctions $\phi_n(x)$ corresponding to the non- degenerate eigenvalues $E_n^{(o)}$. We now assume that the operator F_n is simply a function of x. Equation 12.60 when written out then reads:

$$V_1(x)\phi_n(x) = -\frac{\hbar^2}{2m}\{F_n(x)\frac{d^2\phi_n}{dx^2} - \frac{d^2}{dx^2}[F_n(x)\phi_n(x)]\}$$

$$= \frac{\hbar^2}{2m}\{\frac{d^2F_n}{dx^2}\phi_n(x) + 2\frac{dF_n}{dx}\frac{d\phi_n}{dx}\} . \qquad 12.67$$

Thus,

$$\frac{d^2F_n}{dx^2} + 2\frac{\phi'_n}{\phi_n}\frac{dF_n}{dx} = \frac{2m}{\hbar^2}\frac{V_1(x)}{\phi_n(x)} \qquad 12.68$$

Calling $\frac{dF_n}{dx} = u$ we get the first order linear equation

$$\frac{du}{dx} + 2\frac{\phi'_n}{\phi_n}u = \frac{2m}{\hbar^2}\frac{V_1(x)}{\phi_n(x)} \qquad 12.69$$

which may be integrated by multiplying by the integrating factor

$$R = \phi_n^2(x) . \qquad 12.70$$

$$\frac{d}{dx}(\phi_n^2 u) = \frac{2m}{\hbar^2}V_1(x)\phi_n(x) . \qquad 12.71$$

Hence, up to an integration constant

$$u = \frac{dF_n}{dx} = \frac{2m}{\hbar^2}\frac{1}{\phi_n^2(x)}\int_a^x V_1(y)\phi_n(y)dy \qquad 12.72$$

Integrating again we find

$$F_n(x) = \frac{2m}{\hbar^2}\int_b^x \frac{dy}{\phi_n^2(y)}\int_a^y V_1(z)\phi_n(z)dz . \qquad 12.73$$

The constants a and b simply determine the value of $F_n(b) = 0$ $\left.\frac{dF_n}{dx}\right|_{x=a} = 0$. Since these values can always be changed by adding to our particular solution 12.73 an arbitrary solution of the homogneous equation

$$\phi_n \frac{d^2F_n}{dx^2} + 2\phi_n' \frac{dF_n}{dx} = 0 , \qquad 12.74$$

they have no effect on the sum we want to evaluate. The reason for this is that solutions of the homogeneous equation correspond to $V_1 = 0$.

This simply means, that, as far as our perturbation sum is concerned, the boundary conditions on F_n are arbitrary and may be chosen for convenience.

We have now succeeded in reducing the sum for the second order energy perturbation to integrals over known functions. In practice these integrals can seldom be done in closed form. However whenever the perturbation sum is infinite, the integrals are easier to evaluate numerically than the sum.

12.9 Linear Stark Effect

A practical example, first worked out by Dalgarno and Lewis (ref. 12.4) yields, to second order, the shift in the ground state energy of an hydrogen atom due to a constant electric field. This is known as the linear Stark effect. In this case the unperturbed hamiltonian H_o is just the hydrogen hamiltonian

$$H_o = \frac{\vec{p}^{\,2}}{2m} - \frac{e^2}{r} . \qquad 10.70$$

The corresponding ground state wave functions is

$$\phi_o = (\pi a^3)^{-1/2} e^{-r/a} \qquad 10.113$$

The perturbation term is

$$H' = e\mathcal{E}z \qquad 12.75$$

where $\mathcal{E}$ is the strength of the electric field.

Thus the operator F_o must satisfy the equation

$$e\mathcal{E}ze^{-r/a} = [F_o,H_o]e^{-r/a} \qquad 12.76$$

We try to solve this equation by choosing

$$F_o = zf(r) = \cos\theta\, rf(r). \qquad 12.77$$

Straightforward algebra then yields the following differential equation for $f(r)$:

$$r\frac{d^2f}{dr^2} + (4-\frac{2r}{a})\frac{df}{dr} - \frac{2}{a}f = \frac{2me\mathcal{E}}{\hbar^2}r\,, \qquad 12.78$$

A particular integral of this equation is

$$f(r) = -\frac{me\mathcal{E}}{\hbar^2}a(r/2 + a) \qquad 12.79$$

Thus

$$F_o = -\frac{me\mathcal{E}}{\hbar^2}az(r/2 + a) \qquad 12.80$$

It is now a simple matter to evaluate the energy shift

$$E_n^{(2)} = \frac{-1}{\pi a^3}\int e^{-2r/a}\frac{m e^2\mathcal{E}^2}{\hbar^2}a^2z^2(r/2 + a)\, r^2dr\sin d\theta d\phi \qquad 12.81$$

Since the directions x,y, z are all equivalent we can replace z^2 by $\frac{1}{3}(x^2+y^2+z^2) = \frac{1}{3}r^2$. The θ and ϕ integrations can then be performed immediately and we arrive at

$$E_n^{(2)} = -\frac{4}{a}\frac{me^2\mathcal{E}^2}{3\hbar^2}\int_0^\infty e^{-2r/a}\left(\frac{r}{2}+a\right)r^4 dr \ . \tag{12.82}$$

so

$$E_n^{(2)} = -\frac{9}{4}\frac{me^2\mathcal{E}^2}{\hbar^2}a^4 = -\frac{9}{4}a^3\mathcal{E}^2 \tag{12.83}$$

This result is the same as the sum plus integral

$$-(e\mathcal{E})^2\left(\sum_{\substack{n\neq 1\\ \ell,m}}\frac{|\langle 100|z|n,\ell,m\rangle|^2}{\frac{1}{2}mc^2\alpha^2(1-\frac{1}{n^2})} + \int d^3k\,\frac{|\langle 100|z|\vec{k}\rangle|^2}{\frac{1}{2}mc^2\alpha^2+\frac{\hbar^2k^2}{2m}}\right)$$

where $|n,\ell,m\rangle$ and $|\vec{k}\rangle$ are hydrogen atom eigenkets.

This completes our treatment of non-degenerate perturbation theory. In the next chapter we develop perturbation theory for states which are degenerate in energy.

Notes, References and Bibliography

One of the original references on perturbation theory and still very readable today is: J.W.S. Rayleigh - The Theory of Sound - Vol. I sect. 90, Dover Publications, New York (1945).

A very concise treatment of Rayleigh-Schrödinger perturbation theory is given in:

E.P. Wigner - Group Theory - Chapter 5, Academic Press, New York (1959).

12.1 A. Unsold, Z. Phys. 43 563 (1967).

12.2 R. Jackiw - Quantum-mechanical sum rules - Phys. Rev. 157 , 1220-1225(1967)

12.3 C. Schwartz - Calculations in Schrödinger Perturbation theory - Annals of Physics N.Y. 2, 156-169, (1959).

12.4 A Dalgarno and J.T. Lewis - The exact calculation of long-range forces between atoms by perturbation - Proc. Roy. Soc. A233, 70-84, (1956).

Chapter 12 Problems

12.1 Verify formula 12.29. HINT: Solve the problem

$$\nabla^2 \psi(\vec{r}) = \delta(\vec{r}-\vec{r}\,')$$

by

a) expanding in spherical harmonics,

b) realizing that $\psi(\vec{r})$ is the potential for a unit charge located at $\vec{r}\,'$,

and compare the two solutions.

12.2 In 12.11 set r=2 and derive equation 12.36 for the second order correction to the wave function.

12.3 Consider the hamiltonian

$$H = \frac{p^2}{2m} + \frac{1}{2}kx^2 + \frac{1}{2}\lambda x^2 \qquad k > 0$$

a) Find the exact energy of the n'th state of this hamiltonian and expand it to order λ^2 assuming $|\lambda| < k$.

b) Use perturbation theory, treating $\frac{1}{2}\lambda x^2$ as a perturbation, and find the energy of the n'th state to order λ^2.

12.4 Find the approximate ground state energy to second order for the hamiltonian

$$H = \frac{p^2}{2m} + \frac{1}{2} kx^2 + \frac{1}{4} \lambda x^4$$

using the Rayleigh-Schrödinger perturbation theory.

12.5 Repeat problem 12.4 using Brillouin-Wigner perturbation theory

12.6 Consider the hamiltonian

$$H = H_o + \lambda H'$$

where

$$H_o = \begin{pmatrix} E_1 & 0 \\ 0 & E_2 \end{pmatrix}$$

$$H' = \begin{pmatrix} 0 & ia \\ -ia & 0 \end{pmatrix}$$

a) solve for the exact eigenvalues and eigenfunctions

b) Solve for both eigenvalues and eigenfunctions to 2nd order using Rayleigh-Schrödinger perturbation theory.

12.7 A particle of mass m moves in a potential

$$V = \frac{1}{2} k|x|^{2+\varepsilon} \qquad |\varepsilon| < 1.$$

Using appropriate harmonic oscillator wave functions estimate the energies of the ground state and first excited state. Find an approximation for the ground state wave function.

HINT:

a) $\frac{1}{2} k|x|^{2+\varepsilon} = \frac{1}{2} kx^2 - \frac{1}{2} k(x^2 - |x|^{2+\varepsilon})$

$$\simeq \frac{1}{2} kx^2 + \frac{1}{2} k\varepsilon x^2 \ \ln|x| \ . \qquad \varepsilon \ll 1$$

b) $\int_0^\infty e^{-\alpha x^2} x^2 \ln x \, dx = \frac{\sqrt{\pi}}{4} \alpha^{-3/2} [1 - \frac{1}{2}(c + \ln 4\alpha)]$ where $c = 0.577\ 216$ ... = Euler's constant.

12.8 For a particle of mass m moving in the potential

$$V = \frac{1}{2} k_1 x^2 + \frac{1}{2} k_2 y^2 + \lambda xy$$

$$|k_1 - k_2| \gg 2\lambda$$

a) find the exact ground state energy.

b) Find the ground state energy to order λ^2 without diagonalizing the hamiltonian

c) Find to order λ^2 the energy of the first excited state both exactly and by using perturbation theory.

Chapter 13

Time-Independent Degenerate Perturbation Theory

In many cases of physical interest the energy levels of a simple hamiltonian are degenerate. This is always the case if the simple hamiltonian is invariant under some symmetry operation such as a spatial rotation or reflection. Thus for example all levels except the ground state of the three-dimensional simple harmonic oscillator show the $2\ell+1$ fold degeneracy arising from rotation symmetry. A similar statement is true for the hydrogen atom although in this case the degeneracy is larger since an even larger symmetry group than just rotations leaves the hydrogen hamiltonian invariant.

Now in order that the perturbation expansions 12.5, 12.6 be meaningful and useful, it is necessary that the series converge relatively fast. In fact to be useful we should be able to truncate the series after the first few terms for otherwise the computations become excessive. This further requires that the successive terms in the series decrease rapidly. On the other hand if we have degenerate or even just close-lying energy levels, the second order terms

$$\lambda^2 \sum_{\ell \neq n} \frac{|H'_{n,\ell}|^2}{E_n^{(o)} - E_\ell^{(o)}}$$

will give an excessively large contribution whenever the ℓ and n levels are degenerate or close. This is due to our treatment and we shall now develop techniques to handle this situation.

13.1 Two Levels: Rayleigh-Schrödinger Method

As discussed above, for ordinary perturbation theory as described, to be applicable we require that

$$\lambda|H'_{\ell n}| < |E_\ell^{(o)}-E_n^{(o)}| \qquad \ell \neq n \tag{13.1}$$

Now suppose that two eigenvalues say $E_1^{(o)}$ and $E_2^{(o)}$ of H_o are degenerate. In that case condition 13.1 will be violated for these two levels. Suppose further that all other energy levels are non-degenerate and widely spaced. In that case the sum

$$\lambda^2 \sum_{\ell\neq n} \frac{|H'_{n,\ell}|^2}{E_n^{(o)} - E_\ell^{(o)}} = \lambda^2 E_n^{(2)} \tag{13.2}$$

will diverge for $\ell = 1$, $n = 2$ and vice versa unless the matrix element H'_{12} (and hence H'_{21}) vanishes. This vanishing can indeed be accomplished due to the fact that in the subspace corresponding to the degenerate eigenvalues any linear combination of the eigenfunctions is also an eigenfunction.

Thus instead of the basis set $|1\rangle^{(o)}$, $|2\rangle^{(o)}$, $|3\rangle^{(o)}$, $|4\rangle^{(o)}$,... we could also take the set $a_{11}|1\rangle^{(o)} + a_{12}|2\rangle^{(o)}$, $a_{21}|1\rangle^{(o)} + a_{22}|2\rangle^{(o)}$, $|3\rangle^{(o)}$, $|4\rangle^{(o)}$,... where the matrix

$$A = \begin{pmatrix} a_{11} & a_{12} \\ a_{21} & a_{22} \end{pmatrix} \tag{13.3}$$

is unitary. We further specify the matrix A by requiring that it diagonalize H' in the degenerate subspace. Thus defining the matrix

$$V = \begin{pmatrix} H'_{11} & H'_{12} \\ H'_{21} & H'_{22} \end{pmatrix} \tag{13.4}$$

where

$$H'_{ij} = {}^{(o)}\langle i|H'|j\rangle^{(o)} \qquad i = 1,2 \qquad j = 1,2$$

we require that $AVA^{\dagger}$ be diagonal. This is accomplished by solving the eigenvalue problem

$$V\binom{\alpha}{\beta} = v\binom{\alpha}{\beta} . \tag{13.5}$$

For a non-trivial solution we require

$$\det(V - v1) = 0 \tag{13.6}$$

giving us the characteristic equation

$$(H'_{11} - v)(H'_{22} - v) - |H'_{12}|^2 = 0 . \tag{13.7}$$

The roots of this equation are

$$v_{\pm} = \frac{H'_{11}+H'_{22}}{2} \pm \frac{1}{2}\,[(H'_{11}-H'_{22})^2 + 4|H'_{12}|^2]^{1/2} . \tag{13.8}$$

Defining,

$$\tan\theta = \frac{2|H'_{12}|}{H'_{11} - H'_{22}} . \tag{13.9}$$

we get that the corresponding eigenvectors are

$$\frac{\alpha_+}{\beta_+} = \cot\frac{\theta}{2}$$

$$\frac{\alpha_-}{\beta_-} = -\tan\frac{\theta}{2}$$

so that

$$|\phi_1\rangle = \cos\frac{\theta}{2}\,|1\rangle^{(o)} + \sin\frac{\theta}{2}\,|2\rangle^{(o)} \qquad 13.10a$$

$$|\phi_2\rangle = -\sin\frac{\theta}{2}\,|1\rangle^{(o)} + \cos\frac{\theta}{2}\,|2\rangle^{(o)} \qquad 13.10b$$

The use of the basis set with $|\phi_1\rangle$, $|\phi_2\rangle$ as the first two members now ensures that in the degenerate subspace H' is diagonal so that all its off-diagonal elements vanish. Thus, in the perturbation to second order the two divergent terms do not appear.

We can now proceed with the standard perturbation theory. However we use the set $\{|\phi_n\rangle\} = \{|\phi_1\rangle , |\phi_2\rangle , |3\rangle^{(o)} , |4\rangle^{(o)} ,\ldots\}$ as a basis set from the start.

Thus, we let

$$E_n = E_n^{(o)} + \lambda E_n^{(1)} + \lambda^2 E_n^{(2)} + \ldots \; (n=1,2) \,. \qquad 13.11$$

Also the exact wavefunctions $|n\rangle$ $(n=1,2)$ are given by

$$|n\rangle = |\phi_n\rangle + \lambda(a_1^n|\phi_1\rangle + a_2^n|\phi_2\rangle) + \lambda \sum_{k\neq 1,2} a_k^n|k\rangle^{(o)}$$

$$+ \lambda^2(b_1^n|\phi_1\rangle + b_2^n|\phi_2\rangle) + \lambda^2 \sum_{k\neq 1,2} b_k^n|k\rangle^{o} + \ldots\ldots \qquad 13.12$$

for $(n = 1,2)$

Substituting these expansions into the Schrödinger equation

$$(H_o + \lambda H')|n\rangle = E_n|n\rangle \qquad 13.13$$

and proceeding to second order in λ for the energy and first order in λ for the wavefunction, we arrive at

$$H_o|\phi_n\rangle = E_n^{(o)}|\phi_n\rangle \qquad 13.14^{(o)}$$

$$(H_o-E_n^{(o)})|n\rangle^{(1)} + (H'-E_n^{(1)})|\phi_n\rangle = 0 \qquad 13.14^{(1)}$$

$$(H_o-E_n^{(o)})|n\rangle^{(2)}+(H'-E_n^{(1)})|n\rangle^{(1)}-E_n^{(2)}|\phi_n\rangle = 0 \qquad 13.14^{(2)}$$

These equations may now be solved just as in the non-degenerate case. For $n \neq 1,2$ (12.9) still holds. Thus, we get

$$E_n^{(1)} = \langle\phi_n|H'|\phi_n\rangle \qquad 13.15$$

and (see problem 13.7)

$$|1\rangle^{(1)} = \sum_{\ell\neq 1,2}\left\{\frac{\langle\phi_2|H'|\ell\rangle\langle\ell|H'|\phi_1\rangle}{(E_1^{(o)}-E_\ell^{(o)})(v_+-v_-)}|\phi_2\rangle + \frac{|\ell\rangle^{(o)(o)}\langle\ell|H'|\phi_1\rangle}{E_1^{(o)}-E_\ell^{(o)}}\right\}$$

$$|2\rangle^{(1)} = \sum_{\ell\neq 1,2}\left\{\frac{\langle\phi_1|H'|\ell\rangle\langle\ell|H'|\phi_2\rangle}{(E_2^{(o)}-E_\ell^{(o)})(v_--v_+)}|\phi_1\rangle + \frac{|\ell\rangle^{(o)(o)}\langle\ell|H'|\phi_2\rangle}{E_2^{(o)}-E_\ell^{(o)}}\right\} \qquad 13.16$$

as well as

$$|n\rangle^{(1)} = \sum_{\ell\neq n}\frac{|\ell\rangle^{(o)(o)}\langle\ell|H'|n\rangle^{(o)}}{E_n^{(o)}-E_\ell^{(o)}} \qquad n \neq 1,2 \qquad 13.17$$

The second order term for the energy is then given by

$$E_n^{(2)} = \sum_{\ell=3}^{\infty}\frac{|\langle\phi_n|H'_{n,\ell}|\ell\rangle^{(o)}|^2}{E_n^{(o)}-E_\ell^{(o)}} \qquad n = 1,2 \qquad 13.18a$$

and

$$E_n^{(2)} = \sum_{\ell \neq n} \frac{|H'_{n,\ell}|^2}{E_n^{(o)} - E_\ell^{(o)}} \qquad n \neq 1,2 \tag{13.18b}$$

This technique will of course only work if the perturbation H' removes the degeneracy in first order, that is, if $v_\pm$ are distinct. Otherwise one must diagonalize exactly the matrix corresponding to second order in the degenerate subspace. The case discussed is the most important in practice and we shall not discuss the more complicated situation where one must go to second order to remove the degeneracy.

13.2 General Rayleigh-Schrödinger Method for Degenerate Levels

Again the hamiltonian is

$$H = H_o + \lambda H' \ .$$

The eigenvalues $E_\ell^{(o)}$ of H_o are respectively g_ℓ-fold degenerate with corresponding eigenfunctions $|\ell,1\rangle^{(o)}$, $|\ell,2\rangle^{(o)}$,... $|\ell,g_\ell\rangle^{(o)}$. As discussed in the previous section, the difficulty due to degeneracy which is obvious in second order perturbation for the energy will not occur if the degeneracy can be lifted in first order by diagonalizing H' in each of the degenerate subspaces. To do this we take a new basis set $\{|\psi_{\ell,k}\rangle \quad k = 1,\ldots,g_\ell\}$ obtained from the original set $\{|\ell,k\rangle^{(o)}\}$ by a unitary transformation. Thus,

$$|\psi_{\ell,k}\rangle = \sum_{m=1}^{g_\ell} a_{km}^\ell |\ell,m\rangle^{(o)} \ . \tag{13.19}$$

We further choose the set $\{|\psi_{\ell k}\rangle\}$ so that using it as a basis diagonalizes H' in the corresponding g_ℓ-dimensional subspaces. We therefore have to solve the following set of eigenvalue equations

$$\langle\psi_{\ell k}|H'|\psi_{\ell j}\rangle = E^{(1)}_{\ell,k}\delta_{kj} \tag{13.20}$$

where $E^{(1)}_{\ell,k}$ are the resultant eigenvalues. This then reads

$$\sum_{m=1}^{g_\ell} [\langle\ell,m|H'|\ell,n\rangle - E^{(1)}_{\ell,k}\,\delta_{mn}]a^{\ell}_{km} = 0.$$

We now proceed with the usual perturbation theory using the $|\psi_{\ell,m}\rangle$ from the start. Thus we let

$$E_{\ell,m} = E^{(o)}_{\ell} + \lambda E^{(1)}_{\ell,m} + \lambda^2 E^{(2)}_{\ell,m} + \ldots$$

Substituting this into the Schrödinger equation

$$H|\ell,m\rangle = E_{\ell,m}|\ell,m\rangle$$

we obtain the equations corresponding to (12.8). Thus we have

$$(H_o - E^{(o)}_{\ell})|\psi_{\ell m}\rangle = 0 \tag{13.21$^{(o)}$}$$

$$(H_o - E^{(o)}_{\ell})|\ell,m\rangle^{(1)} + (H'-E^{(1)}_{\ell,m})|\psi_{\ell,m}\rangle = 0 \tag{13.21$^{(1)}$}$$

$$(H_o - E^{(o)}_{\ell})|\ell,m\rangle^{(2)} + (H'-E^{(1)}_{\ell,m})|\ell,m\rangle^{(1)} - E^{(2)}_{\ell,m}|\psi_{\ell,m}\rangle = 0. \tag{13.21$^{(2)}$}$$

As for the case of two degenerate levels, we again expand in terms of the unperturbed eigenkets $|\psi_{\ell,m}\rangle$.

$$|\ell,m\rangle^{(1)} = \sum_{n=1}^{g_\ell} a_{\ell m,\ell n}|\psi_{\ell,n}\rangle + \sum_{n=1}^{g_k}\sum_{k\neq\ell} a_{\ell m,kn}|\psi_{k,n}\rangle \tag{13.22}$$

with a similar expression for $|\ell,m\rangle^{(2)}$.

The first order perturbations are now obtained from $(13.21^{(1)})$. Thus,

$$E^{(1)}_{\ell,m} = \langle\phi_{\ell m}|H'|\phi_{\ell m}\rangle \tag{13.23}$$

and

$$|\ell,m\rangle^{(1)} = \sum_{p\neq m}^{g_\ell}\sum_{n=1}^{g_k}\sum_{k\neq\ell} \frac{|\phi_{\ell p}\rangle\langle\phi_{\ell p}|H'|\phi_{kn}\rangle\langle\phi_{kn}|H'|\phi_{\ell m}\rangle}{(E_\ell^{(o)} - E_k^{(o)})(\langle\phi_{\ell m}|H'|\phi_{\ell m}\rangle-\langle\phi_{\ell p}|H'|\phi_{\ell p}\rangle)}$$

$$+ \sum_{n=1}^{g_k}\sum_{k\neq\ell} \frac{|\phi_{kn}\rangle\langle\phi_{kn}|H'|\phi_{\ell m}\rangle}{E_\ell^{(o)} - E_k^{(o)}} \tag{13.24}$$

To second order, the energy is now given by

$$E^{(2)}_{\ell,m} = \sum_{\substack{k\neq\ell \\ n}} \frac{|\langle\phi_{\ell m}|H'||\phi_{kn}\rangle|^2}{E_\ell^{(o)} - E_k^{(o)}} \quad . \tag{13.25}$$

We now illustrate this technique with some simple examples.

13.3 Example: Spin Hamiltonian

We want to consider the following simple "hamiltonian"

$$H = AS_z^2 + b(S_x^2 - S_y^2) \tag{13.26}$$

for a system with a total spin $S = 1$. This hamiltonian arises if one considers an ion with spin 1 located in a crystal at a point where the effective potential has rhombic symmetry and one only considers this perturbation on the background of large kinetic and Coulomb energies.

The constants A and b are determined by the ionic crystal properties. In general $|b| < |A|$ so that we can treat the term $b|S_x^2 - S_y^2|$ as a perturbation on the term AS_z^2.

The eigenfunctions of the unperturbed hamiltonian AS_z^2 are

$$|1\rangle^{(o)} = \begin{pmatrix} 1 \\ 0 \\ 0 \end{pmatrix} \tag{13.27}$$

$$|2\rangle^{(o)} = \begin{pmatrix} 0 \\ 1 \\ 0 \end{pmatrix} \tag{13.28}$$

$$|3\rangle^{(o)} = \begin{pmatrix} 0 \\ 0 \\ 1 \end{pmatrix} \tag{13.29}$$

with the corresponding energies

$$E_1^{(o)} = A\hbar^2 \ , \ E_2^{(o)} = 0 \ , \ E_3^{(o)} = A\hbar^2 \tag{13.30}$$

$E_1^{(o)}$ and $E_3^{(o)}$ are degenerate. In fact on the basis set above the hamiltonian becomes

$$H = \hbar^2 \begin{pmatrix} A & 0 & b \\ 0 & 0 & 0 \\ b & 0 & A \end{pmatrix} . \tag{13.31}$$

We shall first give an exact solution and then use Rayleigh-Schrödinger perturbation theory.

13.3.1 Exact Solution

To diagonalize H we simply solve

$$H|n\rangle = E_n|n\rangle \ . \tag{13.32}$$

This requires with $|n\rangle = a_{1n}|1\rangle^{(o)} + a_{2n}|2\rangle^{(o)} + a_{3n}|3\rangle^{(o)}$ that

$$\det(H - E_n 1) = 0$$

giving the characteristic equation

$$(A\hbar^2 - E_n)^2 E_n - (b\hbar^2)^2 E_n = 0. \tag{13.33}$$

Hence one value of E_n is $E_2 = 0$ and E_1, E_3 are given by

$$E_n^2 - 2A\hbar^2 E_n + (A\hbar^2)^2 - (b\hbar^2)^2 = 0 \ , \ n = 1,3$$

or

$$\begin{aligned} E_1 &= A\hbar^2 + b\hbar^2 \\ E_2 &= 0 \\ E_3 &= A\hbar^2 - b\hbar^2 \quad . \end{aligned} \tag{13.34}$$

As is easily checked the corresponding normalized eigenkets are

$$\begin{aligned} |1\rangle &= \frac{1}{\sqrt{2}}\left(|1\rangle^{(o)} + |3\rangle^{(o)}\right) \\ |2\rangle &= |2\rangle^{(o)} \\ |3\rangle &= \frac{1}{\sqrt{2}}\left(|1\rangle^{(o)} - |3\rangle^{(o)}\right) . \end{aligned} \tag{13.35}$$

13.3.2 Rayleigh-Schrödinger Solution

Since $E_1^{(o)}$ and $E_3^{(o)}$ are degenerate we must use the degenerate perturbation theory approach. This requires that we diagonalize the perturbation part H' of the hamiltonian in the degenerate subspace. On the basis set of unperturbed states this portion of the hamiltonian is

$$H'_{degen} = \hbar^2 \begin{pmatrix} 0 & b \\ b & 0 \end{pmatrix} \tag{13.36}$$

The corresponding eigenvalues are

$$E_1^{(1)} = b\hbar^2$$

$$E_3^{(1)} = -b\hbar^2 \tag{13.37}$$

Also the corresponding eigenkets which must be linear combinations of $|1\rangle^{(o)}$ and $|3\rangle^{(o)}$ are easily found to be $\frac{1}{\sqrt{2}}(|1\rangle^{(o)} \pm |3\rangle^{(o)})$ respectively. These are the kets $|\phi_1\rangle$ and $|\phi_3\rangle$ which we use to do further perturbation calculations. To first order we then have

$$E_1^{(1)} = \langle\phi_1|H'|\phi_1\rangle = b\hbar^2$$

$$E_2^{(1)} = {}^{(o)}\langle 2|H'|2\rangle^{(o)} = 0 \tag{13.38}$$

$$E_3^{(1)} = \langle\phi_3|H'|\phi_3\rangle = -b\hbar^2$$

and to this order the eigenkets are $|\phi_1\rangle$, $|2\rangle^{(o)}$, $|\phi_3\rangle$. These results coincide with the exact solution. It is easily checked that the second order perturbations vanish. We do this only for E_1. Thus,

$$E_1^{(2)} = \sum_{k\neq 1,3} \frac{|\langle\phi_1|H'|^2 k\rangle^{(o)}|^2}{E_1^{(o)} - E_k^{(o)}} = \frac{|\langle\phi_1|H'|2\rangle^{(o)}|^2}{E_1^{(o)} - E_2^{(o)}} = 0$$

as stated.

Notes, References and Bibliography

The book by Wigner (ref. 12.2) provides a very detailed treatment of degenerate Rayleigh-Schrödinger perturbation theory and shows how group-theoretic methods may be used to simplify the computations.

Chapter 13 Problems

13.1 A system has three energy levels as shown (fig. 13.1). E_1 and E_2 are almost degenerate. A perturbation of the system is then performed so that transitions between 1 and 3 and 2 and 3 become possible. If the exact matrix elements of the perturbation $\langle 1|V|3\rangle = a$ and $\langle 2|V|3\rangle = b$, compute the new energy levels

a) exactly

b) using an appropriate form of perturbation theory to second order.

13.2 Find the shift in the energy of the $n = 1$ levels of an hydrogen atom, to second order due to a constant electric field (linear Stark effect). The potential is

$$V' = e\vec{E}\cdot\vec{r} = e\mathcal{E}z$$

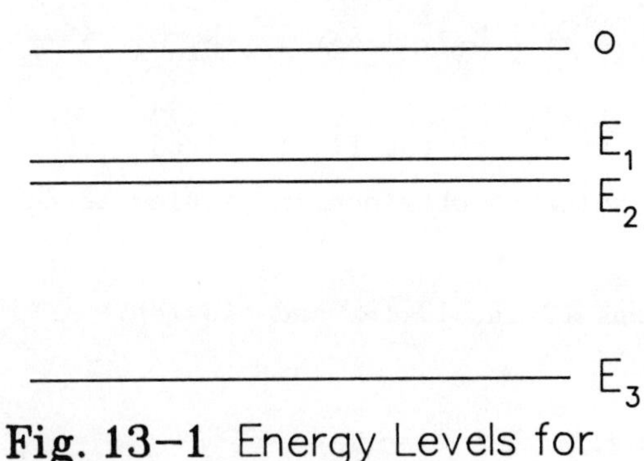

Fig. 13–1 Energy Levels for Problem 13.1

13.3 A particle is in a 2-dimensional box of sides a. If a perturbation

$$V' = \lambda \sin \frac{m\pi x}{a} \sin \frac{n\pi y}{a}$$

is applied. Calculate to second order the shifts in the energy levels of the ground state and first excited state.

13.4 For the two-dimensional simple harmonic oscillator with

$$H_o = h\omega(a_1^\dagger a_1 + a_2^\dagger a_2)$$

calculate the effect of the perturbation

$$H' = \lambda(a_1^\dagger a_1^\dagger a_2 a_2 + a_2^\dagger a_2^\dagger a_1 a_1)$$

on the second and third excited states.

13.5 Repeat 13.4 with

$$H' = \lambda(a_1^\dagger a_2 + a_2^\dagger a_1).$$

This problem can also be solved exactly by introducing

$$A_1 = \cos\theta\ a_1 + \sin\theta\ a_2$$

$$A_2 = -\sin\theta\ a_1 + \cos\theta\ a_2$$

and choosing θ appropriately.

13.6 For a particle of mass m moving in the potential

$$V = \frac{1}{2} k_1 x^2 + \frac{1}{2} k_2 y^2 + \lambda xy \ , \quad |k_1 - k_2| \ll 2\lambda$$

Find to order λ^2 the energy of the first exited state. Compare your answer with the exact solution obtained in problem 12.8c).

13.7 Obtain equations 13.16, 13.18a and 13.18b.

Chapter 14

Further Approximation Methods

The approximation methods for the state vectors and energy levels of a given system that we discussed in the two previous chapters work very well indeed when the perturbation part H' of the hamiltonian is small. The techniques were based on truncating the infinite series expansions for the state vector and energy. In practice it is often simpler to estimate the energy levels by assuming some approximate wavefunction with a reasonable shape. In this case one has to optimize the fit by varying the parameters that determine things like the width of the approximate wavefunction. In a case like this, the computations are greatly facilitated by the use of an expression that is stationary since the result will then be fairly insensitive to the details of the wavefunction.

The use of a stationary expression comes even more to the fore when one picks a set $\{\psi_\alpha\}$ of wavefunctions. Here α is a generic label for a set of parameters. Since the expression used is to be stationary in the vincinity of the exact value we clearly expect the approximation to optimize by choosing those values of the parameters that do in fact make the expression stationary. In general, however, it is not enough to make the expression stationary. What we need, in fact, is an extremum principle. We shall now elaborate these statements somewhat.

Suppose for example that a quantity to be calculated, say Q, is some functional of the wavefunction

$$Q = F(\psi). \tag{14.1}$$

An example of such an expression would be the formula for the energy E given by

$$E = (\psi, H\psi)/(\psi,\psi). \tag{14.2}$$

Suppose now that the expression (14.1) for Q is stationary when ψ satisfies the correct Schrödinger equation i.e. when we have the exact ψ. We then take our family $\{\psi_\alpha\}$ of wavefunctions and compute the quantities

$$Q(\alpha) = F(\psi_\alpha).$$

By varying the parameters α we now make $Q(\alpha)$ stationary. Thus we compute $\delta F/\delta\psi$ and form the set of equations

$$\frac{\delta F}{\delta \psi} = \frac{\partial Q}{\partial \alpha} = 0. \tag{14.3}$$

This will give one (or several) values α' for the parameters such that Q is stationary. One would then expect these values of α to optimize the value for Q. However, unless one has an extremum principle rather than just a stationary pinciple this will not necessarily be the case.

For example, let $F(\psi)$ be maximized by the correct ψ. Then choosing a ψ which satisfies 14.3 could in fact give the worst value for Q if it minimizes F. This shows the necessity for an extremum principle if one wants to be sure of optimizing the approximation.

In this chapter we first develop and then demonstrate the use of an extremum principle for the ground state energy of a system. The technique can also be extended to higher energy levels and we shall indicate how this is done. In general, however, the method is most useful for the ground state energy. That the ground state energy should satisfy a minimum principle is not at all surprising since it is after all the minimum energy that a system can possess.

In the last half of this chapter we shall then introduce and discuss the "geometrical optics" approximation for quantum mechanics. This is also known as the WKB approximation (see problems 2.7, 2.8). This will then conclude our treatment of time independent approximation methods.

14.1 Rayleigh-Ritz Method

We shall not be concerned with the formal structure of the variational formulas obtained, nor shall we be concerned with their possible deeper interpretations. To us, the purpose of variational formulas, will be simply a means for facilitating approximations. As a first step we shall derive a minimum principle for the ground state energy for a system.

Theorem 1 Let H be the hamiltonian of a system, then the functional,

$$E(\psi) = \langle\psi|H|\psi\rangle/\langle|\psi|\psi\rangle \qquad 14.4$$

is minimized when $|\psi\rangle$ is the ground state wavefunction.

Proof

Let $\{|n\rangle\}$ be the eigenfunctions of H corresponding to the eigenvalues $\{E_n\}$ where E_o is the ground state energy, E_1 is the energy of the first excited state and so on. Now consider any normalized state

$$|\psi\rangle = \sum_n a_n|n\rangle \quad . \qquad 14.5$$

Then $E(\psi) = \sum |a_n|^2 E_n$

or $E(\psi) \geq \sum |a_n|^2 E_o = E_o$ 14.6

where the equal sign holds only if all $a_n = 0$ for $n \neq 0$. In that case, however, $|\psi\rangle = |0\rangle$ is the eigenfunction corresponding to E_o as required.

Theorem 2

Let $E(\psi)$ be defined as in (14.4). Then any $|\psi\rangle$, for which $E(\psi)$ is stationary, is an eigenfunction of H.

Proof

Rewrite (14.4) in the form

$$E\langle\psi|\psi\rangle = \langle\psi|H|\psi\rangle \ . \qquad 14.7$$

Varying both sides with respect to ψ yields

$$\delta E\langle\psi|\psi\rangle + E(\langle\psi|\delta\psi\rangle + \langle\delta\psi|\psi\rangle) = \langle\psi|H|\delta\psi\rangle + \langle\delta\psi|H|\psi\rangle$$

and if E is stationary (i.e $\delta E = 0$) this yields

$$\langle\psi|H-E|\delta\psi\rangle + \langle\delta\psi|H-E|\psi\rangle = 0 \ . \qquad 14.8$$

Now although the variations $|\delta\psi\rangle$ and $\langle\delta\psi|$ are not independent they may be treated as independent since (14.8) is valid for arbitrary variations. In that case replacing $|\delta\psi\rangle$ by $i|\delta\psi\rangle$ in 14.8 yields

$$i\langle\psi|H-E|\delta\psi\rangle - i\langle\delta\psi|H-E|\psi\rangle = 0. \qquad 14.9$$

Multiplying (14.9) by i and adding the result to and subtracting it from (14.8) yields

$$\langle\delta\psi|H-E|\psi\rangle = 0 \quad \text{and} \quad \langle\psi|H-E|\delta\psi\rangle = 0. \qquad 14.10$$

These two equations are equivalent to (14.8) if the variations $|\delta\psi\rangle$ and $\langle\delta\psi|$ are treated as independent and arbitrary. In that case, however, they imply

$$(H-E)|\psi\rangle = 0 \tag{14.11a}$$

$$\langle\psi|(H^{\dagger}-E^{*}) = 0. \tag{14.11b}$$

On the other hand, H is an observable and therefore self-adjoint so that E is also real. Thus (14.11b) reads

$$\langle\psi|(H-E) = 0. \tag{14.11c}$$

This is simply the hermitean adjoint of (14.11a). Thus, as stated, if E is stationary $|\psi\rangle$ satisfies the Schrödinger equation for the energy.

This now provides a means for obtaining a minimum principle for the higher energy levels. The technique presupposes that the wavefunctions for the lower levels are known. Thus, for example, if we are interested in finding the energy of the first excited state and if $|0\rangle$ is the exact ground state wavefunction then it is a simple matter to show that

$$E_1(\psi) = \frac{(\langle\psi|-\langle\psi|0\rangle\langle 0|)|H|(|\psi\rangle-\langle 0|\psi\rangle|0\rangle)}{(\langle\psi-\langle\psi|0\rangle\langle 0|)\;(|\psi\rangle-\langle 0|\psi\rangle|0\rangle)} \tag{14.12}$$

is minimized by the wavefunction for the first excited state.

Similarly if $\{|k\rangle,\ k = 0,\ldots,n-1\}$ are the wavefunctions corresponding to the first n levels and

$$P_{n-1} = \sum_{k=0}^{n-1} |k\rangle\langle k|$$

then

$$E_n(\psi) = \langle\psi|(1-P_{n-1})H(1-P_{n-1})|\psi\rangle/\langle\psi|1-P_{n-1}|\psi\rangle \tag{14.13}$$

is minimized by $|\psi\rangle = |n\rangle$, the wavefunction for the n'th excited state. As we stated before, this is not generally a useful method for obtaining the energy of excited states.

To use the method one now picks a set of wavefunctions $|\psi(\alpha_1 \ldots \alpha_n)\rangle$ depending on a set of parameters $\{\alpha_1, \ldots, \alpha_n\}$ and computes

$$\langle H\rangle = \langle\psi|H|\psi\rangle / \langle\psi|\psi\rangle \ .$$

We then minimize $\langle H\rangle$ with respect to these parameters. In choosing these trial wavefunctions we must insure that they satisfy the correct boundary conditions. We then have to solve the n equations

$$\frac{\partial\langle H\rangle}{\partial\alpha_j} = 0 \qquad j = 1, \ldots, n \tag{14.14}$$

for the n α_j's. If several sets of solutions are obtained one takes that set which minimizes $\langle H\rangle$. We now give some illustrations.

14.2 Example 1 - Simple Harmonic Oscillator

This first example is simply to illustrate the technique before we consider a more realistic problem. We have

$$H = \frac{p^2}{2m} + \frac{1}{2} m\omega^2 x^2 . \tag{14.15}$$

As a trial wavefunction we choose

$$\psi(x) = \left(\frac{2\alpha}{\pi}\right)^{1/4} e^{-\alpha x^2} \ . \tag{14.16}$$

Even if we did not know the answer this choice would be reasonable since the ground state must be an even function of x and should decay rapidly at $\pm\infty$. Evaluating $\langle H\rangle$ we get

$$\langle H\rangle = \frac{\hbar^2\alpha}{2m} + \frac{m\omega^2}{8\alpha} \ . \tag{14.17}$$

Solving $\partial\langle H\rangle/\partial\alpha = 0$ yields

$$\alpha = \frac{m\omega}{2\hbar} \tag{14.18}$$

and hence the ground state energy is approximated by

$$\langle H\rangle\Big|_{\alpha=\frac{m\omega}{2\hbar}} = \frac{1}{2}\hbar\omega .$$

This result coincides of course with the exact result as given by (14.20). We next consider a more realistic example.

14.3 Example 2 - He Ground State

We discussed this problem before. The hamiltonian is given by

$$H = \frac{p_1^2}{2m} + \frac{p_2^2}{2m} - 2e^2\left(\frac{1}{r_1} + \frac{1}{r_2}\right) + \frac{e^2}{r_{12}} . \tag{14.19}$$

As we saw previously, neglecting the term e^2/r_{12} arising from the interaction of the two electrons gives an approximate wavefunction

$$\psi(\vec{r}_1,\vec{r}_2) = \frac{8}{\pi a_o^3} e^{-\frac{2}{a_o}(r_1+r_2)} . \tag{14.20}$$

In fact we used this wavefunction as a basis for our perturbation calculation. Now the effect of the interaction of the two electrons is to repel each other. This can be thought of as a mutual screening of the nucleus. In that case it becomes reasonable to think of the range of the wavefunction as a parameter. Thus we shall take as a normalized trial wavefunction

$$\psi(\vec{r}_1,\vec{r}_2) = \frac{Z^3}{\pi a_o^3} e^{-\frac{Z}{a_o}(r_1+r_2)} . \tag{14.21}$$

where Z is a parameter. With this wavefunction we evaluate

$$\langle H \rangle = \langle \psi | H | \psi \rangle$$

just as we did in the perturbation calculation to get (see problem (14.1))

$$\langle H \rangle = \frac{e^2}{a_o} (Z^2 - \frac{27}{8} Z). \qquad 14.22$$

Solving $\partial \langle H \rangle / \partial Z = 0$ yields

$$Z = \frac{27}{16} . \qquad 14.23$$

Thus our physical intuition was correct and Z is indeed less than 2, indicating a screened nucleus. Substituting this result back in (14.22) we get an estimate for the ground state energy

$$E_o = - \frac{e^2}{a_o} (\frac{27}{16})^2 = -76.6 \text{ eV}. \qquad 14.24$$

This is closer to the experimental values of -78.6 eV than the value of -74.0 eV previously obtained from perturbation theory. Furthermore both approximate values are higher than the experimental result. This is simply in conformity with the minimum principle previously enunciated.

The Rayleigh-Ritz variational principle has found a tremendous amount of application in molecular physics. The use of computers has facilitated the use of trial functions with very large numbers of parameters. Although one can thus improve the fits to experimental data, the physical interpretation of many of the parameters is frequently obscure. This situation is particularly accute in these cases since one is not only trying to obtain the best fit for the energy eigenvalue but is actually interested in the wavefunction itself. Nevertheless such unaesthetic procedures are sometimes unavoidable.

14.4 The W.K.B. Approximation

In chapter 2, part 4 we gave a heuristic derivation of the Schrödinger equation by postulating that the Hamilton-Jacobi equation is a geometrical optics approximation for some wave equation. This was examined a little further in problems 2.7 and 2.8. We shall now study this question more systematically to develop a useful approximation technique.

It is useful to recall the geometrical optics approximation and work at first in analogy to it. We shall then justify our procedure later. Now, in general, geometrical optics is valid if the index of refraction n varies little in a distance of one wavelength. This means we need

$$\lambda \left|\frac{dn}{dx}\right| < 1. \tag{14.25}$$

Actually a number of step discontinuities in n are permitted as long as the distances between these discontinuities are also large compared to the wavelength. This occurs for instance at the surfaces of mirrors or lenses in optics. On the other hand, if the jump discontinuities are separated by distances comparable to the wavelength, as in the case of gratings, physical optics becomes necessary.

In Chapter 1 we saw that if an index of refraction is to be associated with a given medium for matter waves then $n \propto p$ and since $\lambda \propto 1/p$ condition (14.25) reduces to

$$\lambda \left|\frac{d}{dx}\frac{\lambda_o}{\lambda}\right| = \frac{\lambda_o}{\lambda}\left|\frac{d\lambda}{dx}\right| < 1$$

where we have written $n = \lambda_o/\lambda$. In general n is of the order of unity so this condition reduces to

$$\left|\frac{d\lambda}{dx}\right| < 1. \tag{14.26}$$

Thus we expect a geometrical optics approximation for quantum mechanics to be valid when (14.26) is satisfied. Now

$$\lambda = \frac{h}{p} = \frac{h}{\sqrt{2m(E-V)}}$$

$$\therefore \qquad \left|\frac{d\lambda}{dx}\right| = \frac{mh\left|\frac{dV}{dx}\right|}{[2m(E-V)]^{3/2}} \qquad 14.27$$

So (14.26) becomes

$$\frac{mh\left|\frac{dV}{dx}\right|}{[2m(E-V)]^{3/2}} < 1 . \qquad 14.28$$

Having given a heuristic argument for the validity of a geometrical optics approximation we now derive the approximation as well as its region of validity.

In the transition from physical to geometrical optics we ignore the wave nature of light and follow the trajectories of light rays. Thus to make the analogous transition for particles is to ignore their wave nature and follow their trajectories. We accomplish this by expanding in powers of $\hbar$ and keeping only the lowest order terms.

Our starting point is the time-dependent Schrödinger equation

$$i\hbar\frac{\partial\psi}{\partial t} = -\frac{\hbar^2}{2m}\nabla^2\Psi + V\Psi. \qquad 14.29$$

First we write

$$\Psi = A_o e^{iw/\hbar} \qquad 14.30$$

with A_o a constant. The Schrödinger equation (14.29) then becomes

$$\frac{\partial w}{\partial t} + \frac{1}{2m}(\nabla w)^2 + V - \frac{i\hbar}{2m}\nabla^2 w = 0 . \qquad 14.31$$

Letting $\hbar \to 0$ we get

$$\frac{\partial w}{\partial t} + H(\vec{r},\vec{\nabla}w) = 0. \qquad 14.32$$

This is just one form of the classical Hamilton-Jacobi equations (equation 2.61) where w is Hamilton's principal function. Furthermore, just as in classical mechanics one can get a separation of the time variable. Thus if Ψ is a stationary state, we have

$$\Psi = \psi(\vec{r})e^{-iEt/\hbar}. \qquad 14.33$$

In that case

$$w(\vec{r},t) = S(\vec{r}) - Et \qquad 14.34$$

where $S(\vec{r})$ corresponds to Hamilton's characteristic function. We now have

$$\psi(\vec{r}) = A_o e^{iS(\vec{r})/\hbar} \qquad 14.35$$

and

$$\frac{1}{2m}(\nabla S)^2 - (E-V) - \frac{i\hbar}{2m}\nabla^2 S = 0. \qquad 14.36$$

The W.K.B. appoximation uses the last term in (14.36) to get one term beyond the classical expression.

We shall now show how to do this quite generally for the one-dimensional Schrödinger equation. In one dimension we start with

$$\frac{d^2u}{dx^2} + \frac{2m}{\hbar^2}(E-V)u = 0 \qquad 14.37$$

and put

$$u = Ae^{iS/\hbar} \qquad 14.38$$

as before. However, in order that A and S be real we also let A depend on x. Substituting (14.38) into (14.37) we get on separating real and imaginary parts

$$\left(\frac{dS}{dx}\right)^2 - 2m(E-V) = \hbar^2 \frac{1}{A}\frac{d^2A}{dx^2} \qquad 14.39$$

and

$$2\,\frac{dA}{dx}\frac{dS}{dx} + A\,\frac{d^2S}{dx^2} = 0. \qquad 14.40$$

From (14.40) we get, using primes to indicate differentiation, that

$$\frac{A'}{A} = -\frac{1}{2}\frac{S''}{S'} \qquad 14.41$$

so that

$$A = A_o(S')^{-1/2}\ . \qquad 14.42$$

Substituting this back in (14.39) we get

$$(S')^2 = 2m(E-V) + \hbar^2\left[\frac{3}{4}\left(\frac{S''}{S'}\right)^2 - \frac{1}{2}\frac{S'''}{S'}\right]. \qquad 14.43$$

Now writing S as a power-series expansion in $\hbar^2$

$$S = S_o + \hbar^2 S_1 + \hbar^4 S_2 + \ldots$$

and collecting the coefficients of equal powers of $\hbar^2$ we get to 0'th order in $\hbar^2$

$$(S_o')^2 = 2m(E-V) \tag{14.44}$$

and to first order in $\hbar^2$

$$2S_o'S_1' = \frac{3}{4}\left(\frac{S_o''}{S_o'}\right)^2 - \frac{1}{2}\frac{S_o''}{S_o'} \quad . \tag{14.45}$$

The S_1 term will give a correction to the S_o term. We shall use it to estimate the region of validity of the W.K.B. approximation, but first we solve (14.44) for S_o . To integrate this equation we must distinguish between two regions.

1. The classically allowed region $E > V$

and

2. the classically forbidden region $E < V$.

1. In the classically allowed region we define

$$k(x) = \frac{1}{\hbar}\sqrt{2m(E-V(x))} \quad . \tag{14.46}$$

Then

$$S_o' = \pm\hbar\, k(x) \tag{14.47}$$

giving

$$S_o = \pm\hbar \int_{x_1}^{x} k(x')dx' + \phi \tag{14.48}$$

and combining this with (14.42) we get the approximate wave function

$$u_o = \frac{A_o}{\sqrt{k(x)}} \cos\left[\int_{x_1}^{x} k(x')dx' + \phi\right] \tag{14.49}$$

where A_o and ϕ are constants and the lower limit x_1 of the integral may be picked to be any convenient number.

2. In the classically forbidden region we define

$$\kappa(x) = \frac{1}{\hbar}\sqrt{2m(V-E)} \tag{14.50}$$

leading eventually to the approximate wave function

$$u_o = \frac{1}{\sqrt{\kappa(x)}}[A_o \exp\int_{x_1}^{x}\kappa(x')dx'' + B_o \exp-\int_{x_1}^{x}\kappa(x')dx']. \tag{14.51}$$

Before proceeding with the discussion of how to match the solutions (14.49), (14.51) at the classical turning points given by $E = V(x)$ we shall examine conditions for the validity of this approximation. We shall in fact only derive a necessary condition, but in practice this condition is also usually sufficient. Recall that by analogy with optics we arrived at the condition

$$\frac{m\hbar V'}{[2m(E-V)]^{3/2}} < 1. \tag{14.28}$$

Now in order that u_o be a good approximation it is necessary that the next correction to u_o be small. The next correction gives $e^{i\hbar S_1}$ and this will be small if $\hbar S_1$ is small. Thus a necessary condition is that

$$\hbar S_1 < 1 . \tag{14.52}$$

We therefore look at (14.45), the equation for S_1 in more detail. Now in the classically allowed region $(E > V)$, $S_o' = \pm\hbar k(x)$ so that substituting this result into (14.45) yields

$$S_1' = \pm\frac{1}{4\hbar}[\frac{3}{2}\frac{k'^2}{k^3} - \frac{k''}{k^2}]$$

$$= \mp\frac{1}{4\hbar}[(\frac{k'}{k^2})' + \frac{1}{2}k(\frac{k'}{k^2})^2] . \tag{14.53}$$

Hence,

$$\hbar S_1 = \mp \frac{1}{4}\left[\frac{k'}{k^2} + \frac{1}{2}\int_{x_1}^{x} \left(\frac{k'}{k^2}\right)^2 k\, dx'\right] . \qquad 14.54$$

Clearly if $k'/k^2 < 1$ then $\hbar S_1 < 1$ as required. This, however, is precisely the condition (14.28). Thus our intuitive formula is verified. For the classically forbidden region $(E < V)$ simply replace k by $i\kappa$ to get a similar result. Both results can be included in the single condition

$$\frac{m\hbar|V'|}{[2m|E - V|]^{3/2}} < 1 \qquad 14.55$$

Although the condition (14.55) is necessary in order that the first-order W.K.B. approximation be useful, it tells us nothing about the convergence of the power series for S. Detailed examination shows that this series is asymptotic and therefore provides a good approximation if $\hbar$ is small. This simply means that, in examining the series, $\hbar$ is treated as a parameter and not that $\hbar$ actually varies. In practice, if (14.55) is satisfied, $\hbar$ is sufficiently small so that the first term yields a good approximation.

14.4.1 Turning points.

Those points x at which

$$E = V(x) \qquad 14.56$$

are where the kinetic energy changes sign. Classically of course, the kinetic energy is always non-negative and therefore the classical motion reverses at these points. These points are known, therefore, as turning points. Furthermore, the approximate wave function u_o given by (14.49) and (14.51) changes at a turning point. This "discontinuity" is not a property of the solutions of the original Schrödinger equation (14.37) but is rather a consequence of the approximations made. In fact at the turning point $k = 0$, so that the condition $k'/k^2 < 1$ necessary for the validity of the approximation is violated. Thus, the approximate solutions u_o are valid only up to some distance (several wavelengths) from the turning points. It is therefore necessary to find a means of connecting an approximate solution in the classically forbidden region with an approximate solution in the classically allowed region. The difficulty just mentioned can be traced back to the effective potential for the approximate solutions. By straightforward differentiation we find that in the classically allowed region u_o satisfies the Schrödinger equation

$$\frac{d^2u_o}{dx^2} + (k^2 - \frac{3k'^2}{4k^2} + \frac{k''}{2k})u_o = 0. \qquad 14.57$$

Thus we have in fact introduced a singular effective potential. The singularities, being due to the vanishing of k, occur precisely at the turning points.

Now suppose that x_1 is a turning point with the allowed region $x > x_1$ (see fig. 14.1). In the region of the turning point we can approximate V(x) by the tangent to V(x) at $x = x_1$. Thus we have

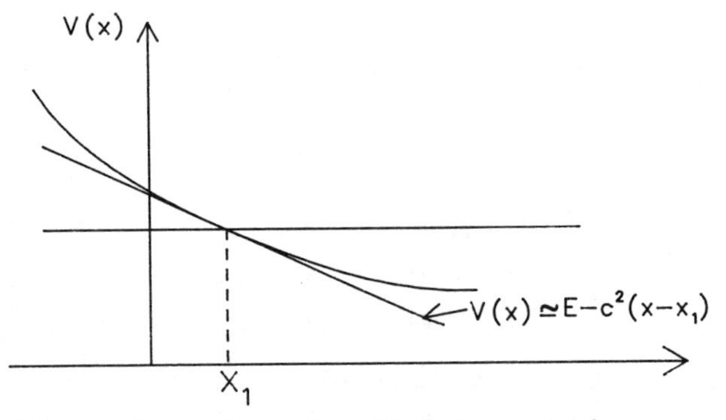

Fig. 14–1 Turning Point and Linear Approximation for Potential

$$V(x) \simeq E - \frac{\hbar^2}{2m} c^2(x-x_1) \qquad \text{near} \quad x = x_1 \ . \tag{14.58}$$

This leads to

$$\begin{aligned} k^2(x) &= \frac{2m}{\hbar^2}(E-V) = c^2(x-x_1) && x > x_1 \\ \kappa^2(x) &= \frac{2m}{\hbar^2}(V-E) = -c^2(x-x_1) && x < x_1 \ . \end{aligned} \tag{14.59}$$

Substituting these linear approximations for k^2 and κ^2 into the Schrödinger equation we find

$$\begin{aligned} \frac{d^2u}{dx^2} + c^2(x-x_1)u &= 0 && x > x_1 \\ \frac{d^2u}{dx^2} - c^2(x-x_1)u &= 0 && x < x_1 \ . \end{aligned} \tag{14.60}$$

The solutions of these equation can be expressed in terms of Bessel functions.

$$\begin{aligned} u_a^{\pm}(x) &= A_{\pm} y^{1/3} J_{\pm 1/3}(y) \\ u_f^{\pm}(x) &= B_{\pm} z^{1/3} I_{\pm 1/3}(z) \end{aligned} \tag{14.61}$$

where "a" stands for "allowed" ($x > x_1$ in this case) and "f" stands for forbidden ($x < x_1$ in this case), and

$$\begin{aligned} y &= \int_{x_1}^{x} k(x')dx = \frac{2}{3} c(x-x_1)^{3/2} && x > x_1 \\ z &= \int_{x}^{x_1} \kappa(x')dx' = \frac{2}{3} c(x_1-x)^{3/2} = \frac{2}{3} c|x-x_1|^{3/2} && x < x_1 . \end{aligned} \tag{14.62}$$

As stated, $J_n(x)$ is an ordinary Bessel function satisfying equation (10.28) and $I_n(x)$ is a modified Bessel function corresponding to a $J_n(x)$ with imaginary agreement.

These, equations (14.61), are exact solutions of a differential equation that coincides with the exact Schrödinger equation at a turning point whenever (14.58) is a good approximation. Thus we can use these solutions to match the W.K.B. solutions across a turning point. Away from a turning point we have the W.K.B. solutions. Also away from the turning point we can use the asymptotic forms of the solutions (14.61). Notice that both y(x) and z(x) were defined so as to increase as we move away from the turning point.

The asymptotic forms of these solutions can be found in any book on Bessel or transcendental functions (ref. 14.1) and are given by

$$J_{\pm 1/3}(y) \underset{y\to\infty}{\longrightarrow} \left(\frac{2}{\pi y}\right)^{1/2} \cos\left(y - \frac{\pi}{4} \mp \frac{\pi}{6}\right) \qquad 14.63$$

$$I_{\pm 1/3}(z) \underset{z\to\infty}{\longrightarrow} (2\pi z)^{-1/2}\left[e^{z} + e^{-z} e^{-i\pi(1/2 \pm 1/3)}\right] \qquad 14.64$$

The exponentially damped term e^{-z} in (14.64) is only meaningful if we take a linear combination $I_{+1/3}(z) - I_{-1/3}(z)$ so that the e^{z} terms cancel, since in writing (14.64) we have dropped terms or order e^{z}/z and these are large compared to e^{-z}. For small $|x-x_1|$ we also have the asymptotic forms

$$J_{\pm 1/3}(y) \underset{y\to 0}{\longrightarrow} \frac{(y/2)^{\pm 1/3}}{\Gamma(1 \pm \frac{1}{3})} \qquad 14.65$$

$$I_{\pm 1/3}(z) \underset{z\to 0}{\longrightarrow} \frac{(z/2)^{\pm 1/3}}{\Gamma(1 \pm \frac{1}{3})} \quad . \qquad 14.66$$

Thus in the vicinity of a turning point the solutions (14.61) become

$$u_a^{+} \simeq A_{+} \frac{2^{-1/3}(2c/3)^{2/3}}{\Gamma(4/3)} (x-x_1) \qquad 14.67$$

$$u_a^- = A_- \frac{2^{1/3}}{\Gamma(2/3)} \tag{14.68}$$

$$u_f^+ = B_+ \frac{2^{-1/3}(2c/3)^{2/3}}{\Gamma(4/3)} |x-x_1| \tag{14.69}$$

$$u_f^- = B_- \frac{2^{1/3}}{\Gamma(2/3)} \tag{14.70}$$

So we find that u_a^+ joins smoothly to u_f^+ if $A_+ = -B_+$, and u_a^- joins smoothly to u_f^- if $A_- = B_-$. So we choose the constants

$$A_+ = -B_+ = A_- = B_- = A. \tag{14.71}$$

The solutions then divide into two kinds: u^+ and u^-. They have the asymptotic forms

$$\begin{aligned} u^+ &\xrightarrow[|x-x_1|\to\infty]{} A\left(\frac{2}{\pi k(x)}\right)^{1/2} \cos\left(y-\frac{5\pi}{12}\right) \quad \text{allowed region} \\ u^+ &\xrightarrow[|x-x_1|\to\infty]{} -A(2\pi\kappa(x))^{-1/2}[e^z+e^{-z}e^{-i\pi\, 5/6}] \quad \text{forbidden region} \end{aligned} \tag{14.72}$$

and

$$\begin{aligned} u^- &\xrightarrow[|x-x_1|\to\infty]{} A\left(\frac{2}{\pi k}\right)^{1/2} \cos\left(y-\frac{\pi}{12}\right) \quad \text{allowed region} \\ u^- &\xrightarrow[|x-x_1|\to\infty]{} A(2\pi\kappa(x))^{-1/2}[e^z+e^{-z}e^{-i\pi\,/6}] \quad \text{forbidden region} \end{aligned} \tag{14.73}$$

Thus for example if we want an asymptotically damped solution in the classically forbidden region we take u^+ plus u^-. In the classically forbidden region this has the asymptotic form

$$3^{-1/2}A(2\pi\kappa)^{-1/2}(e^{-i\pi/6}-e^{-i\pi\, 5/6})e^{-z}$$

and joins smoothly onto the solution $u^+ + u^-$ which in the classically allowed region has the asymptotic behaviour

$$A(\frac{2}{\pi k})^{1/2}[\cos(y-\frac{5\pi}{12}) + \cos(y-\frac{\pi}{12})] = A(\frac{2}{\pi k})^{1/2}\cos(y-\frac{\pi}{4}).$$

Simplifying this expression we obtain the first of our connection formulas

$$\kappa^{-1/2}\, e^{-z} \rightarrow 2k^{-1/2} \cos(y-\frac{\pi}{4}) \ . \tag{14.74}$$

The arrow means that the asymptotic solution in the classically forbidden region appearing on the left of the arrow goes over into the solution with asymptotic behaviour appearing on the right of the arrow. The converse is false since a very small error in the phase of the cosine term would integrate back to produce a small admixture of the dominating exponentially growing term on the left. In fact in computer integration of the Schrödinger equation one frequently integrates "in from infinity" precisely because one wants to avoid the exponentially growing term due to round-off error.

On the other hand if we want a solution that in the classically allowed region has a definite phase, as for example $k^{-1/2} \cos(y-\frac{\pi}{4}+\phi)$ where $\phi \neq 0$ or a multiple of π we start by realizing that

$$\cos(y-\frac{5\pi}{12}) + \cos(y-\frac{\pi}{12}) = 3^{1/2}\cos(y-\frac{\pi}{4})$$

$$\cos(y-\frac{5\pi}{12}) - \cos(y-\frac{\pi}{12}) = \sin(y-\frac{\pi}{4})$$

and we further recall that

$$\cos(y-\frac{\pi}{4})\cos\phi - \sin(y-\frac{\pi}{4})\sin\phi = \cos(y-\frac{\pi}{4}+\phi).$$

Thus the solution in the classically forbidden region matching onto the one with the above asymptotic behaviour is given by

$$3^{-1/2}\cos\phi(u^{+} + u^{-}) - \sin\phi(u^{+} - u^{-})$$

and assumes after some simplification the form

$$(2\pi\kappa)^{-1/2}[\sin\phi\, e^{z} + \frac{1}{2}\, e^{-z-i\phi}]$$

The exponentially damped term on the right is again only meaningful for $\phi = 0$, otherwise it is to be dropped. Thus combining these results we get two more connection formulas:

$$k^{-1/2}\cos(y-\frac{\pi}{4}+\phi) \rightarrow \kappa^{-1/2}\sin\phi e^{z} \quad , \quad \phi \neq 0 \qquad 14.75$$

$$k^{-1/2}\cos(y-\frac{\pi}{4}) \quad \rightarrow \frac{1}{2}\kappa^{-1/2}e^{-z} \; . \qquad 14.76$$

These three formulas, (14.74), (14.75) and (14.76), suffice for handling all problems. In using them it is important to remember two things, namely that y and z are so defined as to increase as we move <u>away</u> from the turning point and secondly that the formulas may only be used to connect the solution on the left of the arrow to the solution on the right of the arrow and never in the reverse direction.

Next we illustrate the use of this method with some examples that frequently occur in practice.

14.5 Example 3 - W.K.B. Applied to a Potential Well

Consider a potential well as shown in fig. 14.2 rising to infinity for both very large x and $-x$. For the energy E shown there are clearly two turning points indicated by x_1 and x_2. Regions 1 and 3 are classically forbidden and region 2 is classically allowed. Thus we want exponentially damped solutions in regions 1 and 3. As usual this should, and will, lead to an energy quantization.

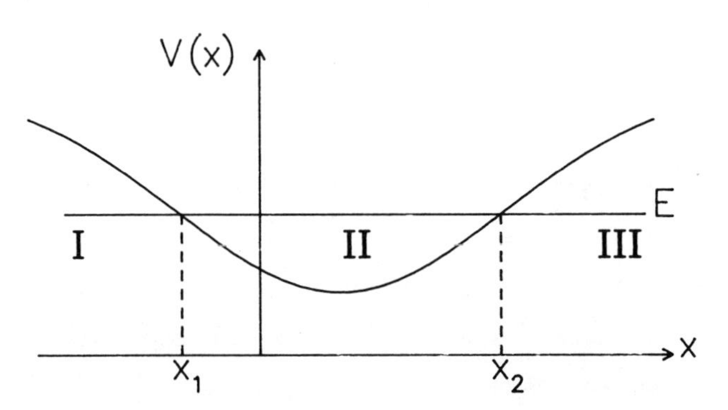

Fig. 14–2 Bound State Problem in W.K.B Approximation

Considering first the point x_1 we have using (14.49), (14.51) that

$$u_o = \frac{A}{\sqrt{\kappa(x)}} \exp - \int_x^{x_1} \kappa(x')dx' \quad x < x_1 \qquad 14.77$$

$$u_o = \frac{2A}{\sqrt{k(x)}} \cos\left[\int_{x_1}^{x} k(x')dx' - \frac{\pi}{4}\right] \quad x_1 < x < x_2 \qquad 14.78$$

Similarly we have that

$$u_o = \frac{B}{\sqrt{\kappa(x)}} \exp - \int_{x_2}^{x} \kappa(x')dx' \quad x > x_2 \qquad 14.79$$

and hence by applying the connection formula (14.74) at x_2 we get that for $x_1 < x < x_2$

$$\begin{aligned} u_o &= \frac{2B}{\sqrt{k(x)}} \cos\left[\int_x^{x_2} k(x')dx' - \frac{\pi}{4}\right] \\ &= \frac{2B}{\sqrt{k(x)}} \cos\left[\int_x^{x_1} k(x')dx' - \frac{\pi}{4} + \int_{x_1}^{x_2} k(x')dx'\right]. \end{aligned} \qquad 14.80$$

This last solution can be further rewritten as

$$u_o = \frac{2B}{\sqrt{k(x)}} \cos\left[-\int_{x_1}^{x} k(x')dx' + \frac{\pi}{4} + \int_{x_1}^{x_2} k(x')dx' - \frac{\pi}{2}\right]$$

$$= \frac{2B}{\sqrt{k(x)}} \cos\left[\int_{x_1}^{x} k(x')dx' - \frac{\pi}{4} - \alpha\right] \quad x_1 < x < x_2$$

where

$$\alpha = \int_{x_1}^{x_2} k(x')dx' - \frac{\pi}{2} . \qquad 14.81$$

Since this solution must coincide with the solution (14.78) for this region we require that

$$\alpha = n\pi \qquad n = 0,1,2,... \qquad 14.82$$

and

$$B = (-1)^n A . \qquad 14.83$$

Hence we get the energy quantization

$$\int_{x_1}^{x_2} k(x')dx' = (n+\frac{1}{2})\pi \qquad n = 0,1,2,... \qquad 14.84$$

But

$$k(x) = \frac{1}{\hbar}\sqrt{2m(E-V)} = \frac{2\pi}{h}\, p$$

so that (14.82) reads

$$2\int_{x_1}^{x_2} p\, dx = (n+\frac{1}{2})h$$

or more compactly

$$\oint p\, dx = (n+\frac{1}{2})h \ . \qquad 14.85$$

Except for the term $\frac{1}{2}h$ on the right this is just the Bohr-Sommerfeld quantization rule which now emerges quite naturally as a semiclassical approximation to the Schrödinger equation.

14.6 Special Boundaries

So far we have only discussed turning points occurring at points where $V(x)$ is continuous. If $V(x)$ experiences a finite jump discontinuity then one knows that the exact solution of the Schrödinger equation as well as the first derivative are continuous. In this case one simply matches the function and its derivative at the point of discontinuity regardless of whether or not this point is also a turning point. As long as V varies slowly on either side of this point, the asymptotic W.K.B. solutions may be used right up to it. In general this simplifies the calculation since it eliminates the necessity for explicit use of the connection formulas.

If $V(x)$ experiences an infinite discontinuity at the point $x = a$ so that say $V = \infty$ for $x < a$, then the exact wavefunction vanishes there and the appropriate W.K.B. solution to use for $x > a$, a classically allowed region, is

$$u = k(x)^{-1/2} \sin(\int_a^x k(x')dx').$$

This assumes of course that $k(x)$ is slowly varying so that the whole W.K.B. procedure is valid.

This type of situation occurs in practice for the s-wave ($\ell = 0$) radial wavefunction $R_o(r)$. One has an equation for $u(r) = rR_o(r)$ so that $u(r)$ must vanish at the origin.

14.7 Example 4-W.K.B. Approximation for Tunneling

As a second example to illustrate the use of the connection formulas we consider tunneling through a smooth potential barrier as shown in fig. 14.3. The turning points are x_1 and x_2 with $x_1 < x_2$. The classically allowed regions are

region I : $x < x_1$

and region III : $x > x_2$.

The classically forbidden region is

region II : $x_1 < x < x_2$.

We are interested in the case of a particle with energy E incident from the left. Thus in region I we have the W.K.B. solution

$$\phi_I(x) = k^{-1/2}\exp -i\left[\int_x^{x_1} k(x')dx' - \frac{\pi}{4}\right] + Rk^{-1/2}\exp i\left[\int_x^{x_1} k(x')dx' - \frac{\pi}{4}\right] \qquad 14.86$$

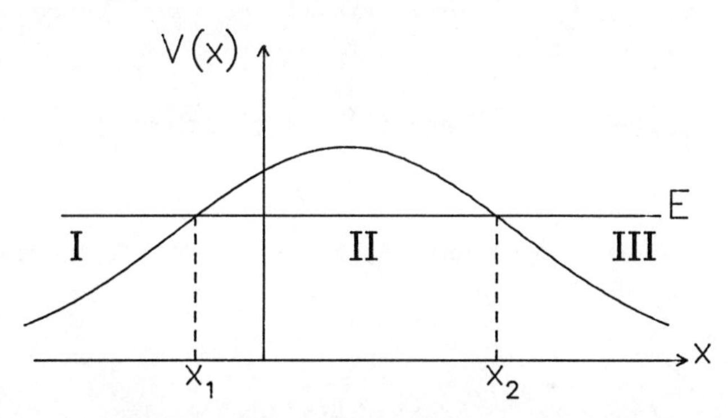

Fig. 14–3 Tunnelling in W.K.B. Approximation

We have arbitrarily chosen the amplitude of the incoming wave as 1 and called the amplitude of the reflected wave R. Also the phase factors of $e^{\pm i\pi/4}$ have been included to facilitate the use of the connection formulas. Rewriting (14.86) in terms of trigonometric functions we obtain

$$\Psi_I(x) = k^{-1/2}\{(1+R)\cos[\int_x^{x_1} k(x')dx' - \frac{\pi}{4}] - i(1-R)\sin[\int_x^{x_1} k(x')dx' - \frac{\pi}{4}] \tag{14.87}$$

Now using the connection formula (14.75) with $\phi = -\pi/2$ as well as the connection formula (14.76) we find that Ψ_I connects to the solution in region II given by

$$\Psi_{II}(x) = \kappa^{-1/2}\{\frac{1}{2}(1+R)\exp - \int_{x_1}^{x} \kappa(x')dx' + i(1-R)\exp \int_{x_1}^{x} \kappa(x')dx'\} \tag{14.88}$$

In region III we must have a purely transmitted wave. Hence we get

$$\Psi_{III}(x) = Tk^{-1/2}\exp i[\int_{x_2}^{x} k(x')dx' - \frac{\pi}{4}] \tag{14.89}$$

where we have called the amplitude of this wave T. This wave can be rewritten as

$$\Psi_{III}(x) = Tk^{-1/2}\{\cos[\int_{x_2}^{x} k(x')dx' - \frac{\pi}{4}] + i\sin[\int_{x_2}^{x} k(x')dx' - \frac{\pi}{4}]\}. \tag{14.90}$$

Again using (14.75) with $\phi = -\pi/2$ and (14.76) we see that the corresponding W.K.B. wave function in region II is given by

$$\Psi_{II}(x) = T\kappa^{-1/2}\{\frac{1}{2}\exp -\int_x^{x_2} \kappa(x')dx' - i\exp \int_x^{x_2} \kappa(x')dx'\} \qquad 14.91$$

To compare this with the expression given by (14.88) we rewrite the exponentials as follows:

$$-\int_x^{x_2} \kappa(x')dx' = -\int_{x_1}^{x_2} \kappa(x')dx' + \int_{x_1}^{x} \kappa(x')dx'$$

and define

$$S = \exp -\int_{x_1}^{x_2} \kappa(x')dx' \quad . \qquad 14.92$$

Then (14.91) can be written to read

$$\Psi_{II}(x) = T\kappa^{-1/2}\{\frac{1}{2} S \exp \int_x^{x_1} \kappa(x')dx' -i S^{-1}\exp -\int_{x_1}^{x} \kappa(x')dx'\} \quad . \qquad 14.93$$

Since this expression must coincide with the expression given by (14.87) we find that

$$\frac{1}{2}(1+R) = -iTS^{-1}$$
$$i(1-R) = \frac{1}{2}ST \;. \qquad 14.94$$

Solving these equations for T and R we find

$$T = \frac{iS}{1+S^2/4} \simeq iS \qquad 14.95$$

$$R = \frac{1-S^2/4}{1+S^2/4} \simeq 1 - \frac{S^2}{2} \;. \qquad 14.96$$

The last two approximations are in keeping with the spirit of the W.K.B. approximation which requires that $S \ll 1$. The quantities T and R have the obvious interpretation of transmission and reflection amplitude respectively.

Two more comments are in order regarding this example. The first deals with the region of validity of our approach. Clearly if $x_2 - x_1 = L$ is too small then the wavefunction ψ_{II} never assumes its asymptotic form and our connection formulas are not valid. This occurs if the energy E is too close to a maximum of the tunneling potential.

The second point is that we do not have connection formulas connecting an exponentially growing solution in the forbidden region to solutions in the allowed region. The reason is that the phase in the allowed region depends crucially on the admixture of the neglegible, damped solution. Thus if the forbidden region is of finite extent so that both exponentially growing and damped solutions can exist, then we must always connect from the classically allowed regions to the classically forbidden region. That is the reason we made the conections from regions I and III to region II and did not try to connect I → II → III or III → II → I.

Finally it is to be remembered that the W.K.B. approximation is just that, an approximation and as such the results obtained may sometimes have only qualitative validity.

Notes, References and Bibliography

14.1 It is again possible to find several of the techniques discussed in this chapter, in the book by Rayleigh ref. 13.1

The W.K.B. approximation was first applied to wave propagation by Rayleigh in 1872. It was then applied to quantum mechanics by:

H. Jeffreys - Proc. London Math. Soc. 23 428(1923).

Later it was simultaneously rediscovered by:

L. Brillouin - Comptes rendus 183 24(1926).

H.A. Kramers - Zeits. f. Physik 39 828 (1926).

G. Wentzel - Zeits. f. Physik 38 518 (1926).

There have also been many papers devoted to the connection formulae for the W.K.B. solutions. Two representative papers are:

R.E. Langer - On the Connection Formulas and the Solutions of the Wave Equation - Phys. Rev. 51 669 (1937).

W.H. Furry - Two Notes on Phase-Integral Methods - Phys. Rev. 71 360(1947).

See also the lectures by W.H. Furry - pages 34-67 in W.E. Brittin, W.B. Downs and J. Downs (editors) - Lectures in Theoretical Physics Vol. V - Summer Institute for Theoretical Physics, University of Colorado, Boulder (1962).

Chapter 14 Problems

14.1 Use the trial wave function (14.21) to evaluate the expectation value of the hamiltonian (14.19).

14.2 Use the connection formulae to solve the tunneling problem for a square barrier with the W.K.B. approximation. This result differs from the exact result in the text. However, they agree to the order to which the approximation has been carried.

14.3 Use a variational approach to find the ground state energy for a particle in the potential

$$V(r) = -V_o e^{-\alpha r^2}$$

14.5 A particle is in a potential

$$V = \frac{1}{4}\lambda^2 x^4 - \frac{1}{2}\mu^2 x^2 \ .$$

Sketch the potential and guess two possible ground state wave-functions. Use a variational approach to estimate the ground state energies for $m = 5\times10^{-27}$kg $\lambda = 10^{18}$kg$^{1/2}$/ms, $\mu = 10^{21}$ kgm/s^2. What happens as $\mu \to 0$?

14.5 A particle is in a potential

$$\begin{aligned} V &= -\ mg\ z \qquad && z > 0 \\ &= \infty && z = 0. \end{aligned}$$

This corresponds to a perfectly elastic ball bouncing on a floor. Find the W.K.B. solution for the ground state of this particle.

14.6 Estimate the ground state energy of an ionized hydrogen molecule. Use the experience obtained in solving (14.4).

14.7 Estimate the ground state energy of an hydrogen molecule if

a) the wavefunction is symmetric in the electron coordinates,

b) the wavefunction is antisymmetric in the electron coordinates.

Hint: Treat the separation of the two nuclei as a parameter and assume atomic hydrogen wavefunctions for the electrons.

Chapter 15

Time-Dependent Perturbation Theory

In the previous chapters we discussed techniques that are applicable if the perturbation is considered as causing a change in the states of the unperturbed system leading to shifts in the energy. This required that the perturbation be time-independent.

In the present chapter we are not interested in the modifications of the states of the unperturbed system, rather we are interested in the transitions occurring between levels of the unperturbed system under the influence of the perturbation. Thus we are interested in computing transition rates. In general we thus have time-dependent perturbations.

15.1 Formal Considerations

For the sake of concreteness consider a hamiltonian

$$H = H_o + \lambda V(t) \qquad 15.1$$

where the perturbation $V(t)$ is written as explicitly time-dependent. Let $|n\rangle$ be a complete set of eigenkets (in the energy representation) of H_o so that

$$H_o|n\rangle = E_n|n\rangle \quad . \qquad 15.2$$

If λ were zero these would be the stationary states of the system, but due to the perturbation transitions between these states occur. Now suppose that at some definite time $t = 0$ the system is known (prepared) to be in the state $|n\rangle$. Then at some later time t the system will be in a state $U(t)|n\rangle$ where $U(t)$ is the evolution operator and is given by solving (7.26):

$$i\hbar \frac{\partial U(t)}{\partial t} = HU(t) \qquad 15.3$$

with

$$U(0) = 1 \quad . \qquad 15.4$$

The probability of finding the system at time t in the state $|m\rangle$ is then given by

$$P_{n,m} = |\langle m|U(t)|n\rangle|^2 \quad . \qquad 15.5$$

If $n = m$, this gives the probability that the system has not changed in the time t. Thus the main problem in computing transitions reduces to calculating the amplitudes $\langle m|U(t)|n\rangle$. We, therefore, reconsider the equations for $U(t)$ in more detail.

The equation 15.3 may be combined with the initial condition (15.4) in a single integral equation

$$U(t) = 1 - \frac{i}{\hbar}\int_0^t H(t')U(t')dt' \quad . \qquad 15.6$$

We can also extract the behaviour due to H_o by going to the interaction picture. Thus, let $U^{(o)}(t)$ be the evolution operator corresponding to H_o , that is

$$i\hbar \frac{\partial U^{(o)}(t)}{\partial t} = H_o U^{(o)}(t) \tag{15.7}$$

with

$$U^{(o)}(0) = 1. \tag{15.8}$$

We then look for a $U_I(t)$ such that

$$U(t) = U^{(o)}(t) U_I(t) \tag{15.9}$$

and since $U^{(o)}$ is unitary we have

$$U_I(t) = U^{(o)\dagger}(t) U(t) . \tag{15.10}$$

Substituting this into (15.3), (15.4), and using (15.7), (15.8) we obtain the equation for $U_I(t)$.

$$i\hbar \frac{\partial U_I(t)}{\partial t} = \lambda H_I(t) U_I(t) \tag{15.11}$$

with

$$U_I(0) = 1 \tag{15.12}$$

where

$$H_I(t) = U^{(o)\dagger}(t) V(t) U^{(o)}(t) . \tag{15.13}$$

This again gives rise to an equivalent integral equation

$$U_I(t) = 1 - \frac{i\lambda}{\hbar} \int_0^t H_I(t') U_I(t') dt' . \tag{15.14}$$

A formal solution, which is also the basis of an approximation procedure, is now obtained by iterating 15.14. Thus we get

$$U_I(t) = 1 + \left(\frac{\lambda}{i\hbar}\right)\int_0^t H_I(t')dt' + \left(\frac{\lambda}{i\hbar}\right)^2 \int_0^t dt' \int_0^{t'} dt'' H_I(t')H_I(t'')U_I(t'') \qquad 15.15$$

or continuing the procedure

$$U_I(t) = 1 + \sum_{n=1}^{\infty} U_I^{(n)}(t) \qquad 15.16$$

where

$$U_I^{(n)}(t) = \left(\frac{\lambda}{i\hbar}\right)^n \int_0^t dt_1 \int_0^{t_1} dt_2 \cdots \int_0^{t_{n-1}} dt_n H_I(t_1)H_I(t_2)\ldots H_I(t_n) \qquad 15.17$$

Combining this with the definition of U_I in terms of $U^{(o)}$ we get

$$U(t) = U^{(o)}(t) + \sum_{n=1}^{\infty} U^{(n)}(t) \qquad 15.18$$

where

$$U^{(n)}(t) = U^{(o)}(t)\left(\frac{\lambda}{i\hbar}\right)^n \int_0^t dt_1 \int_0^{t_1} dt_2 \ldots \int_0^{t_{n-1}} dt_n U^{(o)\dagger}(t_1)V(t_1)U^{(o)}(t_1) U^{(o)\dagger}(t_2)V(t_2)U^{(o)}(t_2)\ldots U^{(o)\dagger}(t_n)V(t_n)U^{(o)}(t_n) \; . \qquad 15.19$$

If we now explicitly use the fact that H_o is time-independent we have

that

$$U^{(o)}(t) = \exp - \frac{iH_o t}{\hbar} . \tag{15.20}$$

In that case the transition probability can be written to different orders in λ explicitly in terms of the transition amplitudes

$$\langle m'|U(t)|m\rangle = \sum_{n=1}^{\infty} \langle m'|U^{(n)}(t)|m\rangle \tag{15.21}$$

where $U^{(n)}$ is given by (15.19). Writing out the first few orders explicitly we have

$$\langle m'|U^{(o)}(t)|m\rangle = e^{-i \frac{E_{m'}}{\hbar} t} \delta_{mm'} \tag{15.22^o}$$

$$\langle m'|U^{(1)}(t)|m\rangle = \frac{\lambda}{i\hbar}\int_0^t dt' e^{-i \frac{E_{m'}}{\hbar}(t-t')} \langle m'|V(t)|m\rangle e^{-i \frac{E_m}{\hbar} t'} \tag{15.22^1}$$

$$\langle m'|U^{(2)}(t)|m\rangle = \left(\frac{\lambda}{i\hbar}\right)^2 \sum_n \int_0^t dt_1 \int_0^{t_1} dt_2 e^{-i \frac{E_{m'}}{\hbar}(t-t_1)}$$
$$\times \langle m'|V(t_1)|n\rangle e^{-i \frac{E_n}{\hbar}(t_1-t_2)} \langle n|V(t_2)|m\rangle e^{-i \frac{E_m}{\hbar} t_2} \tag{15.22^2}$$

where the sum over n runs over the complete set of kets $|n\rangle$.

$$\langle m'|U^{(3)}(t)|m\rangle = \left(\frac{\lambda}{i\hbar}\right)^3 \sum_{n,\ell} \int_0^t dt_1 \int_0^{t_1} dt_2 \int_0^{t_2} dt_3 e^{-i\frac{E_{m'}}{\hbar}(t-t_1)}$$

$$\times \langle m'|V(t_1)|n\rangle e^{-i\frac{E_n}{\hbar}(t_1-t_2)} \langle n|V(t_2)|\ell\rangle e^{-i\frac{E_\ell}{\hbar}(t_2-t_3)} \qquad 15.22^3$$

$$\times \langle \ell|V(t_3)|m\rangle e^{-i\frac{E_m}{\hbar}t_3}$$

where as before, the sum n,ℓ is over complete sets of states.

15.2 Direct Computation of Transition Amplitudes

This whole procedure may also be obtained in a much less formal and more direct manner, which we now describe. We start with the exact Schrödinger equation

$$i\hbar \frac{\partial}{\partial t} |\Psi(t)\rangle = H|\Psi(t)\rangle \qquad 15.23$$

and expand $|\Psi(t)\rangle$ in the terms of the stationary states of H_o . Thus

$$|\Psi(t)\rangle = \sum_n a_n(t)\, e^{-iE_n t/\hbar} |n\rangle . \qquad 15.24$$

Note that the coefficients $a_n(t)$ have to be explicitly time dependent. Substituting this in the Schrödinger equation 15.23 we get

$$i\hbar \sum \dot{a}_n e^{-iE_n t/\hbar} |n\rangle + \sum a_n E_n e^{-iE_n t/\hbar} |n\rangle$$

$$= \sum a_n (H_o + \lambda V) e^{-iE_n t/\hbar} |n\rangle$$

$$= \sum a_n \lambda V e^{-iE_n t/\hbar} |n\rangle + \sum a_n E_n e^{-iE_n t/\hbar} |n\rangle .$$

Hence forming the inner product with $\langle k|$ and calling

$$E_k - E_n = \hbar\omega_{kn} \tag{15.25}$$

we get

$$\dot{a}_k = \frac{\lambda}{i\hbar} \sum_n \langle k|V(t)|n\rangle a_n e^{i\omega_{kn} t} . \tag{15.26}$$

This differential equation is exact and corresponds to equation 15.3. If furthermore in the past, say for $t < 0$, $V(t)$ vanishes and at that time the system was in the state $|m\rangle$ then we have $\dot{a}_k = 0$ for $t < 0$ so that a_k = constant. This constant is either 0 or 1 depending on whether $k \neq m$ or $k = m$. Thus we have an initial condition for (15.26), namely

$$a_k(0) = \delta_{km} . \tag{15.27}$$

We can now rewrite (15.26) as an integral equation incorporating (15.27)

$$a_k(t) = \delta_{km} + \frac{\lambda}{i\hbar} \sum_n \int_0^t dt' \langle k|V(t')|n\rangle a_n(t') e^{i\omega_{kn} t'} \tag{15.28}$$

This is analogous to (15.6). It is now a simple matter to iterate this equation to obtain equations analogous to (15.22). Thus the zeroth order term in λ is

$$a_k^{(o)}(t) = \delta_{km} \tag{15.29^o}$$

and the first order term is

$$a_k^{(1)}(t) = \frac{\lambda}{i\hbar} \int_0^t \langle k|V(t')|m\rangle e^{i\omega_{km} t'} dt' . \tag{15.29^1}$$

The second order term is

$$a_k^{(2)}(t) = \left(\frac{\lambda}{i\hbar}\right)^2 \int_0^t dt_1 \int_0^{t_1} dt_2 \sum_n e^{-i\omega_{kn}t_1} \langle k|V(t_1)|n\rangle e^{i\omega_{nm}t_2}$$

$$\times \langle n|V(t_2)|m\rangle \ . \qquad 15.29^2$$

We shall now apply this procedure to several special cases which are of particular interest in practice.

15.3 Periodic Perturbation of Finite Duration

The perturbation part of the hamiltonian is in this case assumed to have the special form

$$\langle k|V(t)|m\rangle = \begin{cases} 0 & t < 0 \quad \text{or} \quad t > t_o \\ 2\langle k|v|m\rangle \sin \omega t & 0 < t < t_o \end{cases} \ . \qquad 15.30$$

Here v is assumed to be some time-independent operator. We can immediately compute the first order term $a_k^{(1)}(t)$. For $t \geq t_o$ we then have

$$a_k^{(1)}(t) = \frac{\lambda}{i\hbar} \langle k|v|m\rangle \int_o^{t_o} 2 \sin \omega t' e^{i\omega_{km}t'} dt'$$

or

$$a_k^{(1)}(t) = \frac{\lambda\langle k|v|m\rangle}{i\hbar} \left(\frac{e^{i(\omega_{km}-\omega)t_o}-1}{\omega_{km}-\omega} - \frac{e^{i(\omega_{km}+\omega)t_o}-1}{\omega_{km}+\omega}\right) \ . \qquad 15.31$$

Since λ is assumed small, the transition probability will be appreciable only if one of the denominators almost vanishes. This leads to two cases.

<u>Case 1</u> $\omega_{km} - \omega \simeq 0$

or

$$E_k \simeq E_m + \hbar\omega \tag{15.32}$$

<u>Case 2</u> $\omega_{km} + \omega \simeq 0$

or

$$E_k \simeq E_m - \hbar\omega. \tag{15.33}$$

Thus the effect of the perturbation is to transmit to or absorb from the system one quantum $\hbar\omega$ of energy. For a perturbation with only one sharp frequency, as given, the probability is appreciable only if one of the "resonance" conditions (15.32) or (15.33) is satisfied

We now specialize this problem even further and consider the transition probability from a bound state $|m\rangle$ to a continuum state $|k\rangle$. In that case $E_k > E_m$ and only case 1 applies. The probability of finding the system in the state $|k\rangle$ if it was initially in the state $|m\rangle$ is then to lowest order in λ given by

$$|a_k^{(1)}(t)|^2 = |\frac{\lambda}{\hbar}\langle k|v|m\rangle|^2 \left| \frac{e^{i(\omega_{km}-\omega)t_o}-1}{\omega_{km}-\omega} \right|^2$$

or

$$|a_k^{(1)}(t)|^2 = \frac{4\lambda^2|\langle k|v|m\rangle|^2}{\hbar^2(\omega_{km}-\omega)^2} \sin^2 \frac{1}{2}(\omega_{km}-\omega)t_0 . \tag{15.34}$$

Intuitively one would expect this probability to be proportional to t_o, the length of time the interaction was on. If one considers the function

$$\frac{\sin^2 \frac{1}{2}(\omega_{km}-\omega)t_o}{(\omega_{km}-\omega)^2} = \frac{\sin^2 x}{4x^2} t_o^2 \tag{15.35}$$

where

$$x = \frac{1}{2}(\omega_{km}-\omega)t_o \tag{15.36}$$

we see that for reasonably small t_o and due to the resonance condition $x \simeq 0$, so that $|a_k^{(1)}(t)|^2$ is essentially proportional to t_o^2. This seems to contradict our intuition since we would expect the number of quanta inducing the transition to be proportional to the length of time the interaction is on. The answer lies, in fact, in the finite time over which the interaction is effective. Since a sinusoidal signal lasting a time t_o can be thought of as a pulse, of length t_o, such a signal is not monochromatic with frequency ω, but has in fact its energy distributed over a band width proportional to $\frac{1}{t_o}$. In particular the Fourier transform $F(\omega')$ of the function

$$f(t) = \begin{cases} 0 & t \leq 0 \quad \text{or} \quad t > t_o \\ \sin \omega t & 0 \leq t \leq t_o \end{cases} \tag{15.37}$$

is

$$F(\omega') = \frac{1}{2\pi i}\left\{ e^{i(\omega'+\omega)t_o/2} \frac{\sin(\omega'+\omega)t_o/2}{(\omega'+\omega)} - e^{i(\omega'-\omega)t_o/2} \frac{\sin(\omega'-\omega)t_o/2}{(\omega'-\omega)} \right\} \tag{15.38}$$

For the region of interest to us, namely $\omega' > 0$, $\omega > 0$ this function peaks at $\omega' = \omega$ and simplifies to

$$F(\omega') \simeq \frac{i}{2\pi} e^{i(\omega'-\omega)t_o/2} \frac{\sin(\omega'-\omega)t_o/2}{(\omega'-\omega)} \tag{15.39}$$

The points ω'' at which $|F(\omega'')| = \frac{1}{2}|F(\omega)|$ determine the "width" of this function. They are approximately given by

$$\omega'' \simeq \omega \pm \frac{2}{t_o} \times 1.9 \qquad 15.40$$

Thus

$$\Delta\omega' \simeq 2\times \frac{3.8}{t_o} = \frac{7.6}{t_o} \qquad 15.41$$

as stated. That is why for a unit frequency interval centered about the peak of this spectral distribution the energy of this pulse is proportional to the square of the duration of the pulse.

In practice one does not usually have such short pulses of radiation and one does not usually measure the transition to exactly one final state $|k\rangle$ but rather to a set of such states all with approximately the same energy and hence all approximately satisfying the "resonance" condition. In such a case one is interested in the rate at which such transitions occur. Thus one calculates W, the transition probability per unit time. This is given to lowest order in λ by

$$W = \frac{1}{t_o} \int |a_k^{(1)}(t)|^2 \rho(k)dE_k \qquad 15.42$$

where $\rho(k)\, dE_k$ is the number of states with energy between E_k and $E_k + dE_k$. Changing the variable of integration from E_k to

$$x = \frac{1}{2\hbar}(E_k - E_m - \hbar\omega)t_o$$

so that $dx = \frac{t_o}{2\hbar} dE_k$, and substituting from (15.36) we get

$$W = \frac{2\lambda^2}{\hbar} \int |\langle k|v|m\rangle|^2 \rho(k) \frac{\sin^2 x}{x^2} dx\ . \qquad 15.43$$

Now in order to evaluate this integral we make some further approximations. If t_o becomes large, the function $\sin^2 x/x^2$ has a high narrow peak with respect to the k variable. Thus the density of final states $\rho(k)$ and the matrix element $\langle k|v|m\rangle$ are essentially constant over the range where this function is large and may be taken outside the integral. Furthermore for very large t_o we may extend the limits of integration from $-\infty$ to ∞. Thus

$$W \simeq \frac{2}{\hbar} |\lambda\langle k|v|m\rangle|^2 \rho(k) \int_{-\infty}^{\infty} \frac{\sin^2 x}{x^2}\, dx$$

so that to a good approximation

$$W = \frac{2\pi}{\hbar} |\lambda\langle k|v|m\rangle|^2 \rho(k). \qquad 15.44$$

This formula is the celebrated Fermi's Golden Rule. It will crop up again in several more examples. To illustrate the use of this first-order theory we apply it to a physically interesting problem.

15.4 Photo-Ionization of Hydrogen Atom

The problem we are interested in consists of calculating the transition probability per unit time that a hydrogen atom in its ground state placed in a high frequency electromagnetic field ejects an electron into a solid angle lying between Ω and $\Omega + d\Omega$.

The hamiltonian for the electron, assuming it is bound in a fixed Coulomb potential, is then given by

$$H = H_o + V \qquad 15.45$$

where H_o is the hydrogen atom hamiltonian (10.51), and

$$V = e\, E_o \vec{r} \cdot \vec{e}\; 2 \sin \omega t\ . \qquad 15.46$$

E_o has the obvious interpretation as the strength of the applied electric field and $\vec{e}$ is a unit vector in the direction of the electric field. The initial state ψ_{in} of the electron is the ground state of an electron in the hydrogen atom

$$\psi_{in} = \psi_{1oo}(\vec{r}) = \pi^{-1/2} a_o^{-3/2} e^{-r/a_o} . \qquad 15.47$$

The final state ψ_{out} is a positive energy (scattering state) of an electron in the presence of a point charge. The resulting wave function is very complicated due to the very long range of the Coulomb potential. We, therefore, <u>assume</u> that the Coulomb potential is effectively screened by surrounding matter so that the ejected electron is <u>free</u>. Then the final electron state is approximated by a plane wave

$$\psi_{out} = (2\pi)^{-3/2} e^{i\vec{k}\cdot\vec{r}} . \qquad 15.48$$

We also need to calculate the density of final states. For this purpose it is simpler first to consider the electron placed in a very large cube of sides L and then later take the limit $L \to \infty$. At the surface of the box we impose periodic boundary conditions since these are the simplest boundary conditions yielding a self-adjoint extension of the momentum operator. In that case we have

$$\psi_{out} = L^{-3/2} e^{i\vec{k}\cdot\vec{r}} \qquad 15.49$$

with

$$\vec{k} = \frac{2\pi}{L} (n_x, n_y, n_z) \qquad 15.50$$

where n_x, n_y, n_z are integers. The counting now proceeds just as for the black body radiation (section 1.1). Thus the number of modes in the range between k_x and $k_x + dk_x$ etc. is

$$dN = \Delta n_x \Delta n_y \Delta n_z = \frac{L}{2\pi} dk_x \frac{L}{2\pi} dk_y \frac{L}{2\pi} dk_z$$
$$= (\frac{L}{2\pi})^3 d^3k = (\frac{L}{2\pi})^3 k^2 dk \sin\theta d\theta d\phi . \qquad 15.51$$

The use of spherical coordinates in k space is convenient since energy conservation fixes $|\vec{k}| = k$ according to

$$E_k = E_o + \hbar\omega$$

or

$$\frac{(\hbar k)^2}{2m} = -\frac{1}{2} mc^2\alpha^2 + \hbar\omega. \qquad 15.52$$

From equation (15.46) we get that

$$\rho(k)\, dE_k = \frac{1}{L^3} dN = \frac{1}{(2\pi)^3} k^2 dk \sin\theta\, d\theta d\phi \qquad 15.53$$

However $E_k = \frac{\hbar^2 k^2}{2m}$ so that

$$dE_k = \frac{\hbar^2}{m} kdk \qquad 15.54$$

and hence

$$\rho(k) = \frac{m}{8\pi^3\hbar^2} k \sin\theta\, d\theta d\phi \qquad 15.55$$

Next we need to evaluate the matrix element

$$\langle\vec{k}|v|1,0,0\rangle = \int \psi^*_{out}(r)\, eE_o\, \vec{\varepsilon}\cdot\vec{r}\, \psi_{in}(\vec{r})d^3r$$
$$= eE_o(\pi a_o^3)^{-1/2}(2\pi)^{-3/2}\int e^{i\vec{k}.\vec{r}}\, \vec{\varepsilon}\cdot\vec{r} e^{-r/a_o}\, d^3r \qquad 15.56$$

The integral may be written as $\vec{\varepsilon}.\vec{I}$ where

$$\vec{I} = \int e^{i\vec{k}.\vec{r}}\, \vec{r}e^{-r/a_o} d^3r \qquad 15.57$$

Such vector integrals may be more easily evaluated by using the following symmetry argument. The direction of $\vec{I}$ must be the same as $\vec{k}$ since that is the only vector occuring in the integrand besides the integration variable itself. We can therefore write

$$\vec{I} = \vec{k}I \qquad 15.58$$

and take the inner product of this equation with $\vec{k}$ to get

$$k^2I = \int e^{i\vec{k}.\vec{r}}\, \vec{k}.\vec{r}\, e^{-r/a_o} d^3r \ . \qquad 15.59$$

This integral is easier to evaluate than the original expression (15.56) since it involves only the angle between $\vec{k}$ and $\vec{r}$, rather than two angles.

We now choose the z axis in the integrand to be parallel to $\vec{k}$. Then

$$k^2I = \int_0^\infty r^2 dr\, e^{-r/a_o} \int_0^\pi e^{ikr\cos\theta'} kr\cos\theta' \sin\theta' d\theta' \int_0^{2\pi} d\phi' \qquad 15.60$$

where $\vec{k}.\vec{r} = kr\cos\theta'$. The ϕ' integration is immediate and the θ' integration is easily performed by introducing $u = \cos\theta'$. Then we get:

$$k^2I = 2\pi k\int_0^\infty r^3 dr\, e^{-r/a_o} \int_{-1}^{1} ue^{ikru} du \qquad 15.61$$

These integrals are standard and yield

$$k^2 I = \frac{32\pi i a_o^3 (ka_o)^2}{[1 + (ka_o)^2]^3} . \qquad 15.62$$

Hence we have the matrix element

$$\langle \vec{k}|v|1,0,0\rangle = \frac{32\pi i e\mathcal{E}_o a_o^5 \vec{\varepsilon}.\vec{k}}{(2\pi)^{3/2}(\pi a_o^3)^{1/2}[1+(ka_o)^2]^3} . \qquad 15.63$$

Combining this with our expression for $\rho(k)$, the density of final states, we can use Fermi's golden rule to get

$$dW = \frac{32(e\mathcal{E}_o a_0^2)^2 m}{\pi^4 \hbar^3} \frac{(ka_o)^3}{[1 + (ka_o)^2]^6} \cos^2\theta \sin\theta \, d\theta d\phi \qquad 15.64$$

where

$$k\cos\theta = \vec{\varepsilon}.\vec{k} \qquad 15.65$$

With $d\Omega = \sin\theta \, d\theta d\phi$ we see that $dW/d\Omega \propto \cos^2\theta$. This simply means that the electron is most likely to be ejected in the direction in which the incident photons are travelling. Also the energy for which $dW/d\Omega$ is maximum is given by setting $d^2W/dkd\Omega = 0$ and solving for k. The result is

$$k = \frac{2}{a_o} = \frac{2me^2}{\hbar^2} . \qquad 15.66$$

Hence the energy is

$$E = \frac{(\hbar k)^2}{2m} = 2mc^2\alpha^2 . \qquad 15.67$$

This is in fact four times the ground state energy of the hydrogen atom.

We next consider two other types of approximations that are frequently used in practice.

15.5 The Adiabatic Approximation

The perturbation method developed above is generally valid if $\lambda V(t)$ is small for all t, the criterion of smallness being the same as for time-independent perturbation theory. It is also possible to develop approximation methods based on how rapidly $V(t)$ varies with time. Thus we have an approximation based on very slow time variation, called the adiabatic approximation, which we now discuss and we also have an approximation based on very rapid time variation, called the sudden approximation which we discuss in the next section.

In the adiabatic approximation $\lambda \frac{\partial V}{\partial t}$ is assumed to be small, so we try to arrange the computation in such a manner that this term will appear. We start with

$$H(t) = H_o + \lambda V(t) \qquad 15.1$$

and using the slow variation of $H(t)$ write the "instantaneous" Schrödinger equation

$$H(t)u_n(t) = E_n(t)u_n(t). \qquad 15.68$$

In this equation t is treated simply as a parameter. What we are tacitly assuming in writing (15.68) is that if at $t = 0$ the system is in the state $u_n(0)$ with energy $E_n(0)$ then at a slightly later time the system will be in the state approximated by

$$u_n(t)e^{-\frac{i}{\hbar}\int_0^t E_n(t')dt'}$$

with energy $E_n(t)$ where the energy $E_n(t)$ varies very slowly. Thus if we write the time dependent Schrödinger equation

$$i\hbar \frac{\partial \Psi}{\partial t} = H(t)\Psi \qquad 15.69$$

and expand Ψ in terms of the slowly varying "eigenstates" $u_n(t)$ so that

$$\Psi = \sum_k a_k(t)u_k(t)e^{-\frac{i}{\hbar}\int_0^t E_n(t')dt'} \qquad 15.70$$

then the $a_n(t)$ should be approximately constant and equal to δ_{kn}. We further assume that the $u_n(t)$ are orthonormal so that

$$(u_n(t),u_k(t)) = \delta_{nk} . \qquad 15.71$$

They are of course complete since $H(t)$ is assumed self- adjoint. Substituting (15.70) into (15.69) yields

$$i\hbar \sum_k (\dot{a}_k u_k + a_k \dot{u}_k - \frac{i}{\hbar} a_k E_k u_k)e^{-\frac{i}{\hbar}\int_0^t E_k(t')dt'} = \sum_k a_k H u_k e^{-\frac{i}{\hbar}\int_0^t E_k(t')dt'} . \qquad 15.72$$

Using (15.68) then yields

$$\sum_k (\dot{a}_k u_k + a_k \dot{u}_k)e^{-\frac{i}{\hbar}\int_0^t E_k(t')dt'} = 0. \qquad 15.73$$

Taking the inner product of this equation with

$$u_m e^{-\frac{i}{\hbar}\int_0^t E_m(t')dt'}$$

yields

$$\dot{a}_m = -\sum_k a_k(u_m(t),\dot{u}_k(t))e^{i\int_0^t \omega_{mk}(t')dt'} \qquad 15.74$$

where we have defined

$$\hbar\omega_{mk}(t) = E_m(t) - E_k(t) \ . \qquad 15.75$$

In order to solve (15.74) we need to compute $(u_m,\dot{u}_k)$. It is at this point that the slow variation of $\lambda V(t)$ appears. From (15.68) we get by differentiation

$$\frac{\partial H}{\partial t} u_k + H \frac{\partial u_k}{\partial t} = \frac{\partial E_k}{\partial t} u_k + E_k \frac{\partial u_k}{\partial t} \ .$$

Taking the inner product with u_m we get

$$(u_m, \frac{\partial H}{\partial t} u_k) + E_m(u_m, \frac{\partial u_k}{\partial t}) = \frac{\partial E_k}{\partial t} (u_m,u_k) + E_k(u_m, \frac{\partial u_k}{\partial t})$$

and hence for $m \neq k$

$$(u_m,\dot{u}_k)(E_k-E_m) = (u_m, \frac{\partial H}{\partial t} u_k) \qquad 15.76$$

or more compactly,

$$(u_m,\dot{u}_k) = -\frac{\lambda}{\hbar\omega_{mk}}(u_m, \frac{\partial V(t)}{\partial t} u_k) \qquad k \neq m \ . \qquad 15.77$$

We also need $(u_m, \dot{u}_m)$. From $(u_m, u_m) = 1$ we get

$$(u_m, \dot{u}_m) + (\dot{u}_m, u_m) = 0$$

or

$$(u_m, \dot{u}_m) + (u_m, \dot{u}_m)^* = 0 \ .$$

Thus $(u_m, \dot{u}_m) = i\alpha(t)$ where α is real. If we now consider changing the phase of $u_m(t)$ by an amount $\gamma(t)$ to

$$u_m'(t) = u_m e^{i\gamma(t)} \ .$$

Then,

$$(u_m', \dot{u}_m') = (u_m, \dot{u}_m) + i\dot{\gamma}$$

$$= i(\alpha + \dot{\gamma}).$$

Therefore by choosing

$$\gamma = - \int_0^t \alpha(t')dt'$$

so that $\alpha + \dot{\gamma} = 0$ we get that

$$(u_m', \dot{u}_m') = 0.$$

We henceforth assume that the phase of u_m has been chosen in this way and therefore drop the primes. Then we have

$$(u_m, \dot{u}_m) = 0. \qquad 15.78$$

In that case we arrive at the following <u>exact</u> equation replacing the Schrödinger equation (15.54)

$$\dot{a}_m = \sum_{k \neq m} a_k \frac{\lambda}{\hbar\omega_{mk}} (u_m, \frac{\partial V}{\partial t} u_k) e^{i\int_0^t \omega_{mk}(t')dt'} \quad . \qquad 15.79$$

Our approximation now consists of recalling that $V(t)$ and hence $E_n(t)$ and $u_n(t)$ vary slowly with t. Thus, as a first approximation for a_k we choose these quantities constant. As initial condition we further assume that at $t = 0$ the system is in the state u_n so that

$$a_k(t) = a_k(0) = \delta_{kn} \quad . \qquad 15.80$$

With the approximations above (15.79) becomes

$$\dot{a}_k = \frac{\lambda}{\hbar\omega_{kn}} (u_k, \frac{\partial V}{\partial t} u_n) \; e^{i\omega_{kn} t} \quad , \quad k \neq n \quad . \qquad 15.81$$

This combined with the initial condition yields

$$a_k(t) = \frac{\lambda}{i\hbar\omega_{kn}^2} (u_k, \frac{\partial V}{\partial t} u_n)(e^{i\omega_{kn} t} - 1), \quad k \neq n \; . \qquad 15.82$$

This is then the solution in the case of the adiabatic approximation. It follows from this equation that for $k \neq n$, a_k oscillates and does not increase monotonely with t. In fact

$$|a_k(t)| \propto \left| \frac{1}{E_k - E_n} \cdot (\frac{\partial V}{\partial t} \cdot T) \right|$$

where $T = 2\pi/\omega_{kn}$. However if V is itself oscillatory with a frequency ω comparable to ω_{kn} then $V(t)$ can no longer be considered to vary slowly and, in fact, the approximation breaks down since we get "resonance". In this case small changes in V can cause large changes in $|a_k|$ as we saw in the perturbation treatment. This means we can no longer neglect $\partial V/\partial t$.

To be specific, suppose we have

$$H = H_o + 2V \sin \omega t. \tag{15.83}$$

Then

$$\frac{\partial V}{\partial t} = 2\omega V \cos \omega t. \tag{15.84}$$

We assume $\partial V/\partial t$ and V are small so that $a_n(t), u_n(t)$ and $\omega_{kn}(t)$ still depend only weakly on t. Thus in (15.63) we neglect their time dependence and as before put

$$a_k = \delta_{kn} . \tag{15.85}$$

Then,

$$\begin{aligned} \dot{a}_k &\simeq \frac{2\omega}{\hbar\omega_{kn}} (u_k, Vu_n) \cos\omega t \, e^{i\omega_{kn}t} \\ &= \frac{\omega}{\hbar\omega_{kn}} (u_k, Vu_n)[e^{i(\omega_{kn}-\omega)t} + e^{i(\omega_{kn}+\omega)t}] \end{aligned} \tag{15.86}$$

or integrating

$$a_k = \frac{\omega}{i\hbar\omega_{kn}} (u_k, Vu_n) \left[\frac{e^{i(\omega_{kn}-\omega)t} - 1}{\omega_{kn}-\omega} + \frac{e^{i(\omega_{kn}+\omega)t} - 1}{\omega_{kn}+\omega} \right] \tag{15.87}$$

Clearly for $\omega_{kn} \simeq \omega$ the adiabatic approximation breaks down since then for $t < 1/(\omega_{kn}-\omega)$ we have

$$\frac{e^{i(\omega_{kn}-\omega)t} - 1}{\omega_{kn}-\omega} \simeq t$$

so that $|a_k| \propto t$. The same thing happens of course for $\omega_{kn} \simeq -\omega$. In fact in either case we obtain the same result as in the perturbative treatment. Thus we have for $\omega \simeq \omega_{kn}$

$$a_k \simeq \frac{(u_k, Vu_m)}{i\hbar} \frac{e^{i(\omega_{kn}-\omega)t}-1}{\omega_{kn}-\omega}$$

and for $\omega \simeq -\omega_{kn}$

$$a_k \simeq \frac{-(u_k, Vu_m)}{i\hbar} \frac{e^{i(\omega_{kn}-\omega)t}-1}{\omega_{kn}+\omega}$$

These are precisely the results obtained from the perturbative treatment.

15.6 The Sudden Approximation

As outlined in the previous section the sudden approximation is based on the fact that $H = H_o + \lambda V(t)$ changes rapidly. In fact, it is complementary to the perturbative treatment in the following sense. The perturbative treatment is valid if the time t_o over which the interaction is on is relatively long compared to $1/\omega_{kn}$. The sudden approximation is valid when the time dependence of the hamiltonian is on for a very short time t_o. In general the hamiltonian will be of the type

$$H = \begin{cases} H_o & t < 0 \\ H_i & 0 < t < t_o \\ H_1 & t > t_0 \end{cases} . \qquad 15.88$$

This type of behaviour will occur, for example, if electrons are bound to an atom whose nucleus at $t = 0$ beta- decays so that suddenly the nuclear charge is increased by one. In this case for $t < 0$ the nuclear charge is Z and for $t > t_o$ it is $Z+1$. During the interval

$0 < t < t_o$ the hamiltonian changes rapidly and in a very complicated manner. The advantage of the sudden approximation is that if t_o is sufficiently small one need not even know H_i.

The approximation consists, in fact, in replacing (15.88) by

$$H' = \begin{matrix} H_o & t < 0 \\ H_1 & t > 0 \end{matrix} \quad . \tag{15.89}$$

In this case we solve the two time independent Schrödinger equations

$$H_o u_n = E_n u_n \tag{15.90}$$

and

$$H_1 v_\mu = \varepsilon_\mu v_\mu \ . \tag{15.91}$$

Then $\{u_n\}$ and $\{v_\mu\}$ both form complete (not necessarily discrete) sets. We assume that both sets are orthonormalized. Then the general solution of the time dependent Schrödinger equation

$$i\hbar \frac{\partial \Psi}{\partial t} = H\,\Psi \tag{15.92}$$

is

$$\begin{aligned} \Psi &= \sum_n a_n u_n e^{-iE_n t/\hbar} \qquad & t < 0 \\ \Psi &= \sum_\mu b_\mu v_\mu e^{-i\varepsilon_\mu t/\hbar} \qquad & t > 0 \end{aligned} \tag{15.93}$$

where the sum is to be interpreted as a sum over the discrete and an integral over the continuous energy eigenvalues. Furthermore since Ψ satisfies a first order differential equation in time with a simple jump discontinuity in $\partial\Psi/\partial t$, it follows that Ψ is continuous. Hence we get that

$$\sum_n a_n u_n = \sum_\mu b_\mu v_\mu \quad . \tag{15.94}$$

Using the orthonormality of the $\{v_\mu\}$ we then get that

$$b_\lambda = \sum_n a_n (v_\lambda, u_n) \quad . \tag{15.95}$$

Typically we start with the system in a pure state say u_m. Then for $t < 0$, $a_n = \delta_{n,m}$ so that

$$b_\lambda = (v_\lambda, u_m) \quad , \tag{15.96}$$

Thus, starting with a pure state with energy E_m we wind up after the interaction in a superposition of states with energy ε_μ. Actually this is only due to our mode of description. In fact, as stated, Ψ is continuous so the state of the system has not changed and remains an eigenstate of H_o. In a sense, the hamiltonian changes too rapidly for the system to follow and thus it remains unchanged in an eigenstate of H_o. However, it does evolve for $t > 0$ according to the new hamiltonian. This is to be contrasted with the adiabatic case where the system evolves from an eigenstate of the original hamiltonian into an eigenstate of the final hamiltonian.

We shall now consider the case of a magnetic dipole in a magnetic field that oscillates or is increased either adiabatically or suddenly to some final value.

15.7 Dipole in a Time-Dependent Magnetic Field

The problem we want to consider is that of an electron in a magnetic field. We shall assume that all other energies of the electron such as its kinetic energy can be neglected. Since the electron has a spin $\vec{S}$ it has associated with it a magnetic dipole moment operator (see section 9.6)

$$\vec{\mu} = -\frac{e}{mc}\vec{S} \ . \tag{15.97}$$

Thus with our approximations, the hamiltonian for this electron is

$$H = -\vec{\mu}.\vec{B} = \frac{e}{mc}\vec{B}.\vec{S} = \frac{e\hbar}{2mc}\vec{B}.\vec{\sigma} \ , \tag{15.98}$$

where $\vec{\sigma}$ are the Pauli matrices (see section 9.6).
We shall consider three separate cases in all of which

$$\vec{B} = \vec{B}_o + \vec{B}_1(t) \tag{15.99}$$

where $\vec{B}_o$ has only a z component B_o and $\vec{B}_1(t)$ has only an x component. The three cases are:

$$1) \quad B_1(t) = \begin{matrix} 0 & t < 0 \\ b \sin \omega t & t > 0 \end{matrix} \tag{15.100}$$

$$2) \quad B_1(t) = \begin{matrix} 0 & t < 0 \\ b(1-e^{-\alpha t}) & t > 0 \end{matrix} \tag{15.101}$$

$$3) \quad B_1(t) = \begin{matrix} 0 & t < 0 \\ b & t > 0 \end{matrix} \tag{15.102}$$

We first solve for the steady states of the unperturbed hamiltonian

$$H_o = \frac{e\hbar B_o}{2mc}\begin{pmatrix} 1 & 0 \\ 0 & -1 \end{pmatrix} . \tag{15.103}$$

The eigenstates are

$$|+\rangle = \begin{pmatrix} 1 \\ 0 \end{pmatrix}$$
$$|-\rangle = \begin{pmatrix} 0 \\ 1 \end{pmatrix} . \tag{15.104}$$

with eigenvalues $\pm\hbar\Omega_o$ respectively, where

$$\Omega_o = \frac{eB_o}{2mc} . \tag{15.105}$$

15.7.1 Oscillatory Perturbation

In this case the hamiltonian is

$$H = \hbar\Omega_o\sigma_3 + \hbar\Omega_1(\sin\omega t)\sigma_1$$

or

$$H = \hbar\Omega_o\begin{pmatrix}1 & 0\\ 0 & -1\end{pmatrix} + \hbar\Omega_1 \sin\omega t \begin{pmatrix}0 & 1\\ 1 & 0\end{pmatrix} \tag{15.106}$$

where

$$\Omega_1 = \frac{eb}{2mc} . \tag{15.107}$$

If we now expand the solutions of

$$i\hbar \frac{\partial}{\partial t} |\Psi(t)\rangle = H|\Psi(t)\rangle \tag{15.108}$$

according to (15.24) with

$$E_\pm = \pm\hbar\Omega_o$$

then we have only one transition frequency

$$E_+ - E_- = 2\hbar\Omega_o \tag{15.109}$$

so that (15.26) reads

$$\dot{a}_k = \frac{\Omega_1}{i} \sum_{n=\pm} \langle k|\sigma_1 \sin\omega t|n\rangle a_n e^{-i2\Omega_o t} \qquad t > 0 \tag{15.110}$$

where $k = \pm$. If furthermore for $t < 0$ the electron is in the state $|+\rangle$ then the integral equations equivalent to (15.28) are

$$a_+(t) = 1 - i\Omega_1\int_0^t \sin\omega t' e^{i2\Omega_o t'} a_-(t')dt' \qquad 15.111$$

$$a_-(t) = -i\Omega_1\int_0^t \sin\omega t' e^{i2\Omega_o t'} a_+(t')dt' \ . \qquad 15.112$$

These equations are still exact. If we now iterate, then the zeroth order term in Ω_1 is

$$a_+^{(o)} = 1 \ , \quad a_-^{(o)} = 0 \qquad 15.113$$

and the first order term in Ω_1 is

$$a_+^{(1)} = 0$$

$$a_-^{(1)} = -i\Omega_1\int_0^t \sin\omega t' e^{i2\Omega_o t'} dt' \ . \qquad 15.114$$

The second order terms are

$$a_+^{(2)} = -\Omega_1^2\int_0^t dt' \sin\omega t' e^{i2\Omega_o t'} \int_0^{t'} \sin\omega t'' e^{i2\Omega_o t''} dt''$$

$$a_-^{(2)} = 0. \qquad 15.115$$

These equations are precisely (15.29) 0), 1) and 2) for the specific hamiltonian (15.98) with $\vec{B}_1(t)$ given by (15.100).

To first order in Ω_1 we then have

$$|a_+^{(1)}(t)|^2 = 1$$

$$|a_-^{(1)}(t)|^2 = \Omega_1^2 \left|\int_0^t e^{i2\Omega_o t'} \sin \omega t' dt'\right|^2 \qquad 15.116$$

or

$$|a_-^{(1)}(t)|^2 = \Omega_1^2 \frac{1}{4} \left|\frac{e^{i(2\Omega_o+\omega)t} - 1}{2\Omega_o+\omega} - \frac{e^{i(2\Omega_o-\omega)t} - 1}{2\Omega_o-\omega}\right|^2 \qquad 15.117$$

Thus to lowest order we find that the system will almost certainly remain in its original state $|+\rangle$ and there is only a second order term giving the probability of finding the system in the state $|-\rangle$ after a time t.

15.7.2 Slowly Varying Perturbation

Here we consider

$$B_1(t) = \begin{cases} 0 & t < 0 \\ b(1-e^{-\alpha t}) & t \geq 0 \end{cases} \qquad 15.101$$

and apply the <u>adiabatic approximation</u>. This will be valid if α is very small. In that case

$$\frac{dB_1}{dt} = 0 \qquad t < 0$$

and

$$\frac{dB_1}{dt} = \alpha b e^{-\alpha t} \ . \qquad t \geq 0 \ .$$

So $dB_1/dt < \alpha b$, which is small for sufficiently small α.

The hamiltonian is

$$H = \hbar\Omega_o\sigma_3 + \hbar\Omega_1(1-e^{-\alpha t})\sigma_1 \qquad t \geq 0 \tag{15.118}$$

$$H = \hbar\Omega_o\begin{pmatrix}1 & 0\\0 & -1\end{pmatrix} + \hbar\Omega_1\ (1-e^{-\alpha t})\begin{pmatrix}0 & 1\\1 & 0\end{pmatrix} \quad t \geq 0 \tag{15.119}$$

and

$$H_o = \hbar\Omega_o\begin{pmatrix}1 & 0\\0 & -1\end{pmatrix} = \hbar\Omega_o\sigma_3 \qquad t < 0 \tag{15.120}$$

where as before

$$\Omega_o = \frac{eB_o}{2mc}\ , \qquad \Omega_1 = \frac{eb}{2mc}\ . \tag{15.121}$$

The time-dependent Schrödinger equation can be rewritten to yield the equations corresponding to (15.79). They are

$$\dot{a}_+ = \frac{\alpha\hbar\Omega_1}{E_+ - E_-}\, e^{-\alpha t}(u_+, \sigma_x\, u_-)e^{i\int_0^t \frac{E_+-E_-}{\hbar}dt'}\, a_- \tag{15.122}$$

$$\dot{a}_- = \frac{\alpha\hbar\Omega_1}{E_+ - E_-}\, e^{-\alpha t}(u_-, \sigma_x\, u_+)e^{i\int_0^t \frac{E_+-E_-}{\hbar}dt'}\, a_+ \tag{15.123}$$

where E_+ , E_- and u_+ , u_- are solution of the full hamiltonian

$$Hu_\pm = E_\pm u_\pm\ . \tag{15.124}$$

These equations are still exact and also far too complicated. The adiabatic approximation consists in choosing the $a_\pm$, $u_\pm$ on the right hand side of (15.122) and (15.123) as well as the $E_\pm$ as constants. Thus if the initial state at $t = 0$ was u_+ then on the right side of (15.123) and (15.122) $a_+ = 1$, $a_- = 0$. This yields,

$$\dot{a}_+ = 0 \tag{15.125}$$

$$\dot{a}_- = -\frac{\alpha\hbar\Omega_1}{2\hbar\Omega_o} e^{-\alpha t} e^{-2i\Omega_o t} \tag{15.126}$$

so

$$a_+ = 1 \tag{15.127}$$

$$\begin{aligned} a_- &= \frac{\alpha}{2(\alpha+2i\Omega_o)} \frac{\Omega_1}{\Omega_o} [e^{-(\alpha+2i\Omega_o)t} - 1] \\ &\simeq -\frac{i\alpha\Omega_1}{4\Omega_o^2} [e^{-(\alpha+2i\Omega_o)t} - 1] \end{aligned} \tag{15.128}$$

Thus to lowest order in α, the state at time t is given by

$$\Psi(t) = \begin{pmatrix} 1 \\ -\frac{i\alpha\Omega_1}{4\Omega_o^2}(e^{-(\alpha+2i\Omega_o)t} - 1) \end{pmatrix} \tag{15.129}$$

This shows that the state evolves slowly from its original state u_+ to a mixture of u_+ and u_- .

15.7.3 Sudden Approximation

In this case

$$H = \hbar\Omega_o \begin{pmatrix} 1 & 0 \\ 0 & -1 \end{pmatrix} = H_o = \hbar\Omega_o\sigma_3 \qquad t < 0 \tag{15.130}$$

and

$$H = \hbar\Omega_o\sigma_3 + \hbar\Omega_1\sigma_1 = H_1 \qquad t \geq 0 \tag{15.131}$$

or

$$H = \hbar\Omega_o \begin{pmatrix} 1 & 0 \\ 0 & -1 \end{pmatrix} + \hbar\Omega_1 \begin{pmatrix} 0 & 1 \\ 1 & 0 \end{pmatrix} = H_1 \quad t \geq 0. \tag{15.132}$$

The eigenstates of H_o are, as before,

$$|+\rangle = \begin{pmatrix} 1 \\ 0 \end{pmatrix} \qquad |-\rangle = \begin{pmatrix} 0 \\ 1 \end{pmatrix} \tag{15.104}$$

with

$$E_{\pm} = \pm\hbar\Omega_o \quad .$$

If we call the eigenfunctions and eigenvalues of H, v_μ and ε_μ respectively we get for

$$v_\mu = \begin{pmatrix} a \\ b \end{pmatrix} \tag{15.133}$$

$$\hbar \begin{pmatrix} \Omega_o & \Omega_1 \\ \Omega_1 & -\Omega_o \end{pmatrix} \begin{pmatrix} a \\ b \end{pmatrix} = \varepsilon_\mu \begin{pmatrix} a \\ b \end{pmatrix} . \tag{15.134}$$

Taking the determinant of $(H_1 - \varepsilon_\mu)$ yields

$$(\hbar\Omega_o - \varepsilon_\mu)(\hbar\Omega_o + \varepsilon_\mu) + \hbar^2\Omega_1^2 = 0$$

$$-\varepsilon_\mu^2 + \hbar^2(\Omega_o^2 + \Omega_1^2) = 0$$

$$\varepsilon_{\pm} = \pm\hbar\sqrt{\Omega_o^2 + \Omega_1^2} \quad . \tag{15.135}$$

Calling

$$\Omega_o^2 + \Omega_1^2 = \Omega^2 \tag{15.136}$$

we have

$$\varepsilon_{\pm} = \pm\hbar\Omega \tag{15.137}$$

We then get

$$v_{\pm} = \frac{1}{\sqrt{2}} (\Omega^2 \mp \Omega_o \Omega)^{-1/2} \begin{pmatrix} \Omega_1 \\ \pm\Omega - \Omega_o \end{pmatrix}. \qquad 15.138$$

If at $t = 0$ the system is in the state u_+, then for $t > 0$ the system will be in the state

$$\Psi(t) = b_+ v_+ e^{-i\Omega t} + b_- v_- e^{i\Omega t} \qquad 15.139$$

where $b_{\pm}$ are determined by the initial condition

$$u_+ = b_+ v_+ + b_- v_-$$

or

$$b_{\pm} = (v_{\pm}, u_+).$$

Thus

$$b_{\pm} = \frac{\Omega_1}{\sqrt{2}} (\Omega^2 \mp \Omega_o \Omega)^{-1/2} . \qquad 15.140$$

Combining these results we get

$$\Psi(t) = \begin{pmatrix} \cos \Omega t - i \frac{\Omega_o}{\Omega} \sin \Omega t \\ -i \frac{\Omega_1}{\Omega} \sin \Omega t \end{pmatrix} \qquad 15.141$$

This Ψ is an <u>exact</u> solution of the hamiltonian (15.130) and (15.131) with the initial condition $\Psi(0) = u_+$. The approximation is made in writing the hamiltonian (15.130), (15.131) in the first place.

Notes, References and Bibliography

15.1 The computation of transition probabilities for atomic systems is carried out in:

E.U. Condon and G.H. Shortley - The Theory of Atomic Spectra - Cambridge University Press (1963).

A very concise, but very readable treatment of time-dependent perturbation theory is to be found in sections 44-46 of Dirac's book (reference 11.2).

Chapter 15 Problems

15.1 A particle is in the ground state of the hamiltonian

$$H = \frac{p^2}{2m} + V$$

where

$$V = \begin{cases} 0 & x < -a\ , \quad x > a \\ -V_0 & -a < x < a \end{cases} .$$

Find the transition probability per unit time to a state of energy $E_k > 0$, due to a perturbation

$$H'(t) = ve^{-x^2/\alpha^2} \sin \omega t$$

where v is a constant and $\alpha \ll a$.

15.2 The deuteron is an s-wave bound state of a proton and neutron with a binding energy of 2.226 MeV. It is well approximated as a bound state in a square well of depth $V_o = 36.2$ MeV and a width $a = 2.02 \times 10^{-13}$ cm. Using these data, compute the probability for photo-disintegration of the deuteron. Assume the incident photon can be approximated by a perturbation

$$\begin{aligned} V &= \vec{A}\cdot\vec{r}\,\sin\omega t \qquad & t > 0 \\ &= 0 & t < 0 \end{aligned}$$

Use whatever other approximations seem reasonable.

15.3 An atom is initially in the ground state of a simple harmonic oscillator

$$H = \hbar\,\omega a^\dagger a.$$

At $t = 0$ a perturbation

$$V' = \hbar\,\Omega(a^\dagger + a)$$

is turned on. Find the transition probability per unit time to any excited state of the system. What is the probability that the atom remains in its ground state?

15.4 Repeat 15.3 with

$$V' = \hbar\,\Omega a^\dagger a.$$

15.5 An atom has two energy levels $\pm\hbar\,\Omega$. A disturbance $V(t)$ connecting these two levels and varying periodically in time such that

$$\langle 1|V(t)|2\rangle = \hbar\,\Omega_1\,\sin\omega t$$

is turned on at $t = 0$.

a) Find a model hamiltonian for this system.

b) If the atom was originally in its ground state, estimate the probability $P(t)$ that it is in its excited state at time t.

15.6 A hydrogen atom in an excited state $|n,\ell,m\rangle$ is perturbed by an electromagnetic field. If the interaction can be written

$$
\begin{aligned}
V(t) &= \vec{E}\cdot\vec{r}\,\sin\omega t \qquad && 0 < t < t_o \\
&= 0 && t < 0\ ,\ t > t_o
\end{aligned}
$$

Find an expression for the transition probabilities per unit time to a lower level. This is how intensities of spectral lines can be computed. See reference 15.1.

Chapter 16

Applications

In this chapter we apply to some specific, interesting problems the techniques developed in the previous chapters. Interactions with the electromagnetic field are of particular interest to us. Since these interactions are written in terms of electromagnetic potentials we study the consequences of choosing different potentials (gauges) that give rise to the same electromagnetic field. We go on to include interactions between the electromagnetic field and the magnetic moment of an electron due to its spin as well as its orbital angular momentum.

In particular, we also study the splitting of spectral lines due to the magnetic field - the Zeeman effect. To carry out computations for this effect requires the study of the addition of angular momenta. This problem constitutes a major portion of this chapter.

16.1 Gauge Transformations

In chapter 3, equation (3.54), we wrote the hamiltonian H for a charged particle in an electromagnetic field. There we found that for a static field decribed by a scalar potential ϕ and a vector potential

$\vec{A}$ such that

$$\vec{E} = -\nabla\phi \tag{16.1}$$

$$\vec{B} = \nabla\times\vec{A} \tag{16.2}$$

the hamiltonian for a particle of mass m and charge q is

$$H = \frac{1}{2m}(\vec{p} - q/c\ \vec{A})^2 + q\phi \tag{16.3}$$

Of course ϕ and $\vec{A}$ do not describe the electromagnetic field $(\vec{E},\vec{B})$ uniquely since both $\vec{E}$ and $\vec{B}$ are left unchanged by gauge transformations. An example of such a transformation is

$$\begin{aligned} \vec{A} \to \vec{A}' &= \vec{A} + \nabla\Lambda \\ \phi \to \phi' &= \phi \end{aligned} \tag{16.4}$$

where Λ is an arbitrary time-independent scalar field. Clearly

$$\vec{\nabla}\times\vec{A}' = \vec{\nabla}\times\vec{A} \tag{16.5}$$

since the curl of a gradient vanishes.

If we now consider the Schrödinger equation with the hamiltonian (16.3) then, to keep this equation invariant under the gauge transformation (16.4), the phase of the wave-function ψ must change

$$\psi \to \psi' = e^{i\chi}\psi\ . \tag{16.6}$$

With a proper choice of χ, the pair of transformations (16.4), (16.6) leave the Schrödinger equation

$$\frac{1}{2m}(\vec{p} - \frac{q}{c}\vec{A})^2\psi + q\phi\psi = E\psi \tag{16.7}$$

unchanged. To see this we replace $\vec{A}$ and ψ by $\vec{A}'$ and ψ' respectively to get

$$\frac{1}{2m}\left(\vec{p}-\frac{q}{c}\vec{A}'+\frac{q}{c}\nabla\Lambda\right)^2 e^{-i\chi}\psi' + q\phi e^{-i\chi}\psi' = Ee^{-i\chi}\psi' \qquad 16.8$$

Now consider

$$\begin{aligned}&\left(\vec{p}-\frac{q}{c}\vec{A}'+\frac{q}{c}\nabla\Lambda\right)e^{-i\chi}\psi' \\ &= e^{-i\chi}\left[-\hbar\nabla\chi+\vec{p}-\frac{q}{c}(\vec{A}-\nabla\Lambda)\right]\psi'\end{aligned} \qquad 16.9$$

Thus if we choose

$$\chi = +\frac{q}{\hbar c}\Lambda \qquad 16.10$$

Then equation (16.9) shows that

$$\left(\vec{p}-\frac{q}{c}\vec{A}+\frac{q}{c}\nabla\Lambda\right)e^{-\frac{iq}{\hbar c}\Lambda}\psi' = e^{-i\frac{q}{\hbar c}\Lambda}\left[\vec{p}-\frac{q}{c}\vec{A}\right]\psi' \qquad 16.11$$

Operating on (16.11) once more with $(\vec{p}-\frac{q}{c}\vec{A}+\frac{q}{c}\nabla\Lambda)$ we find that for χ given by (16.10) the Schrödinger equation (16.8) reduces to

$$e^{-i\frac{q}{\hbar c}\Lambda}\left[\frac{1}{2m}\left(\vec{p}-\frac{q}{c}\vec{A}\right)^2\psi'+q\phi\psi'\right] = e^{-i\frac{q}{\hbar c}\Lambda}E\psi'. \qquad 16.12$$

After cancelling the phase factor $e^{-i\frac{q}{\hbar c}\Lambda}$ we recognize this as the original Schrödinger equation (16.7). Thus we have found that under the <u>local gauge transformations</u>

$$\begin{aligned}\vec{A} &\to \vec{A}' = \vec{A}+\nabla\Lambda \\ \phi &\to \phi' = \phi\end{aligned} \qquad 16.4$$

$$\psi \to \psi' = e^{i\frac{q}{\hbar c}\Lambda}\psi \qquad 16.13$$

the Schrödinger equation

$$\frac{1}{2m}(\vec{p}-\frac{q}{c}\vec{A})^2\psi + q\phi\psi = E\psi \qquad 16.7$$

remains unchanged. This is a very important physical result since it shows that whatever set of potentials $(\phi,\vec{A})$ we choose, the resultant Schrödinger equation does not depend upon our choice as long as these different potentials are connected by local gauge transformations or, what is the same thing, describe the same electromagnetic field $(\vec{E},\vec{B})$. The adjective "local" in the above discussion simply refers to the fact that the gauge field Λ depends on the coordinates $\vec{x}$ in a local manner.

If the electromagnetic field $(\vec{E},\vec{B})$ is time-dependent then the potentials ϕ, $\vec{A}$ are time dependent and

$$\vec{E} = -\nabla\phi - \frac{1}{c}\frac{\partial\vec{A}}{\partial t} \qquad 16.14$$

$$\vec{B} = \nabla\times\vec{A} \qquad 16.15$$

In this case we must consider the invariance of the time- dependent Schrödinger equation

$$i\hbar\frac{\partial\Psi}{\partial t} = \frac{1}{2m}(\vec{p}-\frac{q}{c}\vec{A})^2\Psi + q\phi\Psi\ . \qquad 16.16$$

under time-dependent local gauge transformations. A computation similar to equation (16.9) (see problem 16.8) and (16.11) shows that the set of transformations

$$\vec{A} \to \vec{A}' = \vec{A} + \nabla\Lambda \qquad 16.17$$

$$\phi \to \phi' = \phi - \frac{1}{c}\frac{\partial\Lambda}{\partial t} \qquad 16.18$$

$$\Psi \to \Psi' = e^{i\frac{q}{\hbar c}\Lambda}\Psi \qquad 16.19$$

leaves the time dependent Schrödinger equation 16.10 unchanged or invariant. In this discussion Λ is an arbitrary time-dependent scalar field.

Local gauge invariance, as discussed above, has not only the important physical consequence that the physics of interactions with the electromagnetic fields does not depend on the choice of gauge, but has proved to be an important guiding principle in modern theories of elementary particles.

As an example, that we use in the next section, we consider the vector potential for a <u>constant</u> magnetic field $\vec{B}$. Using the vector identity

$$\nabla\times(\vec{F}\times\vec{G}) = \vec{F}(\nabla.\vec{G}) - \vec{G}(\nabla.\vec{F}) - (\vec{F}.\nabla)\vec{G} + (\vec{G}.\nabla)\vec{F} \qquad 16.20$$

we see that if we choose

$$\vec{A} = -\frac{1}{2}\,\vec{r}\times\vec{B} \qquad 16.21$$

then

$$\vec{\nabla}\times\vec{A} = -\frac{1}{2}\left[\vec{r}(\nabla.\vec{B})-\vec{B}(\nabla.\vec{r})-(\vec{r}.\nabla)\vec{B}+(\vec{B}.\nabla)\vec{r}\right] = \vec{B} \quad . \qquad 16.22$$

Thus a possible choice of vector potentials is given by (16.21). This is sometimes called the "symmetric gauge".

For convenience we now choose our coordinate system so that the z-axis is parallel to $\vec{B}$. Then,

$$\vec{B} = (0,0,B) \qquad 16.23$$

and

$$\vec{A} = -\frac{B}{2}\,(y,-x,0) \qquad 16.24$$

If we now perform a gauge transformation of the type given by equation

16.4 with the gauge function

$$\Lambda = -\frac{B}{2}xy \qquad 16.25$$

then

$$\nabla\Lambda = -\frac{B}{2}(y,x,0) \qquad 16.26$$

and

$$\vec{A}' = \vec{A} + \nabla\Lambda = -B(y,0,0) \qquad 16.27$$

The potential $\vec{A}'$ is just as good as the potential $\vec{A}$ for computing the magnetic field B. The choice of potential $\vec{A}'$ is called the Landau gauge (ref. 16.1). In the next section we consider the motion of an electron in a uniform magnetic field and we shall use both forms of the vector potential to see that they yield the same result.

For computational purposes one should attempt to find a gauge (potentials) that simplify the computations as much as possible.

16.2 Motion in a Uniform Magnetic Field - Landau Levels

The problem of the motion of an electron in a uniform magnetic field is currently a very "hot" topic in solid state physics due to the discovery of the "fractional Hall effect". Much of the physics of the Hall effect can be understood by solving the Schrödinger equation for an electron in a uniform magnetic field. The resultant discrete energy levels are called Landau levels.

The hamiltonian is given by

$$H = \frac{1}{2m}(\vec{p} + \frac{e}{c}\vec{A})^2 \qquad 16.28$$

For convenience we choose the Landau gauge so that

$$\vec{A} = -B(y,0,0). \qquad 16.29$$

The hamiltonian can now be written out explicitly and reads

$$H = \frac{1}{2m}\left(p_x - \frac{eB}{c}y\right)^2 + \frac{1}{2m}p_y^2 + \frac{1}{2m}p_z^2 \qquad 16.30$$

Since H does not contain any function of x or z we see that

$$[H,p_x] = [H,p_z] = 0. \qquad 16.31$$

Since p_x and p_z also commute we see that p_x , p_z and H form a complete set of observables and can be simultaneously diagonalized. To this end we look for a wavefunction of the form

$$\psi = e^{i(k_x x + k_z z)}\phi(y). \qquad 16.32$$

Then,

$$p_x\psi = \hbar k_x \psi \qquad 16.33$$

$$p_z\psi = \hbar k_z \psi \qquad 16.34$$

and the Schrödinger equation 16.30 reduces to

$$\frac{d^2\phi}{dy^2} + \left[\frac{2mE}{\hbar^2} - k_z^2 - \frac{m^2\omega^2}{\hbar^2}(y-y_o)^2\right]\phi = 0 \qquad 16.31$$

Here we have defined the Larmor frequency

$$\omega = \frac{eB}{mc} \qquad 16.32$$

and the parameter

$$y_o = \hbar\frac{ck_x}{eB}. \qquad 16.33$$

Except for the constant k_z^2 , equation 16.31 is just the equation for a simple harmonic oscillator centered at y_o. The energy is therefore given by (section 5.3 and 9.1)

$$E_{n,k_z} = (n+\frac{1}{2})\hbar\omega + \frac{\hbar^2 k_z^2}{2m} . \qquad 16.34$$

The corresponding wavefunction (up to a normalization constant A_n) is

$$\phi_{n,k_x,k_z} = A_n e^{i(k_x x+k_z z)} e^{-\frac{(y-y_o)^2}{2\lambda^2}} H_n\left(\frac{y-y_o}{\lambda}\right) \qquad 16.35$$

Here we have introduced a new parameter

$$\lambda = \left(\frac{\hbar c}{eB}\right)^{1/2} \qquad 16.36$$

called the <u>magnetic length</u>. The physical significance of this parameter and this problem is discussed in section (16.4) where we solve this problem again in a completely different manner.

In the next section we solve a mathematically very similar problem for crossed electric and magnetic fields.

16.3 The Quantum Hall Effect

The Hall effect occurs when uniform magnetic and electric fields perpendicular to each other are applied to a semiconducting medium. We shall model this effect by considering a single electron in empty space. The resultant Hall current is perpendicular to both the electric and magnetic fields.

Our model hamiltonian is

$$H = \frac{1}{2m} (\vec{p}+ \frac{e}{c} \vec{A})^2 - e\phi \qquad 16.37$$

where we again choose the Landau gauge. Thus,

$$\vec{A} = -B(y,0,0) \quad , \quad \vec{B} = (0,0,B) \tag{16.38}$$

$$\phi = -\mathcal{E}y \quad , \quad \vec{E} = (0,\mathcal{E},0) \; . \tag{16.39}$$

Thus, the hamiltonian when written out reads

$$H = \frac{1}{2m}\left(p_x - \frac{eB}{c}\,y\right)^2 + \frac{1}{2m}\,p_y^2 + \frac{1}{2m}\,p_z^2 + e\mathcal{E}y \tag{16.40}$$

Again we find that

$$[p_x,H]=[p_z,H] = 0 \tag{16.41}$$

and we can write

$$\psi = e^{i(k_x x + k_z z)}\,\phi(y) \; . \tag{16.42}$$

The Schrödinger equation for $\phi(y)$ now reads

$$\frac{d^2\phi}{dy^2} - \frac{m^2\omega^2}{\hbar^2}(y-y_0+y_1)^2\phi + \left[\frac{2mE}{\hbar^2} - k_z^2 + \frac{m^2\omega^2}{\hbar^2}(y_1^2 - 2y_1 y_o)\right]\phi = 0 \tag{16.43}$$

We have again introduced the parameters

$$y_o = \hbar\,\frac{ck_x}{eB} \quad , \quad \omega = \frac{eB}{mc}$$

as well as the parameter

$$y_1 = \frac{e\mathcal{E}}{m\omega^2} \; . \tag{16.44}$$

Calling $a = y_o - y_1$ we can now write the solution immediately as

$$\psi_{n,k_x,k_z} = A_n e^{i(k_x x + k_z z)} e^{-\frac{(y-a)^2}{2\lambda^2}} H_n\left(\frac{y-a}{\lambda}\right) . \qquad 16.45$$

The energy is given by

$$E_n = (n + \frac{1}{2})\hbar\omega + \frac{\hbar^2 k_z^2}{2m} - \frac{m\omega^2}{2}(y_1^2 - 2y_1 y_o) . \qquad 16.46$$

The quantity of interest to us is the electric current density

$$\vec{J}_{n,k_x,k_z} = \frac{-e\hbar}{2im}(\psi^*\nabla\psi - \nabla\psi^*\psi) \qquad 16.47$$

The y-component of the current density clearly vanishes. The other two components are given by

$$J_x = -\frac{e\hbar k_x}{m}|\phi_n|^2 \qquad 16.48$$

$$J_z = -\frac{e\hbar k_z}{m}|\phi_n|^2 . \qquad 16.49$$

If we integrate over the y coordinate to get the net current and use

$$\int_{-\infty}^{\infty} |\phi_n(y)|^2 dy = 1$$

then we find

$$\int_{-\infty}^{\infty} J_x(y)dy = -\frac{e\hbar k_x}{m} \qquad 16.50$$

$$\int_{-\infty}^{\infty} J_z(y)dy = -\frac{e\hbar k_z}{m} \qquad 16.51$$

If we now recall that $\hbar k_x$ and $\hbar k_z$ are just respectively the x and z components of momentum then we see that these results coincide with the classical results.

16.4 Motion in a Uniform Magnetic Field - Heisenberg Equations

We now return to the problem discussed in section 16.2 and solve it again. To illustrate how similar the classical equations of motion and the Heisenberg equations are, we solve the Heisenberg equations. To further illustrate the use of other gauges we this time employ the symmetric gauge so that

$$\vec{A} = -\frac{B}{2}(-y,x,0). \tag{16.24}$$

Finally to obtain the wavefunctions we use an algebraic technique similar to the method used in section 9.1 to solve the harmonic oscillator. The hamiltonian is again

$$H = \frac{1}{2m}(\vec{p}+\frac{e}{c}\vec{A})^2 \equiv \frac{1}{2m}\vec{\pi}^2 \tag{16.52}$$

where we have introduced the <u>mechanical momentum</u> $\vec{\pi}$ related to the <u>canonical momentum</u> $\vec{p}$ by

$$\vec{\pi} = \vec{p} + \frac{e}{c}\vec{A} \tag{16.53}$$

If we were doing classical mechanics $\vec{\pi}$ would just be equal to $m\vec{v}$. In quantum mechanics it is, however, the canonical momentum components p_j that satisfy the canonical commutation relations with the coordinate components x_k. In other words

$$[p_j , x_k] = -i\hbar\delta_{jk} . \tag{16.54}$$

Using this fact, and the explicit form for $\vec{A}$ given by (16.24) we can compute the various commutators among the mechanical momenta. Thus,

$$[\pi_x,\pi_y] = \frac{e}{c}[p_x,A_y] + \frac{e}{c}[A_x,p_y]$$

$$= \frac{eB}{2c}[p_x,x] + \frac{eB}{2c}[-y,p_y] \qquad 16.55$$

$$= -\frac{ie\hbar}{c}B.$$

Similarly, we find

$$[\pi_x,\pi_z] = [\pi_y,\pi_z] = 0. \qquad 16.56$$

It is now straightforward to obtain the Heisenberg equations of motion:

$$\dot{\pi}_x = \frac{1}{i\hbar}[\pi_x,H] = -\frac{eB}{mc}\pi_y = -\omega\pi_y \qquad 16.57$$

$$\dot{\pi}_y = \frac{1}{i\hbar}[\pi_y,H] = \frac{eB}{mc}\pi_x = \omega\pi_x \qquad 16.58$$

$$\dot{\pi}_z = \frac{1}{i\hbar}[\pi_z,H] = 0 \qquad 16.59$$

and

$$m\dot{\vec{r}} = \frac{m}{i\hbar}[\vec{r},H] = \vec{\pi} \qquad 16.60$$

Equation 16.59 shows that π_z is a constant of the motion. Furthermore writing out π_z explicitly we see that it coincides with the canonical momentum p_z, thus

$$\pi_z = p_z = \text{constant.} \qquad 16.61$$

This result is exactly the same as the result obtained above and also coincides with the classical result.

We next use equations 16.60 to replace π_y and π_x by $m\dot{y}$ and $m\dot{x}$ respectively on the right hand side of equations 16.57 and 16.58 to get

$$\dot{\pi}_x = -m\omega\dot{y} \tag{16.62}$$

$$\dot{\pi}_y = m\omega\dot{x} \quad . \tag{16.63}$$

These equations can be integrated immediately to yield:

$$\pi_x + m\omega y = m\omega y_o \tag{16.64}$$

$$\pi_y - m\omega x = -m\omega x_o \tag{16.65}$$

The constants of integration are $m\omega y_o$ and $-m\omega x_o$. Thus x_o and y_o are also constants of the motion and commute with the hamiltonian H. They also commute with both π_x and π_y. On the other hand they do not commute with each other. In fact.

$$[x_o,y_o] = [x-\frac{\pi_y}{m\omega}\ , \ y+\frac{\pi_x}{m\omega}]$$

$$= -\frac{1}{m\omega}[\pi_y,y] + \frac{1}{m\omega}[x,\pi_x] - \frac{1}{m^2\omega^2}[\pi_y,\pi_x]$$

so

$$[x_o,y_o] = \frac{i\hbar}{m\omega} = i\,\frac{\hbar c}{eB} = i\lambda^2 \ . \tag{16.66}$$

It is interesting to note that the equations of motion (16.57) to (16.60), as well as the first integrals (16.64) and (16.65) coincide with the classical results. For the <u>classical motion</u> we have free motion along the z-direction and circular motion in the x-y plane with the centers of the circles located at (x_o,y_o). In fact for constant energy we have

$$H = \frac{1}{2m}(\pi_x^2+\pi_y^2) + \frac{p_z^2}{2m} = E_t + \frac{p_z^2}{2m} \qquad 16.67$$

where we have introduced the conserved "transverse energy" E_t. If we now consider $2E_t/m\omega^2$ and write this out we get classically

$$\frac{2E_t}{m\omega^2} = (m\omega)^{-2}(\pi_x^2+\pi_y^2) = (x-x_o)^2 + (y-y_o)^2 \, . \qquad 16.68$$

These are the equations of circular orbits with radius

$$R^2 = 2E_t/m\omega^2 \qquad 16.69$$

and center (x_o,y_o) . The same result holds quantum mechanically. This does not mean that the trajectory of such an electron in a fixed energy eigenstate of

$$H_t = \frac{1}{2m}(\pi_x^2+\pi_y^2) \qquad 16.70$$

is exactly measurable; only the radius R is measurable. The center of the circle is uncertain because x_o and y_o do not commute but satisfy instead the commutation relation given by equation 16.66. Thus they also satisfy (according to (7.12)) the uncertainty relation

$$\Delta x_o \Delta y_o \geq \lambda^2/2 \qquad 16.71$$

So if, in conformity with the classical results, we interpret (x_o,y_o) as the operators whose eigenvalues yield the coordinates of the center of the circular motion, then the center of the circle is not exactly measurable. This is why a trajectory is not observable.

Writing out the hamiltonian (16.70), we find

$$H_t = \frac{1}{2m}(p_x^2+p_y^2) + \frac{e^2B^2}{8mc^2}(x^2+y^2) + \frac{eB}{2mc}(xp_y-yp_x)$$
$$= \frac{1}{2m}(p_x^2+p_y^2) + \frac{1}{8}m\omega^2(x^2+y^2) + \frac{1}{2}\omega L_z \ . \qquad 16.72$$

We can easily check that L_z is a constant of the motion since it commutes with both H_t and $\frac{p_z^2}{2m}$. It is, however, not an independent constant of the motion since we can write

$$L_z = xp_y - yp_z = x\pi_y - yp_z - \frac{m\omega}{2}(x^2+y^2)$$

and then use equations 16.64, 16.45 and 16.68 to rewrite this as

$$L_z = m\omega[x(x-x_o)+y(y-y_o) - \frac{1}{2}(x^2+y^2)]$$
$$= \frac{m\omega}{2}(R^2-x_o^2-y_o^2). \qquad 16.73$$

This concludes our discussion of constants of the motion and solutions of the Heisenberg equations. The physical interpretation is now clear. The motion of the electron is as in the classical case, there is linear motion parallel to the magnetic field and circular motion about the magnetic field. Only the location of the center of the orbit is indeterminate.

16.4.1 Energy Eigenfunctions

We can also obtain the energy eigenfunctions using the Heisenberg operators. In fact the procedure is similar to what we did for the simple harmonic oscillator when we solved that problem algebraically. However, because there are additional conserved quantum numbers the problem is somewhat more complicated.

To start we introduce the operators

$$\pi_\pm = \pi_x \pm i\pi_y \qquad 16.74$$

The hamiltonian H_t can now be written as

$$
\begin{aligned}
H_t &= \frac{1}{4m}(\pi_+\pi_- + \pi_-\pi_+) \\
&= \frac{1}{2m}(\pi_+\pi_- + \frac{1}{2}[\pi_-,\pi_+]) \qquad 16.75 \\
&= \frac{1}{2m}\pi_+\pi_- + \frac{1}{2}\hbar\omega
\end{aligned}
$$

We also find that

$$[\pi_\pm, H_t] = \mp\hbar\omega\pi_\pm \qquad 16.76$$

Thus if we have an eigenstate ϕ_n of H_t

$$H_t\phi_n = E_n\phi_n \qquad 16.77$$

then using (16.76) we get

$$\pi_\pm H_t\phi_n = H_t\pi_\pm\phi_n \mp \hbar\omega\pi_\pm\phi_n = E_n\pi_\pm\phi_n \;.$$

Hence

$$H_t\pi_\pm\phi_n = (E_n \pm \hbar\omega)\pi_\pm\phi_n \;. \qquad 16.78$$

This shows that $\pi_\pm$ are step-up and step-down operators for eigenstates of H_t. We can therefore write

$$\phi_n = A_n\pi_+\phi_{n-1} \qquad 16.79$$

where A_n is a normalization constant. Since ϕ_n and ϕ_{n-1} are assumed normalized we can write

$$
\begin{aligned}
1 = (\phi_n,\phi_n) &= |A_n|^2(\pi_+\phi_{n-1},\pi_+\phi_{n-1}) \\
&= |A_n|^2(\phi_{n-1},\pi_-\pi_+\phi_{n-1}).
\end{aligned} \qquad 16.80
$$

But

$$\pi_-\pi_+ = \pi_+\pi_- + 2m\hbar\omega = 2m(H_t - \frac{1}{2}\hbar\omega) + 2m\hbar\omega$$
$$= 2mH_t + m\hbar\omega. \tag{16.81}$$

So we find

$$1 = |A_n|^2 2m(\phi_{n-1}, (H_t + \frac{1}{2}\hbar\omega)\phi_{n-1})$$
$$= |A_n|^2 2m\cdot n\hbar\omega$$

$$A_n = (2n)^{-1/2}(m\hbar\omega)^{-1/2} = (2n)^{-1/2}\hbar/\lambda. \tag{16.82}$$

Thus proceeding in the now familiar fashion we define the ground state ϕ_o by

$$\pi_-\phi_o = 0 \tag{16.83}$$

$$\pi_z\phi_o = \hbar k\phi_o \tag{16.84}$$

The normalized state ϕ_n is then given by

$$\phi_n = (2^n n!)^{-1/2}(\lambda\pi_+/\hbar)^n\phi_o . \tag{16.85}$$

To find ϕ_o we first write out $\pi_\pm$ to get

$$\pi_\pm = -i\hbar[\frac{\partial}{\partial x} \pm i\frac{\partial}{\partial y} \mp \frac{1}{2\lambda^2}(x\pm iy)]. \tag{16.86}$$

Since $\pi_z = -i\hbar\frac{\partial}{\partial z}$ we can solve equations 16.83 and 16.84 to obtain

$$\phi_o = f(x-iy)e^{ikz-\frac{x^2+y^2}{4\lambda^2}} \tag{16.87}$$

where f is an arbitrary function. This fact reflects the infinite degeneracy of these states. It also means that we can impose another condition on ϕ_o. The condition we impose is that ϕ_o should also be an eigenfunction of x_o. This condition is motivated by the fact that

$$[x_o, \pi_+] = 0 \tag{16.88}$$

and thus if

$$x_o\phi_o = a\phi_o \tag{16.89}$$

then using (16.85) we see that we also have

$$x_o\phi_n = a\phi_n \, . \tag{16.90}$$

Writing out x_o we find

$$x_o = \frac{1}{2}x + \frac{i\hbar}{m\omega}\frac{\partial}{\partial y} = \frac{1}{2}x + i\lambda^2\frac{\partial}{\partial y} \tag{16.91}$$

Thus equation 16.89 becomes

$$\left(\frac{x}{2} + i\lambda^2\frac{\partial}{\partial y}\right)\phi_o = a\phi_o \, . \tag{16.92}$$

Using our explicit form for ϕ_o this equation reduces to

$$f' = -\frac{1}{2\lambda^2}(x-iy-2a)f \tag{16.93}$$

Hence we get

$$f = Ae^{-\frac{1}{4\lambda^2}(x-iy-2a)^2} \qquad 16.94$$

and

$$\phi_{o,a} = Ae^{ikz}\exp -\frac{1}{4\lambda^2}[(x^2+y^2) + (x-iy-2a)^2]. \qquad 16.95$$

We now normalize so that

$$1 = \int_{-\infty}^{\infty} |\phi_{o,a}(x)|^2\, dx$$

$$= |A|^2 \int_{-\infty}^{\infty} e^{-a^2/\lambda^2}\, e^{-(x-a)^2/\lambda^2}\, dx$$

$$= |A|^2\sqrt{\pi\lambda^2}\, e^{-a^2/\lambda^2}$$

Thus

$$A = (\pi\lambda^2)^{-1/4} e^{a^2/2\lambda^2}, \qquad 16.96$$

and

$$\phi_{o,a} = (\pi\lambda^2)^{-1/4} e^{ikz}\exp -\frac{1}{4\lambda^2}[x^2+y^2-2a^2+(x-iy-2a)^2] \qquad 16.97$$

To obtain the higher energy eigenfunctions $\phi_{n,a}$ in terms of hermite polynomials by using the step-up operators as in equation (16.85) we employ the following two identities:

$$\pi_+ e^{\frac{1}{4\lambda^2}(x^2+y^2)} F(x,y)$$

$$= -i\hbar\left[\frac{\partial}{\partial x} + i\,\frac{\partial}{\partial y} - \frac{1}{2\lambda^2}(x+iy)\right] e^{\frac{1}{4\lambda^2}(x^2+y^2)} F(x,y) \qquad 16.98$$

$$= -i\hbar e^{\frac{1}{4\lambda^2}(x^2+y^2)} \left(\frac{\partial}{\partial x} + i\,\frac{\partial}{\partial y}\right) F(x,y)$$

and

$$\left(\frac{\partial}{\partial x} + i\frac{\partial}{\partial y}\right) F(x+iy) G(x,y) = F(x+iy)\left(\frac{\partial}{\partial x} + i\frac{\partial}{\partial y}\right) G(x,y) \qquad 16.99$$

We next compute $\phi_{1,a}$ from (16.70). Thus

$$\phi_{1,a} = (2)^{-1/2}\,\frac{\lambda}{\hbar}\,\pi_+ \phi_{o,a} \qquad 16.100$$

But,

$$\pi_+ \phi_{o,a} = (\sqrt{\pi}\,\lambda)^{-1/2} e^{ikz} \pi_+ e^{-\frac{1}{4\lambda^2}\left[x^2+y^2-2a^2+(x-iy-2a)^2\right]}$$

$$= (\sqrt{\pi}\,\lambda)^{-1/2} e^{ikz} e^{\frac{1}{4\lambda^2}(x^2+y^2+2a^2)} (-i\hbar)\times \qquad 16.101$$

$$\times\left(\frac{\partial}{\partial x} + i\,\frac{\partial}{\partial y}\right) e^{-\frac{1}{2\lambda^2}\left[x^2+y^2+\frac{1}{2}(x-iy-2a)^2\right]}$$

where we have used the identity (16.98) after multiplying to the right of π_+ by $1 = \exp\frac{1}{4\lambda^2}(x^2+y^2)\exp-\frac{1}{4\lambda^2}(x^2+y^2)$. We now multiply on the right of π_+ by

$$1 = \exp\frac{1}{4\lambda^2}(x+iy-2a)^2 \exp-\frac{1}{4\lambda^2}(x+iy-2a)^2$$

and use the identity (16.95) to get

$$\pi_+\phi_{o,a} = (\sqrt{\pi}\,\lambda)^{-1/2} e^{ikz} e^{\frac{1}{4\lambda^2}(x^2+y^2+2a^2)} e^{\frac{1}{4\lambda^2}(x+iy-2a^2)} \cdot$$

$$\cdot(-i\hbar)\left(\frac{\partial}{\partial x} + i\frac{\partial}{\partial y}\right) e^{-\frac{1}{4\lambda^2}(x^2+y^2)} e^{-\frac{1}{4\lambda^2}[(x-iy-2a)^2+(x+iy-2a)^2]}$$

$$= (\sqrt{\pi}\,\lambda)^{-1/2} e^{ikz} e^{\frac{1}{2\lambda^2}[(x-a)^2+iy(x-2a)+2a^2]} \times$$

$$\times(-i\hbar)\left(\frac{\partial}{\partial x} + i\,\frac{\partial}{\partial y}\right) e^{-\frac{1}{\lambda^2}(x-a)^2-\frac{a^2}{\lambda^2}}$$

$$= (\sqrt{\pi}\,\lambda)^{-1/2} e^{ikz} e^{-\frac{1}{2\lambda^2}[(x-a)^2-iy(x-2a)]} \times \qquad 16.102$$

$$\times\left(\frac{-i\hbar}{\lambda}\right) e^{\frac{1}{\lambda^2}(x-a)^2}\,\lambda\,\frac{\partial}{\partial x}\, e^{-\frac{1}{\lambda^2}(x-a)^2}$$

The last term in this formula can now be recognized as the Rodrigues formula for the first hermite polynomial (equation (5.37))

$$H_n(x) = (-1)^n e^{x^2} \frac{d^n}{dx^n} e^{-x^2} . \qquad 16.103$$

Thus we find

$$\pi_+\phi_{o,a} = (\sqrt{\pi}\,\lambda)^{-1/2}\left(\frac{i\hbar}{\lambda}\right) e^{ikz} e^{-\frac{1}{2\lambda^2}[(x-a)^2-iy(x-2a)]} H_1\left(\frac{x-a}{\lambda}\right). \qquad 16.104$$

Combining this result with equation (16.100) we finally get

$$\phi_{1,a} = \frac{i}{\sqrt{2}} (\sqrt{\pi}\,\lambda)^{-1/2} e^{ikz} e^{-\frac{1}{2\lambda^2}[(x-a)^2 - iy(x-2a)]} H_1\left(\frac{x-a}{\lambda}\right) . \tag{16.105}$$

Repeating these same steps n times we get

$$\phi_{n,a} = \left(\frac{i}{\sqrt{2}}\right)^n (\sqrt{\pi}\,\lambda n!)^{-1/2} e^{ikz} e^{-\frac{1}{2\lambda^2}[(x-a)^2 - iy(x-2a)]} H_n\left(\frac{x-a}{\lambda}\right). \tag{16.106}$$

Thus we have again recovered the eigenfunctions for the Landau levels. Comparing them with the solution obtained in the Landau gauge (equation 16.35) we see that the factor $e^{ik_x x}$ has been replaced by a factor $\exp i\,\frac{y(x-2a)}{2\lambda^2}$. Now using the fact that $a = x_o = -\frac{\hbar c}{eB} k_y$ and $\lambda^2 = \frac{\hbar c}{eB}$ we find that this factor is just $\exp\left(i\,\frac{xy}{2\lambda^2} + ik_y y\right)$. The first factor is simply due to the gauge transformation from the Landau to the symmetric gauge with the gauge function

$$\Lambda = \frac{B}{2}\,xy.$$

The factor $\exp\left(-ik_x x + ik_y y\right) = \exp \frac{i}{2\lambda^2}(xy_o - yx_o)$ reflects the arbitrariness in the choice of the center (x_o, y_o) for the orbit.

These eigenfunctions play an important role in understanding the quantum Hall effect as well as the energy levels of atoms in superstrong magnetic fields. By superstrong magnetic fields we mean fields so strong that the "orbits" due to the magnetic field have shrunk to the order of magnitude of Bohr orbits. Thus these magnetic fields must be so strong that

$$R_o = \left(\frac{\hbar\omega}{m\omega^2}\right)^{1/2} \leq a_o = \frac{\hbar^2}{me^2} .$$

Substituting for the Larmor frequency ω and solving for B we find

$$B \geq \frac{m^2 e^3}{\hbar^3} c$$

For such strong fields it is not the magnetic interaction that is the perturbation (section 16.7) but rather the Coulomb term (reference 16.2) and the atoms become more "cylindrical" than spherical in shape.

16.5 Spin and Spin-Orbit Coupling

We now consider an electron interacting with both an electric as well as a constant magnetic field. The "orbital" hamiltonian is then given by (16.3) with $q = -e$,

$$H_o = \frac{1}{2m} (\vec{p} + \frac{e}{c} \vec{A})^2 - e\phi . \tag{16.107}$$

where the subscript "o" stands for "orbital". Choosing the "symmetric gauge" such that

$$\vec{A} = -\frac{1}{2} \vec{r}\times\vec{B} \tag{16.108}$$

we can rewrite this hamiltonian as

$$H_o = \frac{\vec{p}^2}{2m} + V(r) + \frac{e}{2mc} \vec{B}.\vec{L} + \frac{e^2}{2mc^2} \vec{A}^2 \tag{16.109}$$

where $V = -e\phi$. For most cases of weak magnetic fields we drop the last term since it is much smaller than the other terms.

This is not yet the complete hamiltonian for an electron in an electromagnetic field. The fact that the electron has spin adds two more terms to this hamiltonian. The first of these arises from the fact that due to its spin the electron also has a magnetic moment (section 9.6)

$$\vec{\mu} = \frac{ge}{2mc}\vec{S} . \qquad 16.110$$

The factor g is known as the <u>gyromagnetic</u> ratio and was first measured spectroscopically to have the value $g = 2$. This factor arises automatically in Dirac's relativistic equation for an electron and is exactly 2 except for corrections due to quantum electrodynamic effects. The measured and calculated values of g agree and are

$$g = 2(1+1.1596389\times10^{-3}).$$

We shall use the value $g = 2$. Thus the energy due to this magnetic moment is given by

$$H_m = -\vec{\mu}.\vec{B} = \frac{2e}{2mc}\vec{B}.\vec{S}. \qquad 16.111$$

There is a second energy term which arises due to the interaction of the spin magnetic moment of the electron with the magnetic field created by its orbital motion. Crudely speaking, the electric field du to the nucleus as described in (16.26) above is viewed in the rest fram of the nucleus. However to get the same field in the rest frame of the electron we must perform a Lorentz transformation. This causes the originally purely electric field $\vec{E} = -\frac{\partial\phi}{\partial r}\vec{r}$ to acquire a magnetic component

$$\vec{B}_{Lorentz} = -\frac{\vec{v}}{c}\times\vec{E} . \qquad 16.11$$

The interaction of this magnetic field gives rise to the so-called spin-orbit interaction

$$H'_{s.o.} = -\vec{\mu}.\vec{B}_{Lorentz} = \frac{ge}{2mc} \cdot \frac{1}{c} (\vec{v}\times\vec{E})\cdot\vec{S} \qquad 16.113$$

$$= -\frac{e}{m^2c^2} \vec{S}\cdot(\vec{E}\times\vec{p}) \; . \qquad 16.114$$

This term was first proposed by Goudsmit and Uehlenbeck (ref. 16.3). Unfortunately it was too large by a factor of 2. Thomas, (ref. 16.3), however, showed that relativistic effects cause a further precession of the spin vector to effectively reduce $H'_{s.o.}$ by a factor of 2. Thus the spin-orbit interaction is given by

$$H_{s.o.} = \frac{-e}{2m^2c^2} \vec{S}\cdot(\vec{E}\times\vec{p}) \; . \qquad 16.115$$

If the central electrostatic field is described by a potential $\phi(r)$ then

$$\vec{E} = -|\nabla\phi| \frac{\vec{r}}{r} = -\frac{1}{r}\frac{\partial\phi}{\partial r}\vec{r} = -\frac{1}{er}\frac{\partial V}{\partial r}\vec{r} \qquad 16.116$$

Thus finally

$$H_{s.o.} = \frac{1}{2m^2c^2}\frac{1}{r}\frac{\partial V}{\partial r}\vec{S}.\vec{L}. \qquad 16.117$$

Thus the resultant hamiltonian is given by

$$H = \frac{\vec{p}^2}{2m} + V(r) + \frac{e}{2mc}\vec{B}.(\vec{L}+2\vec{S}) + H_{s.o.} \; . \qquad 16.118$$

16.6 Alkali Spectra

Consider any alkali atom in the absence of a magnetic field.

Furthermore restrict your attention to the outer or valence electrons and treat the inner electrons as inert closed shells. Then this is effectively a one electron problem in some central electrostatic potential $V(r)$. According to (16.118) the appropriate corresponding hamiltonian is

$$H = \frac{\vec{p}^2}{2m} + V(r) + \frac{1}{2m^2c^2}\frac{1}{r}\frac{dV}{dr}\vec{L}.\vec{S} \; . \tag{16.119}$$

Now

$$\vec{J}^2 = \vec{L}^2 + \vec{S}^2 + 2\vec{L}\cdot\vec{S} \tag{16.120}$$

so that

$$\vec{L}\cdot\vec{S} = \frac{1}{2}(\vec{J}^2 - \vec{L}^2 - \vec{S}^2) \; . \tag{16.121}$$

Also since H, $\vec{J}^2$, $\vec{L}^2$ and $\vec{S}^2$ are mutually commuting we can label the eigenkets of H by $|n,j,\ell,s\rangle$. Thus fixing j and ℓ the effective potential becomes

$$V_{eff}(r) = V(r) + \frac{\hbar^2}{4m^2c^2}\frac{1}{r}\frac{dV}{dr}\cdot\{j(j+1)-\ell(\ell+1)-\frac{3}{4}\} \; . \tag{16.122}$$

where we have used the fact that $S = 1/2$. Thus, in fact, $j = \ell \pm 1/2$ except for S-waves ($\ell = 0$) in which case $j = 1/2$. The effect of this is to give a different effective potential for levels with the same n and ℓ but different orientations of $\vec{L}$ and $\vec{S}$ or in fact different $\vec{J}$, except for S-waves. This means that all degenerate levels with $\ell \neq 0$ have in fact two different effective potentials depending on the two values of j. Thus all lines with $\ell \neq 0$ are doublets. For example all $\ell = 1$ or P lines split into the doublets $P_{3/2}$ and $P_{1/2}$ where the subscript refers to j.

If we now treat the spin-orbit term in (16.119) as a perturbation then to first order in perturbation theory we can calculate the splitting resulting from the spin-orbit interaction. Thus

$$\Delta E_{doublet} = E_{n,\ell,\ell+1/2} - E_{n,\ell,\ell-1/2}$$

$$= \frac{\hbar^2}{2m^2c^2} \langle n,\ell|\frac{1}{r}\frac{dV}{dr}|n,\ell\rangle\{(\ell+1/2)(\ell+3/2)-\ell(\ell+1) - \frac{3}{4}$$

$$- (\ell-1/2)(\ell+1/2) + \ell(\ell+1) + \frac{3}{4}\} \tag{16.123}$$

or

$$\Delta E_{doublet} = \frac{\hbar^2}{2m^2c^2} \langle n,\ell|\frac{1}{r}\frac{dV}{dr}|n,\ell\rangle(2\ell+1) \tag{16.124}$$

where we have written $|n,\ell\rangle$ for both $|n,\ell,\ell+1/2, 1/2\rangle$ and $|n,\ell,\ell-1/2,1/2\rangle$ since the matrix elements for both these states are identical. It is this interaction that is responsible for the well known doublet in sodium called the sodium D lines. This doublet occurs in the transition from $P_{3/2}$ and $P_{1/2}$ to $S_{1/2}$ states. Here we are using the spectroscopic notation S, P, D, F, G, H, etc. for $\ell = 0,1,2,3$, etc. and the subscript denotes the j value of the level.

The reason that the sodium doublet is more widely split than the hydrogen doublet is due to the more rapid variation of $V(r)$. Thus dV/dr is much larger for sodium than hydrogen since the inner closed shells in the case of sodium screen the valence electron and more of the nuclear charge becomes effective on the valence electron as it penetrates the outer shell.

After considering spin-orbit coupling the obvious next step is to consider an atom in a weak magnetic field as well and to study the hamiltonian (16.118). Unfortunately we must first study another aspect of angular momentum, namely how to add two angular momenta since (16.118) involves the term $\vec{L}+2\vec{S}$. We shall therefore do this first and then return to the hamiltonian given by (16.118).

16.7 Addition of Angular Momenta

Consider a pair of particles with angular momenta $\vec{J}_1$ and $\vec{J}_2$. Then

$$[\vec{J}_1,\vec{J}_2] = 0. \qquad 16.125$$

This will also be the case if $\vec{J}_1$ refers to the orbital angular momentum $\vec{L}$ and $\vec{J}_2$ refers to the spin angular momentum $\vec{S}$ of just one particle. In either case the total angular momentum $\vec{J}$ is given by

$$\vec{J} = \vec{J}_1 + \vec{J}_2 . \qquad 16.126$$

If we now consider commutators, again picking z as the preferred axis, we find two sets of mutually commuting operators, namely

$$1) \quad J^2, J_z, J_1^2, J_2^2$$

and

$$2) \quad J_1^2, J_{1z}, J_2^2, J_{2z} .$$

Thus it is possible to simultaneously diagonalize all the operators in either the first or the second set. We label the corresponding eigenkets as

$$1) \quad |j,m,j_1,j_2\rangle$$

and

$$2) \quad |j_1,m_1,j_2,m_2\rangle .$$

Both sets of eigenkets then form complete orthogonal sets and can be normalized so that both sets may be used as basis sets. Thus the two sets are related by a unitary transformation and we may write

$$|jmj_1j_2\rangle = \sum_{m_1m_2} \langle j_1j_2m_1m_2|jm\rangle|j_1m_1j_2m_2\rangle . \qquad 16.127$$

The coefficients $\langle j_1 j_2 m_1 m_2 | jm\rangle$ are the matrix elements of the unitary transformation connecting the two basis sets and are referred to variously as Clebsch-Gordon, Wigner or Vector-addition coefficients. The inverse of equation 16.43 is written as

$$|j_1 m_1 j_2 m_2\rangle = \sum_{jm} \langle jm | j_1 j_2 m_1 m_2\rangle | jm j_1 j_2\rangle \ . \qquad 16.128$$

Thus we immediately have that

$$\sum_{jm} \langle j_1 j_2 m_1 m_2 | jm\rangle\langle jm | j_1 j_2 m_1' m_2'\rangle = \delta_{m_1 m_1'} \delta_{m_2 m_2'} \qquad 16.129$$

and

$$\sum_{m_1 m_2} \langle jm | j_1 j_2 m_1 m_2\rangle\langle j_1 j_2 m_1 m_2 | j'm'\rangle = \delta_{jj'} \delta_{mm'} \qquad 16.130$$

Clearly the range of summation in (16.127) and (16.130) is over $-j_1 \le m_1 \le j_1$ and $-j_2 \le m_2 \le j_2$ since that is the full range of m_1 and m_2. On the other hand we have yet to find the range for j and m. We shall now do this by a counting procedure as well as by explicitly constructing some of the vectors $|jmj_1 j_2\rangle$ in terms of the vectors $|j_1 m_1 j_2 m_2\rangle$. We first notice that the vector

$$|j_1 m_1 j_2 m_2\rangle = |j_1 m_1\rangle | j_2 m_2\rangle \qquad 16.131$$

where $|j_i m_i\rangle$ are the eigenkets of J_i^2 and J_{iz}. This is easily seen to be the case since J_1^2 , J_{1z} operate only on $|j_1 m_1\rangle$ and J_2^2 , J_{2z} operate only on $|j_2 m_2\rangle$. Furthermore each of these vectors is also an eigenvector of J_z with eigenvalue $(m_1+m_2)\hbar$ since

$$J_z = J_{1z} + J_{2z}$$

and hence

$$J_z|j_1m_1j_2m_2\rangle = J_{1z}|j_1m_1\rangle|j_2m_2\rangle + J_{2z}|j_1m_1|j_2m_2\rangle$$
$$= (m_1+m_2)\hbar|j_1m_1j_2m_2\rangle. \quad 16.132$$

The maximum values of m_1, m_2 are j_1 and j_2 and therefore the maximum eigenvalue m of J_z is

$$m_{max} = j_1 + j_2 \ . \quad 16.133$$

But since J satisfies the standard algebra of an angular momentum operator it follows from this that the maximum value of j is also j_1+j_2. Hence

$$j_{max} = j_1+j_2 \ . \quad 16.134$$

We shall now show that in fact j assumes once and only once all the values

$$j = j_1+j_2,\ j_1+j_2-1,\ j_1+j_2-2\ ,\ldots,|j_1-j_2|. \quad 16.135$$

It is clear that by repeatedly applying the operator J_- to the state

$$|j_{max}m_{max}j_1j_2\rangle = |j_1+j_2,j_1+j_2,j_1,j_2\rangle$$
$$= |j_1,j_1,j_2,j_2\rangle \quad 16.136$$

we get all states $|j_1+j_2\ ,\ m,\ j_1,j_2\rangle$ with m running from $+(j_1+j_2)$ to $-(j_1+j_2)$. Thus there are $2j+1$ different possible values of m for $j = j_1 + j_2$. The same argument holds for any other j value so that there are $2j+1$ different values of m for every state of definite j.

Now consider states with $m = j_1+j_2-1$. These can be obtained in two ways; namely from $m_1 = j_1$, $m_2 = j_2-1$ or $m_1 = j_1-1$, $m_2 = j_2$.

Hence there must be two and only two states of definite j with these m values. One of these is the state with $j = j_1+j_2$, $m = j-1$, the other is the state with $j = j_1+j_2-1$ and $m = j$. Similarly for the case of $m = j_1+j_2-2$ there are exactly three states of definite m_1, m_2 values corresponding to the pairs (j_1,j_2-2), (j_1-1,j_2-1), (j_1-2,j_2). Thus there are also exactly three states of definite j, namely

$$j = j_1 + j_2 , \qquad m = j - 2$$

$$j = j_1 + j_2 - 1, \qquad m = j - 1$$

$$j = j_1 + j_2 - 2 , \qquad m = j.$$

Similarly for $m = j_1+j_2-3$ there correspond exactly four states of definite (m_1,m_2) values, namely, (j_1,j_2-3), (j_1-1,j_2-2) , (j_1-2,j_2-1) , (j_1-3,j_2) and hence also exactly four states of definite j, namely

$$j = j_1 + j_2 \quad , \qquad m = j - 3$$

$$j = j_1 + j_2 - 1 , \quad m = j - 2$$

$$j = j_1 + j_2 - 2 , \quad m = j - 1$$

$$j = j_1 + j_2 - 3 , \quad m = j .$$

Continuing in this way we finaly arrive at states with $m = j_1 - j_2$ where for convenience we assumed $j_1 > j_2$. In this case there are exactly $2j_2+1$ states of definite m_1,m_2 values. They correspond to the pairs

$$(j_1,-j_2),(j_1-1,-j_2+1),\ldots,(j_1-j_2,0),(j_1-j_2-1,1),\ldots,$$

$$(j_1-2j_2+1,j_2-1),\ldots,(j_1-2j_2,j_2).$$

It therefore follows as before that there are also exactly $2j_2+1$ states of definite j, namely

$$\begin{array}{ll} j = j_1 + j_2 & , \ m = j - 2j_2 \\ j = j_1 + j_2 - 1 & , \ m = j - 2j_2 + 1 \\ \vdots & \quad \vdots \\ j = j_1 - j_2 & , \ m = j \ . \end{array}$$

Thus the possible values of j are given in (16.51). As an additional check we can count the number of vectors in the two sets $|j\ m\ j_1 j_2\rangle$ and $|j_1 m_1 j_2 m_2\rangle$. From the form of the states $|j_1 m_1 j_2 m_2\rangle$ as given in (16.131) we see that for fixed (j_1, j_2) there are $(2j_1+1)(2j_2+1)$ states of this type. For the states of type $|j\ m\ j_1 j_2\rangle$ we have that for fixed (j_1, j_2) there are $(2j+1)$ states for each j value over the whole possible range of j values, thus the number of such states is

$$\sum_{j=|j_1-j_2|}^{j_1+j_2} (2j+1) = 2\,\frac{(j_1+j_2+1)(j_1+j_2)}{2} - 2\,\frac{|j_1-j_2|(|j_1-j_2|-1)}{2}$$

$$+ j_1 + j_2 - (|j_1-j_2|-1)$$

$$= (2j_1+1)(2j_2+1)$$

as desired.

We have thus arrived at the

Fundamental Addition Theorem for Angular Momenta.

In the $(2j_1+1)(2j_2+1)$ dimensional space spanned by the basis set $|j\ m\ j_1 j_2\rangle$ with j_1 and j_2 fixed, the possible values of j are $j = j_1+j_2,\ j_1+j_2-1,\ j_1+j_2-2, \ldots, |j_1-j_2|$. Corresponding to each of these j values there are $2j+1$ vectors obtained by repeatedly applying J_- to the state $|j\ j\ j_1 j_2\rangle$.

16.8 Example 1. Two Spin 1/2 States

In this case we have $j_1 = j_2 = 1/2$. The four possible states in the $|j_1,m_1,j_2,m_2>$ representation are

$$|1/2,1/2>|1/2,1/2>, \ |1/2,1/2>|1/2,-1/2>,$$

$$|1/2,-1/2>|1/2,1/2> , \ |1/2,-1/2>|1/2,-1/2>.$$

In the $|J,M;j_1,j_2>$ representation the four possible states are

$$|1,1;1/2,1/2> , \ |1,0;1/2,1/2>, \ |1,-1;1/2,1/2> \text{ and } |0,0;1/2,1/2>.$$

The state of highest weight is given by

$$|1,1;1/2,1/2> = |1/2,1/2>|1/2,1/2> \qquad 16.137$$

The two other states with total $S = 1$ are obtained by applying $S_- = S_{1-} + S_{2-}$ to both sides of equation 16.137. This yields

$$\sqrt{1(1+1) - 1(1-1)}\ \hbar\ |1,0;\ 1/2,1/2>$$

$$= \sqrt{1/2(1/2 + 1) - 1/2(1/2 - 1)}\ \hbar\ (|1/2,-1/2>|1/2,1/2>$$

$$+ |1/2,1/2>\ |1/2,- 1/2>)$$

so that

$$|1,0;1/2,1/2> = \frac{1}{\sqrt{2}}(|1/2,- 1/2>|1/2,1/2> + |1/2,1/2>|1/2,- 1/2>) . \qquad 16.138$$

Similarly applying S_- once more to (16.138) or else realizing that both spins 1/2 must point down to get a total S_z of $-\hbar$ we find that

$$|1,-1;\ 1/2,1/2\ \rangle = |1/2,-\ 1/2\rangle|1/2,-\ 1/2\rangle\ . \qquad 16.139$$

The singlet (S = 0) state must be orthogonal to the above three states. A simple calculation then shows that

$$|0,0;1/2,1/2\rangle = \frac{1}{\sqrt{2}}\,(|1/2,-\ 1/2\rangle|1/2,1/2\rangle - |1/2,1/2\rangle|1/2,-\ 1/2\rangle)\ . \qquad 16.140$$

In writing equation 16.140 we have made an arbitrary choice of phase.

To display the matrix of Clebsch-Gordon coefficients we rewrite equation 16.137 to 16.140 as one matrix equation. Thus,

$$\begin{pmatrix} |1,1;1/2,1/2\rangle \\ |1,0;1/2,1/2\rangle \\ |1,-1;1/2,1/2\rangle \\ |0,0;1/2,1/2\rangle \end{pmatrix} = \begin{pmatrix} 1 & 0 & 0 & 0 \\ 0 & 1/\sqrt{2} & 1/\sqrt{2} & 0 \\ 0 & 0 & 0 & 1 \\ 0 & 1/\sqrt{2} & -1/\sqrt{2} & 0 \end{pmatrix} \begin{pmatrix} |1/2,1/2\rangle|1/2,1/2,\rangle \\ |1/2,-1/2\rangle|1/2,1/2\rangle \\ |1/2,1/2\rangle|1/2,-1/2,\rangle \\ |1/2,-1/2\rangle|1/2,-1/2\rangle \end{pmatrix} \qquad 16.141$$

Clearly the 4×4 matrix $\langle 1/2,1/2,m_1,m_2|jm\rangle$ is unitary.

16.9 Example 2. Spin 1/2 and Orbital Angular Momentum

The only examples of interest to us will be the coupling of orbital and spin angular momentum for an electron. Since S = 1/2 the only possible j values are $\ell\pm1/2$ for $\ell \neq 0$. If $\ell = 0$ only $j = 1/2$ is possible. We shall simplify the notation somewhat in this case and denote the states $|\ell\pm1/2,m;\ell,1/2\rangle$ simply by $|\ell\pm1/2,m,\ell\rangle$. Thus, again the state of maximum weight (maximum J_z) is given by

$$|\ell+1/2,\ell+1/2,\ell\rangle = |\ell,\ell\rangle|1/2,1/2\rangle \; .$$

Now using the matrix representation $\binom{1}{0}$ for $|1/2,1/2\rangle$ and the coordinate representation $Y_{\ell,\ell}$ for $|\ell,\ell\rangle$ we have the representation $\binom{Y_{\ell\ell}}{0}$ for the state $|\ell+1/2,\ell+1/2,\ell\rangle$. All other states $|\ell+1/2,m,\ell\rangle$ are now obtained from this state by repeated application of

$$J_- = L_- + S_- \tag{16.142}$$

Construction of the state $|\ell-1/2,\ell-1/2,\ell\rangle$ from which all states $|\ell-1/2,m,\ell\rangle$ are again obtainable by applying J_- is more difficult. The answer in fact is (see problem 16.3)

$$|\ell-1/2,\ell-1/2,\ell\rangle = \frac{1}{\sqrt{2\ell+1}}\left[\,|\ell,\ell-1\rangle|1/2,1/2\rangle - \sqrt{2\ell}\;|\ell,\ell\rangle|1/2,-1/2\rangle\right] \tag{16.143}$$

16.10 The Weak-Field Zeeman Effect

We are now finally able to consider the hamiltonian (16.118) which reads:

$$H = \frac{p^2}{2m} + V(r) + f(r)\vec{L}\cdot\vec{S} + \frac{e\vec{B}}{2mc}\cdot(\vec{L}+2\vec{S}) \tag{16.118}$$

$$= \frac{p^2}{2m} + V(r) + f(r)\vec{L}\cdot\vec{S} + \frac{e\vec{B}}{2mc}\cdot(\vec{J}+\vec{S})$$

where

$$f(r) = \frac{1}{2m^2c^2}\frac{1}{r}\frac{dV}{dr} \qquad 16.144$$

is the radial part of the spin-orbit term.

We choose the z-direction to be along the direction of the B field which is assumed constant and uniform. Then defining the Larmor frequency

$$\omega = \frac{eB}{2mc} \qquad 16.145$$

the last term becomes $\omega(J_z + S_z)$. The eigenstates of the hamiltonian without the last term may be labelled $|n,\ell\pm1/2,m,\ell\rangle$ where n is the principal quantum number and j is $\ell\pm1/2$.

If we treat the term due to the magnetic field, namely $\omega(J_z + S_z)$ by means of first order perturbation theory we get an energy shift ΔE given by

$$\Delta E = \omega\langle n,\ell\pm1/2,m,\ell|J_z+S_z|n,\ell\pm1/2,m,\ell\rangle \qquad 16.146$$

or

$$\Delta E = \hbar\omega m + \omega\langle S_z\rangle . \qquad 16.147$$

To evaluate this further we must calculate explicitly the spin dependence of the states. Thus we need a formula of the form

$$|n,\ell\pm1/2;m,\ell\rangle = \sum_{m_\ell,m_s} \langle\ell,1/2,m_\ell,m_s|\ell\pm1/2,m\rangle|n,\ell,m_\ell,m_s\rangle. \qquad 16.148$$

In order to determine the Clebsch-Gordon coefficients $\langle\ell,1/2,m_\ell,m_s|\ell\pm1/2,m\rangle$ we use the results of Example 2 above. Thus calling $j = \ell \pm 1/2$ and using that

$$J_-|n,j,m,\ell\rangle = \sqrt{j(j+1)-m(m-1)}\,\hbar\,|n,j,m-1,\ell\rangle$$

we get after some computation that

$$|n,j,m,\ell\rangle = \left(\frac{(j+m)!}{(2j)!(j-m)!}\right)^{1/2} \hbar^{m-j}\, J_-^{j-m}|n,j,j,\ell\rangle \;. \qquad 16.149$$

The coordinate representations of the two states $|n,\ell\pm1/2,\ell\pm1/2,\ell\rangle$ are respectively

$$R_{n,\ell}(r)\begin{pmatrix} Y_{\ell\ell}(\theta,\phi) \\ 0 \end{pmatrix} \qquad 16.150$$

and

$$R_{n,\ell}(r)\begin{pmatrix} \left(\frac{1}{2\ell+1}\right)^{1/2} Y_{\ell,\ell-1}(\theta,\phi) \\ -\left(\frac{2\ell}{2\ell+1}\right)^{1/2} Y_{\ell,\ell}(\theta,\phi) \end{pmatrix} \qquad 16.151$$

Using $J_- = L_- + S_-$ and formula (16.149) we can now obtain all states from these and complete our calculation of ΔE. Thus, for example, for the state $|n,\ell+1/2;\ell+1/2,\ell\rangle$

$$\langle S_z\rangle = \left(\begin{pmatrix} Y_{\ell\ell} \\ 0 \end{pmatrix}, S_z\begin{pmatrix} Y_{\ell\ell} \\ 0 \end{pmatrix}\right) = \frac{\hbar}{2}\,(Y_{\ell\ell},Y_{\ell\ell}) = \frac{\hbar}{2} \qquad 16.152$$

so that in this case

$$\Delta E_{n,\ell+1/2,\ell+1/2,\ell} = \hbar\omega(m + 1/2) \;. \qquad 16.153$$

Similarly for the state $|n,\ell-1/2,\ell-1/2,\ell\rangle$

$$\langle S_z\rangle = \left(\begin{pmatrix} \sqrt{\frac{1}{2\ell+1}} & Y_{\ell,\ell-1} \\ -\sqrt{\frac{2\ell}{2\ell+1}} & Y_{\ell,\ell} \end{pmatrix}, S_z \begin{pmatrix} \sqrt{\frac{1}{2\ell+1}} & Y_{\ell,\ell-1} \\ -\sqrt{\frac{2\ell}{2\ell+1}} & Y_{\ell,\ell} \end{pmatrix}\right) \qquad 16.154$$

so

$$\langle S_z\rangle = \frac{\hbar}{2}\left[\frac{1}{2\ell+1}(Y_{\ell,\ell-1},Y_{\ell,\ell-1}) - \frac{2\ell}{2\ell+1}(Y_{\ell\ell},Y_{\ell\ell})\right]$$

$$= \frac{\hbar}{2}\,\frac{1-2\ell}{2\ell+1}$$

and in this case

$$\Delta E_{n,\ell-1/2,\ell-1/2,\ell} = \hbar\omega\left(m - \frac{1}{2} + \frac{1}{2\ell+1}\right) \qquad 16.155$$

For all other states the calculation proceeds in the same manner and is only complicated by the factors of J_- . However using the fact that

$$J_-^\dagger = J_+$$

and that

$$J_+J_- = J^2 - J_z^2 + \hbar J_z$$

allows the computations to be performed without excessive labour (see problem 16.4).

Notes, References and Bibliography

16.1 L. Landau - Zeitschrift für Physik 64 629 (1930).

16.2 J.E. Skjervold and E. Ostgaard - Physics Scripta 29 543 (1984).

16.3 The original papers on electron spin are:

G.E. Uehlenbeck and S. Goudsmit - Naturwiss. 13 953 (1925); Nature 117 264 (1926).

The Thomas precession was presented in L.H. Thomas Nature 107 514 (1926).

A detailed treatment of atomic spectra is to be found in the book by Condon and Shortley (ref. 15.1).

The addition of angular momenta is simplest when viewed from the point of view of group theory. A representative text for physicists is: M. Hammermesh - Group Theory and its Application to Physical Problems - Addison-Wesley Publishing Co., Inc., Reading, Mass., U.S.A. (1962).

In using tables of Clebsch-Gordon coefficients great care must be exercised since several different phase-conventions are in use. Besides the one used in ref. 16.4 common ones are also found in Wigner's book ref. 12.2 as well as in:

J.M. Blatt and V.F. Weisskopf - Theoretical Nuclear Physics-Wiley, New York (1952).

Chapter 16 Problems

16.1 The strongest static magnetic fields currently achieved in laboratories are of the order of 3×10^5 gauss. For fields of this strength estimate the magnitude of the last terms in (16.109) in eV.

16.2 Solve the eigenvalue problem for $r_0^2 = x_0^2 + y_0^2$ where y_0 and x_0 are given by equations (16.64) and (16.65). It may be useful to write the eigenvalue problem in the form

$$r_0^2 \, f_\ell = \lambda^2 (2\ell+1) f_\ell .$$

Interpret the meaning of this result.

16.3 Show that the operator $T = A[j_2 J_{1-} - j_1 J_{2-}]$, where A is a normalization constant, has the property that

$$T|j\ j\ j_1 j_2\rangle = |j-1, j-1, j_1, j_2\rangle$$

for $j = j_1 + j_2$. Hence show that equation (16.143) is correct.

16.4 Evaluate the expectation value $\langle S_z\rangle$ for the states $|\ell\pm1/2, m, \ell\rangle$ and hence the shift in energy due to a uniform magnetostatic field.

16.5 A particle of angular momentum 1/2 is coupled to a particle of angular momentum 1. List the states that are eigenstates of

$$J^2 = (\vec{J}_1 + \vec{J}_2)^2 \quad \text{and}$$

$$J_z = J_{1z} + J_{2z}$$

and express them in terms of the eigenstates of

$$(J_1^2, J_{1z}) \quad \text{and} \quad (J_2^2, J_{2z}) .$$

16.6 Consider a set of three operators T_m $m = -1,0,1$ such that

$$T_m^\dagger = (-1)^m\, T_{-m}$$

$$[J_\pm, T_m] = \sqrt{2 - m(m\pm1)}\; T_{m\pm1}$$

$$[J_z, T_m] = m\, T_m$$

where J are the total angular momentum operators.
Evaluate the total m', m'' dependence of the matrix elements

$$\langle j, m''|T_m|j', m'\rangle$$

Hint: Express T_m in terms of 3×3 matrices. This is an example of the Wigner-Eckart Theorem.

16.7 Consider the unitary operator

$$R_n(\phi) = e^{i(\vec{J}.\vec{n})\phi/\hbar}$$

where $\vec{J}$ is the angular momentum operator.

a) If $j = 1/2$ expand $R_n(\phi)$ to obtain a simpler expression and apply it to the states $\binom{1}{0}$ and $\binom{0}{1}$. What is the effect of $R_n(\phi)$?

b) If $j = 1$ repeat part a) but consider the states

$$\begin{pmatrix}1\\0\\0\end{pmatrix}, \begin{pmatrix}0\\1\\0\end{pmatrix} \text{ and } \begin{pmatrix}0\\0\\1\end{pmatrix}.$$

Chapter 17

Scattering Theory: Time Dependent Formulation

So far except for a few examples in chapters 4 and 5 we have been concerned almost exclusively with the discrete part of the energy spectrum. In this chapter we commence a discussion of the continuous spectrum. Thus we do not solve for the energy; in fact, we solve for the wave-function corresponding to certain initial conditions. These initial conditions correspond to a current of particles incident on some potential and a current of particles scattered by the same potential.

A very large number of microscopic phenomena have their origin in the collisions of particles. Thus for example such diverse properties as the conductivity of metals and the critical masses of nuclear reactions are ultimately determined by scattering phenomena. Furthermore, almost all nuclear and high energy physics experiments are collision experiments and require some form of scattering theory for their interpretation. Thus scattering theory is one of the most important tools of a modern physicist.

The time-dependent formulation allows a very intuitive approach since the concepts used correspond closely to those used in classical scattering theory. However, for computational purposes, the time-independent formulation is more convenient. Therefore those readers more interested in the applications of scattering theory may omit all of this chapter except section 6. The only other result used later is equation (18.78) of section 8. This section may be read with profit after section 2 of chapter 18.

To bring out the similarity between the time-dependent formulation of scattering theory and classical scattering theory we begin with a quick review of the main physical concepts used in the classical scattering problem.

17.1 Classical Scattering Theory

Consider two particles that may interact with each other. Classically the state of this two-particle system at any time t is given by the four vectors $\vec{x}_1(t)$, $\vec{p}_1(t)$, $\vec{x}_2(t)$, $\vec{p}_2(t)$ specifying the trajectories of both particles.

In practice it is much more convenient to reduce this problem to an equivalent one body problem using the law of conservation of momentum. Thus we define

$$\vec{X}(t) = \frac{m_1\vec{x}_1(t) + m_2\vec{x}_2(t)}{m_1 + m_2} \tag{17.1}$$

$$\vec{x}(t) = \vec{x}_1(t) - \vec{x}_2(t) \tag{17.2}$$

$$\vec{P}(t) = \vec{p}_1(t) + \vec{p}_2(t) \tag{17.3}$$

$$\vec{p}(t) = \frac{m_2}{m_1 + m_2}\,\vec{p}_1(t) - \frac{m_1}{m_1 + m_2}\,\vec{p}_2(t) \tag{17.4}$$

Conservation of momentum then implies that $P(t)$ is a constant independent of time, say $\vec{P}_{in}$, and hence $\vec{X}(t)$ depends linearly on t

so that

$$\vec{X}(t) = \vec{X}_{in} + \frac{\vec{P}_{in}}{m_1 + m_2} t \qquad 17.5$$

Equation 17.5 is simply a statement of the fact that the center of mass moves like a free particle with momentum $\vec{P}_{in}$, mass (m_1+m_2) and position $\vec{X}_{in}$ at $t = 0$. One now has to solve for the trajectory of the equivalent one-body problem described by the variables $\vec{x}(t)$, $\vec{p}(t)$. In a scattering problem involving an interaction of short range the trajectory will have the qualitative features shown in fig. 17.1. Far from the interaction region the fictitious particle corresponding to the reduced mass $m = m_1 m_2/(m_1+m_2)$ moves like a free particle indicated by the two trajectories $\vec{x}_{in}(t)$ and $\vec{x}_{out}(t)$. In the interaction region the trajectory is complicated. We can describe the two asymptotic trajectories by

$$\vec{x}_{in}(t) = \vec{x}_{in}(0) + \frac{\vec{p}_{in}}{m} t \qquad 17.6$$

$$\vec{x}_{out}(t) = \vec{x}_{out}(0) + \frac{\vec{p}_{out}}{m} t \,. \qquad 17.7$$

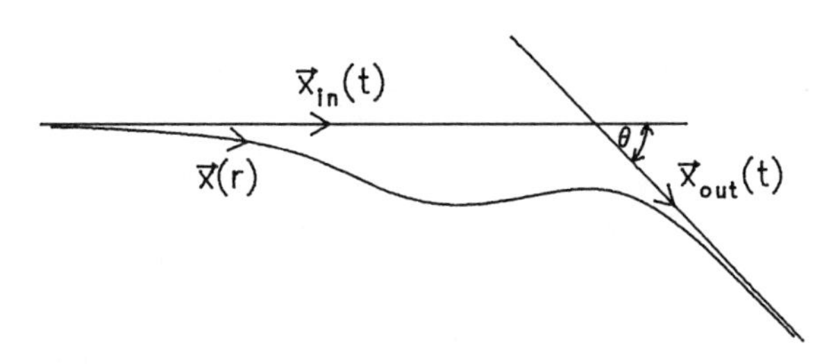

Fig. 17–1 Classical Scattering by Central Force

Due to the assumed short-range* nature of the potential we may also impose the asymptotic conditions

$$\vec{x}(t) \xrightarrow[t\to-\infty]{} \vec{x}_{in}(t) \qquad 17.8$$

and

$$\vec{x}(t) \xrightarrow[t\to\infty]{} \vec{x}_{out}(t) \qquad 17.9$$

This limit has to be defined precisely and is in fact defined by

$$\lim_{t\to-\infty} \|\vec{x}(t) - \vec{x}_{in}(t)\| = 0 \qquad 17.10$$

and

$$\lim_{t\to\infty} \|\vec{x}(t) - \vec{x}_{out}(t)\| = 0 \qquad 17.11$$

where $\|\cdot\|$ here represents the Euclidean norm. From the two asymptotic trajectories we obtain the scattering angle θ by using $\cos\theta = \vec{x}_{in}(t)\cdot\vec{x}_{out}(t)/\|\vec{x}_{in}(t)\|\|\vec{x}_{out}(t)\|$, i.e. the "overlap" between the asymptotic states. We next derive the analogous procedure in quantum mechanics.

17.2 Asymptotic States in the Schrödinger Picture

We start immediately with the hamiltonian for the equivalent one-body problem

$$H = \frac{p^2}{2m} + V(\vec{r}) \qquad 17.12$$

As in the classical case we assume that $V(\vec{r})$ is short range so that for large $\|\vec{r}\|$, $V(\vec{r})$ dies out rapidly and H approaches H_o, the

---

* A precise definition of short-range potentials is not required, but the condition $r^2V(r) \longrightarrow o$ for $r \to \infty$ is sufficient.

hamiltonian for a free particle

$$H_o = \frac{p^2}{2m} . \qquad 17.13$$

In the classical description the trajectories $\vec{x}_{in}(t)$, $\vec{x}_{out}(t)$ were in fact determined by H_o. Thus we assume that there exist states $\Psi_{in}(t)$, $\Psi_{out}(t)$ evolving according to H_o such that

$$\lim_{t\to-\infty} \|\Psi(t) - \Psi_{in}(t)\| = 0 \qquad 17.14$$

$$\lim_{t\to+\infty} \|\Psi(t) - \Psi_{out}(t)\| = 0 . \qquad 17.15$$

These limits, if they exist, define the asymptotic incoming and outgoing states. Here $\Psi(t)$ is a solution of the full Schrödinger equation with the hamiltonian (17.12). The symbol $\|\cdot\|$ here represents the Hilbert space norm and the limits involved are called strong limits. (See problem 6.1). One states this by saying that $\Psi(t)$ converges strongly to $\Psi_{in}(t)$ as $t \to -\infty$ and $\Psi(t)$ converges strongly to $\Psi_{out}(t)$ as $t \to +\infty$.

Since H and H_o are self-adjoint, they determine unitary evolution operators

$$U(t) = e^{-iHt/\hbar} \qquad 17.16$$

and

$$U_o(t) = e^{-iH_ot/\hbar} . \qquad 17.17$$

Thus the requirements for asymptotic states (equations 17.14, 17.15) may be written

$$\lim_{t\to\mp\infty} \|e^{-iHt/\hbar}\Psi(0) - e^{-iH_ot/\hbar}\Psi_{\substack{in\\out}}(0)\| = 0. \qquad 17.18$$

Using the unitarity of these operators we can further write

$$\lim_{t\to\mp\infty} \| e^{iH_o t/\hbar} e^{-iHt/\hbar} \Psi(0) - \Psi_{\substack{in \\ out}}(0)\| = 0. \qquad 17.19$$

We are thus led in a very natural fashion to consider the two operators

$$\Omega_{\pm}^{\dagger} = \lim_{t\to\mp\infty} e^{iH_o t/\hbar} e^{-iHt/\hbar} . \qquad 17.20$$

These operators are known as the Møller wave operators. Just as in the classical case the two asymptotic trajectories gave us the amount of scattering (scattering angle), so also quantum mechanically the amount of scattering is determined by the overlap between the two asymptotic solutions. It is in this respect that the two Møller operators are particularly useful.

17.3 The Møller Wave Operators

We have from the defintion of $\Omega_{\pm}$ that

$$|\Psi_{in}(t)\rangle = \Omega_{+}^{\dagger}|\Psi(t)\rangle \qquad 17.21$$

and

$$|\Psi_{out}(t)\rangle = \Omega_{-}^{\dagger}|\Psi(t)\rangle . \qquad 17.22$$

But from (17.19) we can also deduce that

$$|\Psi(t)\rangle = \Omega_{+}|\Psi_{in}(t)\rangle = \Omega_{-}|\Psi_{out}(t)\rangle . \qquad 17.23$$

Hence combining these with (17.21), (17.22) we get

$$|\Psi_{in}(t)\rangle = \Omega_{+}^{\dagger}\Omega_{+}|\Psi_{in}(t)\rangle \qquad 17.24$$

$$|\Psi_{out}(t)\rangle = \Omega_-^\dagger\Omega_-|\Psi_{out}(t)\rangle \ . \tag{17.25}$$

We assume of course that the "in" and "out" states span the entire Hilbert space of scattering states so that at any time every scattering state can be written as a linear superposition of states $|\Psi_{in}\rangle$ or $|\Psi_{out}\rangle$. From this and (17.24) or (17.25) we conclude that

$$\Omega_\pm^\dagger\Omega_\pm = 1. \tag{17.26}$$

on the space of scattering states. This does not, however, imply that the operators $\Omega_\pm$ are unitary on the whole Hilbert space since one cannot establish the relation $\Omega_\pm\Omega_\pm^\dagger = 1$ for all physical states even though we do have

$$\Omega_\pm\Omega_\pm^\dagger|\Psi(t)\rangle = |\Psi(t)\rangle \ . \tag{17.27}$$

The reason for this is, that if H has bound states then the scattering states $|\Psi(t)\rangle$ do not span the entire Hilbert space. All the scattering states are orthogonal to the bound states and these can therefore not be expanded in terms of scattering states.

If we now return to the Schrödinger equation for $|\Psi(t)\rangle$ or $|\Psi_{\substack{in\\out}}(t)\rangle$ we get

$$i\hbar\frac{\partial}{\partial t}|\Psi(t)\rangle = H|\Psi(t)\rangle \tag{17.28}$$

and

$$i\hbar\frac{\partial}{\partial t}|\Psi_{\substack{in\\out}}(t)\rangle = H_o|\Psi_{\substack{in\\out}}(t)\rangle. \tag{17.29}$$

Substituting (17.23) into (17.28) and comparing with (17.29) we get the so-called inter-twining property of the Moller wave operators

$$H\Omega_\pm = \Omega_\pm H_o \ . \tag{17.30}$$

Combining (17.23) with (17.22) we get the very useful relation that

$$|\Psi_{out}(t)\rangle = \Omega_-^\dagger \Omega_+ |\Psi_{in}(t)\rangle \ . \qquad 17.31$$

This relates the asymptotic states at $t = -\infty$ to those at $t = +\infty$. The combination $\Omega_-^\dagger \Omega_+$ plays an extremely important role in scattering theory and defines the S operator

$$S = \Omega_-^\dagger \Omega_+ \ . \qquad 17.32$$

Evaluated on the basis set of free states this operator is called the S matrix.

If instead of combining (17.23) with (17.22) we had combined (17.23) with (17.21) we would have obtained instead of (17.31) the relation

$$|\Psi_{in}(t)\rangle = \Omega_+^\dagger \Omega_- |\Psi_{out}(t)\rangle \qquad 17.33$$

or

$$|\Psi_{in}(t)\rangle = S^\dagger |\Psi_{out}(t)\rangle \ . \qquad 17.34$$

Since the sets $\{|\Psi_{in}(t)\rangle\}$ and $\{|\Psi_{out}(t)\rangle\}$ are complete we can conclude that S is unitary

$$SS^\dagger = S^\dagger S = 1 \ . \qquad 17.35$$

Using the relation (17.26) and (17.30) as well as the definition of the S operator (equation (17.32)) we find

$$H\Omega_\pm = \Omega_\pm H_o \quad \text{and} \quad H_o \Omega_\pm^\dagger = \Omega_\pm^\dagger H_o \ .$$

Multiplying the first of these by $\Omega_-^\dagger$ on the left and the second by Ω_+ on the right, we get

$$\Omega_-^\dagger H \Omega_+ = \Omega_-^\dagger \Omega_+ H_o = SH_o \qquad 17.36$$

and

$$H_o \Omega_-^\dagger \Omega_+ = H_o S = \Omega_-^\dagger H \Omega_+ \ . \qquad 17.37$$

Thus we conclude that

$$S H_o = H_o S \ . \qquad 17.38$$

Clearly if we take matrix elements of S with eigenstates of H_o (basis of free states) the resultant S matrix elements are independent of time.

17.4 Green's Functions and Propagators

Our hamiltonian is as before

$$H = H_o + V(r) \qquad 17.39$$

with

$$H_o = \vec{p}^{\,2}/2m \ . \qquad 17.40$$

We now define four Green's functions or propagators via the equations

$$\left(i\hbar \frac{\partial}{\partial t} - H_o\right) G_o^{\pm}(t) = \delta(t)\, 1 \qquad 17.41$$

$$\left(i\hbar \frac{\partial}{\partial t} - H\right) G^{\pm}(t) = \delta(t)\, 1 \qquad 17.42$$

and the boundary conditions

$$G_o^+(t) = G^+(t) = 0 \qquad \text{for} \quad t < 0 \qquad 17.43$$

$$G_o^-(t) = G^-(t) = 0 \qquad \text{for} \quad t > 0 \qquad 17.44$$

In the usual terminology of differential equations G_o^+, G^+ are retarded Green's functions, while G_o^-, G^- are advanced Green's functions. They are closely related to the Moller wave operators.

Formal solutions of (17.36), (17.37) incorporating the boundary conditions (17.38) or (17.39) are:

$$G_o^+(t) = -\frac{i}{\hbar} e^{-iH_o t/\hbar} \theta(t) \qquad 17.40$$

$$G_o^-(t) = \frac{i}{\hbar} e^{-iH_o t/\hbar} \theta(-t) \qquad 17.41$$

and

$$G^+(t) = -\frac{i}{\hbar} e^{-iHt/\hbar} \theta(t) \qquad 17.42$$

$$G^-(t) = \frac{i}{\hbar} e^{-iHt/\hbar} \theta(-t) . \qquad 17.43$$

The function $\theta(t)$ is defined by

$$\theta(t) = \begin{matrix} 1 & t > 0 \\ 0 & t < 0 \end{matrix} \qquad 17.44$$

such that

$$\lim_{t \to 0+} \theta(t) = 1$$

$$\lim_{t \to 0-} \theta(t) = 0 . \qquad 17.45$$

These solutions are to be understood in the sense of distributions (see Chapter 9) if matrix elements of these solutions are involved. Clearly they are very closely related to evolution operators and that, in fact, is the reason they are also called propagators. Thus if $|\Psi_o(t)\rangle$ is a solution of the Schrödinger equation

$$(i\hbar \frac{\partial}{\partial t} - H_o)|\Psi_o(t)\rangle = 0$$

then for $t > t'$ we can write

$$|\Psi_o(t)\rangle = i\hbar G_o^+(t-t')|\Psi_o(t')\rangle \tag{17.46}$$

Thus G_o^+ propagates states, evolving according to H_o, from one time to a later time.

Similarly if $|\Psi(t)\rangle$ is a solution of

$$(i\hbar \frac{\partial}{\partial t} - H)|\Psi(t)\rangle = 0$$

then for $t > t'$

$$|\Psi(t)\rangle = i\hbar G^+(t-t')|\Psi(t')\rangle \tag{17.47}$$

so that G^+ propagates states, evolving according to H, from one time to a later time.

In a completely analogous fashion we also obtain the relations

$$|\Psi_o(t)\rangle = -i\hbar G_o^-(t-t')|\Psi_o(t')\rangle \quad \text{for} \quad t < t' \tag{17.48}$$

and

$$|\Psi(t)\rangle = -i\hbar G^-(t-t')|\Psi(t')\rangle \quad \text{for} \quad t < t' . \tag{17.49}$$

We now show that the Moller operators are appropriate limits of these propagators, corresponding to propagation from a time $-\infty$ to a time t or from t to a time $+\infty$.

Consider a state $|\Psi(t')\rangle$, which is a solution of the full Schrödinger equation, and define the state

$$|\Psi_o(t)\rangle = i\hbar G_o^+(t-t')|\Psi(t')\rangle . \tag{17.50}$$

Then for $t > t'$ $|\Psi_o(t)\rangle$ is a solution of the Schrödinger equation with H_o . Furthermore at $t = t'$ we have that

$$|\Psi_o(t')\rangle = |\Psi(t')\rangle$$

$$\text{since } \lim_{t\to 0+} G_o^+(t) = -i/\hbar \qquad 17.51$$

according to the definition (17.45) of the θ function. Now consider taking the limit $t' \to -\infty$. We shall show that this limit yields $|\Psi_{in}(t)\rangle$. To see this we use the formal solutions (17.40) and (17.43) for the Green's function and their properties as propagators given in (17.46) and (17.49).

We have

$$|\Psi(t')\rangle = -i\hbar G^-(t')|\Psi(0)\rangle \qquad t' < 0$$

and from (17.49), and from (17.50)

$$|\Psi_o(t)\rangle = i\hbar G_o^+(t-t')|\Psi(t')\rangle \quad t > t' .$$

Substituting from (17.40) and (17.43) we obtain

$$|\Psi_o(t)\rangle = e^{-iH_o(t-t')/\hbar}\theta(t-t')e^{-iHt'/\hbar}\theta(-t')|\Psi(0)\rangle$$

or since

$$|\Psi_o(t)\rangle = e^{-iH_o t/\hbar}|\Psi_o(0)\rangle$$

we obtain

$$|\Psi_o(0)\rangle = \theta(t-t')\theta(-t')e^{iH_o t'/\hbar}e^{-iHt'/\hbar}|\Psi(0)\rangle .$$

If we now let $t' \to -\infty$ then the two θ functions reduce to 1 and we get

$$|\Psi_o(0)\rangle = \Omega_+^\dagger |\Psi(0)\rangle \ .$$

Thus we have $|\Psi_o(t)\rangle = |\Psi_{in}(t)\rangle$ as required.

In a similar manner one can define a free state $|\Psi_o'(t)\rangle$ by

$$|\Psi_o'(t)\rangle = \lim_{t'\to+\infty} -i\hbar G_o^-(t-t')|\Psi(t')\rangle$$

and show that $|\Psi_o'(t)\rangle$ is in fact $|\Psi_{out}(t)\rangle$. Both results will be used a little later.

17.5 Integral Equations for the Propagators: Asympotic States

We have the equations for the propagators

$$(i\hbar \frac{\partial}{\partial t} - H_o)G_o^\pm(t) = \delta(t)\ 1 \qquad 17.36$$

$$(i\hbar \frac{\partial}{\partial t} - H)G^\pm(t) = \delta(t)\ 1 \qquad 17.37$$

with

$$G_o^+(t) = G^+(t) = 0 \qquad \text{for} \quad t < 0 \qquad 17.38$$

and

$$G_o^-(t) = G^-(t) = 0 \qquad \text{for} \quad t > 0 \qquad 17.39$$

Also $H = H_o + V$.

It is possible to use $G_o^\pm$ to derive integral equations for $G^\pm$. In fact simply using (17.36), (17.37) and the initial conditions (17.38), (17.39) we obtain immediately

$$(i\hbar \frac{\partial}{\partial t} - H_o)G^\pm(t) = \delta(t)\ 1 + VG^\pm(t) \ . \qquad 17.52$$

Hence

$$G^{\pm}(t) = G_o^{\pm}(t) + \int_{-\infty}^{\infty} G_o^{\pm}(t-t')VG^{\pm}(t')dt' \ . \qquad 17.53$$

These equations are integral equations for $G^{\pm}$ in terms of $G_o^{\pm}$. Their Fourier transforms in t play an important role in the time-independent formulation of scattering theory and are known as the Lippmann-Schwinger equations.

It is also possible to derive integral equations for the state $|\Psi(t)\rangle$ by using the results for $|\Psi_{\substack{in \\ out}}(t)\rangle$ and the Green's functions $G_o^{\pm}(t)$. Thus we have

$$(i\hbar \frac{\partial}{\partial t} - H_o)|\Psi\rangle = V|\Psi\rangle \ .$$

Hence

$$|\Psi^{\pm}(t)\rangle = |\Psi_{\substack{in \\ out}}(t)\rangle + \int_{-\infty}^{\infty} G_o^{\pm}(t-t')V|\Psi^{\pm}(t')\rangle dt' \qquad 17.54$$

as is easily checked by differentiation. Also the boundary conditions

$$\lim_{t\to\mp\infty} |\Psi^{\pm}(t)\rangle = |\Psi_{\substack{in \\ out}}(t)\rangle$$

are already included. These equations are sometimes referred to as the Källén-Yang-Feldman equations.

The meaning of these equations is as follows. We prepare a state for a scattering experiment. This would normally be a collimated beam moving towards the scattering target. This prepared state is described by $|\Psi_{in}(t)\rangle$ and contains all the information about the incident beam and target at $t = -\infty$. This state then evolves under the influence of H into the state $|\Psi^{+}(t)\rangle$ which is labelled in exactly the same manner as $|\Psi_{in}(t)\rangle$. In other words, the same set of quantum numbers ($\vec{p}$, s^2, s_z, etc.) may be used to label both. Thus $|\Psi^{+}(t)\rangle$ is a state that arises from a given prepared state $\Psi_{in}(t)$ in the remote past.

As $|\Psi^+(t)\rangle$ continues to evolve, it again becomes a free state $|\Psi_{out}(t)\rangle$ which contains not only the original beam $|\Psi_{in}(t)\rangle$ but also some outgoing scattered wave.

One can in principle also consider a state specified or selected in the remote future, namely $|\Psi_{out}(t)\rangle$. This state would then arise from the state $|\Psi^-(t)\rangle$. In practice this is of course impossible. However, both $|\Psi_{out}(t)\rangle$ and $|\Psi^-(t)\rangle$ would also be labelled by the same set of quantum numbers.

17.6 Cross-Sections

In a scattering experiment the quantity actually measured is the number of particles at a given energy scattered by the target into the element of solid angle between Ω and $\Omega + d\Omega$. This number is proportional to what is known as the differential cross-section $\sigma(\theta,\phi)$.

Alternatively, suppose the incident beam has a current density $\vec{j}_{inc}$ and a fraction of these incident particles are scattered by a potential producing a scattered beam of current density $\vec{j}_{scat}$, then dN the number of scattered particles per unit time that pass through an element of surface area $d\vec{s}$ is $\vec{j}_{scat} \cdot d\vec{s}$. This number is proportional to $|\vec{j}_{inc}|$ and $d\Omega$. The constant of proportionality is defined as the differential cross-section $\sigma(\theta,\phi)$. Thus we have

$$dN = |\vec{j}_{inc}|\sigma(\theta,\phi)d\Omega = \vec{j}_{scat} \cdot d\vec{s} \tag{17.55}$$

where

$$d\Omega = \frac{\vec{n} \cdot d\vec{s}}{r^2} \tag{17.56}$$

and $\vec{n}$ is a unit vector in the direction of the scattered beam. From this discussion it is clear that the differential cross-section has the dimensions of an area.

One can think of $\sigma d\Omega$ as the effective cross-sectional area of the target particles so that incident particles are scattered by them into the element of solid angle $d\Omega$. Thus if the incident current consists of J particles per unit area per unit time and if the target irradiated by the incident beam contains N_o particles (scattering centers) then the number of particles dN scattered into the element of solid angle $d\Omega$ per unit time is given by

$$dN = JN_o\sigma(\theta,\phi)d\Omega. \qquad 17.57$$

In practice the observed quantity is dN. The quantities J and N_o are known from other considerations or measurements and $d\Omega$ is given by the location and effective area of the detector. In this manner one gets a measurement of $\sigma(\theta,\phi)$. The total cross-section σ_t is then defined by

$$\sigma_t = \int \sigma(\theta,\phi)d\Omega \qquad 17.58$$

where the integral extends over the full solid angle 4π.

17.7 The Lippmann-Schwinger Equations

In preparation for the time-independent formulation we now start by Fourier tranforming the Green's functions $G_o^{\pm}(t)$ and $G^{\pm}(t)$. We also call their Fourier transforms $G_o^{\pm}(E)$, $G^{\pm}(E)$ so that the functions are partly defined by their argument. This should not be the source of any confusion.

Thus,

$$G_o^{\pm}(E) = \int_{-\infty}^{\infty} e^{iEt/\hbar}G_o^{\pm}(t)dt \qquad 17.59$$

and

$$G^{\pm}(E) = \int_{-\infty}^{\infty} e^{iEt/\hbar}G^{\pm}(t)dt. \qquad 17.60$$

These integrals as they stand are not well defined due to convergence difficulties for large $|t|$. Since $G_o^+(t)$, $G^+(t)$ both vanish for $t < 0$ we can ensure convergence by giving E a small positive imaginary part $i\varepsilon$. Similarly $G_o^-(t)$, $G^-(t)$ vanish for $t > 0$ and their Fourier transforms are obtained by giving E a small negative imaginary part $-i\varepsilon$.

Inserting (17.59) and (17.60) into the differential equations (17.36), (17.37) we get

$$G_o^{\pm}(E) = (E \pm i\varepsilon - H_o)^{-1} \tag{17.61}$$

$$G^{\pm}(E) = (E \pm i\varepsilon - H)^{-1} \tag{17.62}$$

Thus the $\pm i\varepsilon$'s can be seen to indicate how one is to integrate past the poles in $G_o^{\pm}(E)$, $G^{\pm}(E)$ when transforming back to $G_o^{\pm}(t)$, $G^{\pm}(t)$. They reflect the boundary conditions originally imposed on the time-dependent Green's functions.

One can also Fourier transform the state vectors to get

$$|\Psi(E)\rangle = \int_{-\infty}^{\infty} e^{i\frac{E}{\hbar}t} |\Psi(t)\rangle dt \tag{17.63}$$

as well as

$$|\Psi_{in}(E)\rangle = \int_{-\infty}^{\infty} e^{i\frac{E}{\hbar}t} |\Psi_{in}(t)\rangle dt \tag{17.64}$$

$$|\Psi_{out}(E)\rangle = \int_{-\infty}^{\infty} e^{i\frac{E}{\hbar}t} |\Psi_{out}(t)\rangle dt \tag{17.65}$$

If we now Fourier transform equation (17.54) and use (17.61) we get

$$|\Psi^{(\pm)}(E)\rangle = |\Psi_{\substack{in \\ out}}(E)\rangle + (E \pm i\varepsilon - H_o)^{-1} V |\Psi^{(\pm)}\rangle \tag{17.66}$$

where $|\Psi^{(+)}(E)\rangle$ is a state labelled with the same quantum numbers as $|\Psi_{in}(E)\rangle$ and $|\Psi^{(-)}(E)\rangle$ is a state labelled with the same quantum numbers as $|\Psi_{out}(E)\rangle$.

Equations 17.66 constitute the Lippmann-Schwinger equations for the wave functions. We can also obtain Lippmann-Schwinger equations for the Green's functions by Fourier transforming equations 17.53

$$G^{(\pm)}(E) = (E \pm i\varepsilon - H_o)^{-1} + (E \pm i\varepsilon - H_o)^{-1} V G^{\pm}(E). \qquad 17.67$$

It has become conventional to write

$$(E \pm i\varepsilon - H_o)^{-1} \quad \text{as} \quad \frac{1}{E \pm i\varepsilon - H_o}$$

and we shall do so freely in the future. From our previous discussion of distributions (chapter 9) we know that for such functions

$$\frac{1}{E \pm i\varepsilon - H_o} = P\,\frac{1}{E \pm H_o} \mp i\pi\delta(E-H_o) \quad . \qquad 17.68$$

This is extended to operators by simply defining

$$\delta(E-H_o)|\Psi^{(o)}(E')\rangle = \delta(E-E')|\Psi^{(o)}(E')\rangle \qquad 17.69$$

where

$$H_o|\Psi^{(o)}(E')\rangle = E'|\Psi^{(o)}(E')\rangle. \qquad 17.70$$

Since the $|\Psi^{(o)}(E')\rangle$ form a complete set, this suffices to define $\delta(E-H_o)$. Similarly $P\,\frac{1}{E - H_o}$ is defined by

$$P\,\frac{1}{E - H_o}\,|\Psi^{(o)}(E')\rangle = P\,\frac{1}{E-E'}\,|\Psi^{(o)}(E')\rangle \ . \qquad 17.71$$

17.8 The S Matrix and the Scattering Amplitude

We have already introduced the S operator which connects incoming and outgoing states via

$$|\Psi_{in}\rangle = S^{\dagger}|\Psi_{out}\rangle \tag{17.72}$$

$$|\Psi_{out}\rangle = S|\Psi_{in}\rangle \ . \tag{17.73}$$

Evaluated between free states of definite energy, such as plane wave states, it is called the S matrix and its elements are of considerable interest since they contain all the information for a scattering process. To see this consider a typical scattering experiment. We send a particle down a beam tube. This particle is described by a free state $|\Psi_{in}(t)\rangle = |\Psi_o(\alpha,t)\rangle$ where α is a complete set of labels that are eigenvalues of operators that commute with the free hamiltonian H_o. This state evolves in time into a state $|\Psi^{(+)}(\alpha,t)\rangle$ labelled in exactly the same way. The operators, whose eigenvalues label $|\Psi_o(\alpha,t)\rangle$ and $|\Psi^{(+)}(\alpha,t)\rangle$ do not, however, commute with the full hamiltonian H. They simply state how $|\Psi^{(+)}\rangle$ was prepared in the remote past; they are simply the labels attached to the original particle we sent down the beam tube.

In the remote future this state again evolves into a free state $|\Psi_{out}\rangle$. This state does not carry the same labels as $|\Psi_{in}\rangle$ because it contains an admixture of states due to scattering.

The purpose of any scattering experiment is to measure the probability of finding a given free state $|\Psi_o(\beta,t)\rangle$ in the state that will have evolved in the distant future from the state $|\Psi^{(+)}(\alpha,t)\rangle$. Clearly this is given by

$$\lim_{t\to\infty} \langle\Psi_o(\beta,t)|\Psi^{(+)}(\alpha,t)\rangle = \langle\Psi_o(\beta,t)|\Psi_{out}(t)\rangle$$

and using (17.73) this becomes

$$\langle\Psi_o(\beta,t)|S|\Psi_{in}(t)\rangle = \langle\Psi_o(\beta,t)|S|\Psi_o(\alpha,t)\rangle. \qquad 17.74$$

Thus, as stated, the elements of the S-matrix contain all the information obtained in a scattering experiment.

In order to obtain a useful formula for these matrix elements we view the scattering experiment in a somewhat different fashion.

The state $|\Psi_{out}(t)\rangle$ must arise from the state $|\Psi^{(-)}(\beta,t)\rangle$. Again the labels β are the eigenvalues of a complete set of observables that commute with H_o. It is, of course impossible to prepare such a state $|\Psi_{out}(\beta,t)\rangle$; nevertheless we can now think of a scattering experiment as a means to measure the probability of finding in the state $|\Psi^{(-)}(\beta,t)\rangle$ the particle described by $|\Psi^{(+)}(\alpha,t)\rangle$ and prepared in the remote past as $|\Psi_o(\alpha,t)\rangle$. The corresponding probability amplitude must coincide with the one obtained above. Hence we find

$$\langle\Psi_o(\beta,t)|S|\Psi_o(\alpha,t)\rangle = \langle\Psi^{(-)}(\beta,t)|\Psi^{(+)}(\alpha,t)\rangle. \qquad 17.75$$

This result will be used immediately in conjunction with the Lippmann-Schwinger equations to obtain a neat formula for the S-matrix elements. For this purpose we consider eigenstates of the hamiltonian H_o. Thus our states are labelled $|\Psi_o(E)\rangle$ where temporarily we suppress the additional α labels.

As a first step we write a formal solution of the Lippmann-Schwinger equations

$$(E-H)|\Psi^{(\pm)}\rangle = (E-H_o)|\Psi^{(0)}\rangle = (E-H+V)|\Psi^{(0)}\rangle = (E-H)|\Psi^{(0)}\rangle$$
$$+ V|\Psi^{(0)}\rangle \quad .$$

Thus,

$$|\Psi^{(\pm)}\rangle = |\Psi^{(0)}\rangle + (E-H\pm i\varepsilon)^{-1}V|\Psi^{(0)}\rangle \qquad 17.76$$

where $|\Psi^{(o)}(E)\rangle$ is a free state of energy E corresponding to an incoming or outgoing wave depending on the appropriate initial condition specified by $\pm i\varepsilon$. The states $|\Psi^{(\pm)}\rangle$ are labelled by the same set of quantum numbers as $|\Psi^{(0)}\rangle$. The labels $\pm$ refer to the $\pm i\varepsilon$ and simply mean that for $|\Psi^{(+)}\rangle$, the state $|\Psi^{(0)}\rangle$ is an incoming wave whereas for $|\Psi^{(-)}\rangle$ it is an outgoing wave. The function $|\Psi^{(0)}\rangle$ is the same function in either case; it is just a solution of the free Schrödinger equation. Note that equation 17.76 is purely a formal solution since we must still evaluate the operator $(E+H\pm i\varepsilon)^{-1}$ and from our definition of such expressions this require a knowledge of the eigenfunctions of H. Thus we have not really succeeded in solving the Lippmann- Schwinger equations, we have simply rewritten them.

As we saw (equation (17.75)) the S matrix elements S_{fi} are specified by

$$\begin{aligned} S_{fi} &= \langle\Psi^{(-)}(E_f)|\Psi^{(+)}(E_i)\rangle = \langle\Psi^{(+)}(E_f)|\Psi^{(+)}(E_i)\rangle + \\ &+ (\langle\Psi^{(-)}(E_f)|-\langle\Psi^{(+)}(E_f)|)|\Psi^{(+)}(E_i)\rangle \ . \end{aligned} \qquad 17.77$$

where we have chosen energy eigenstates for $|\Psi^{(0)}\rangle$ and are temporarily suppressing all other labels. But from our formal solutions (17.76) we have

$$|\Psi^{(-)}(E_f)\rangle-|\Psi^{(+)}(E_f)\rangle = \left[\frac{1}{E_f-H-i\varepsilon}V-\frac{1}{E_f-H+i\varepsilon}V\right]|\Psi^{(0)}(E_f)\rangle. \qquad 17.78$$

Also the kets $|\Psi^{(\pm)}(E_f)\rangle$ are assumed orthonormal so that

$$\langle\Psi^{(\pm)}(E_f)|\Psi^{(\pm)}(E_i)\rangle = \delta_{fi} \qquad 17.79$$

where

$$\delta_{fi} = \delta(E_f - E_i)\delta_{\alpha\beta} \ . \qquad 17.80$$

Here α,β specify all the other quantum numbers, besides energy, required to label $|\Psi^{(+)}(E)\rangle$.

We therefore obtain the relation

$$\begin{aligned} S_{fi} &= \delta_{fi} + \langle\Psi^{(o)}(E_f)|V \frac{1}{E_f - H + i\varepsilon} - V \frac{1}{E_f - H - i\varepsilon} |\Psi^{(+)}(E_i)\rangle \\ &= \delta_{fi} + \left[P \frac{1}{E_f - E_i} - i\pi\delta(E_f - E_i) - P \frac{1}{E_f - E_i} - i\pi\delta(E_F - E_i)\right] \\ &\quad \times \langle\Psi^{(o)}(E_f)|V|\Psi^{(+)}(E_i)\rangle \end{aligned} \qquad 17.81$$

where we have used the relations (17.68), (17.69), and (17.71). Thus finally

$$S_{fi} = \delta_{fi} - 2\pi i\delta(E_f - E_i)\langle\Psi^{(o)}(E_f)|V|\Psi^{(+)}(E_i)\rangle \qquad 17.82$$

or somewhat more concisely

$$S_{fi} = \delta_{fi} - 2\pi i\delta(E_f - E_i)T_{fi} \qquad 17.83$$

where

$$T_{fi} = \langle\Psi^{(o)}(E_f)|V|\Psi^{(+)}(E_i)\rangle \qquad E_f = E_i \qquad 17.84$$

is the T matrix element on the energy shell. Multiplied by $\frac{-1}{4\pi}$ it's called the scattering amplitude f . Aside from the conservation of energy and other quantum numbers specified by the two delta functions, T_{fi} contains all the information about a scattering process.

If originally we had replaced $|\Psi^{(+)}(E_i)\rangle$ by

$$|\Psi^{(-)}(E_i)\rangle + (|\Psi^{(+)}(E_i)\rangle - |\Psi^{(-)}(E_i)\rangle)$$

then by exactly similar steps we would have arrived at the relationship

$$S_{fi} = \delta_{fi} - 2\pi i\delta(E_f - E_i)\langle\Psi^{(-)}(E_f)|V|\Psi^{(o)}(E_i)\rangle \qquad 17.85$$

so that we also have

$$T_{fi} = \langle\Psi^{(-)}(E_f)|V|\Psi^{(o)}(E_i)\rangle \quad E_f = E_i . \qquad 17.86$$

In the next chapter we shall develop a systematic formulation of scattering theory starting from the time-independent Schrödinger equation . During this process we shall rederive some of the results obtained here. In this way we hope, not only to emphasize the more important results, but also to elucidate the concepts involved. In practice, computations commence almost always with the time-independent formalism.

Notes, References and Bibliography

There are serveral books devoted exclusively to scattering theory. Two very detailed books are: M.L. Goldberger and K.M. Watson - Collision Theory - John Wiley and Sons, Inc., New York (1964), N.F. Mott and H.S.W. Massey - The Theory of Atomic Collisions - Oxford University Press, London 2nd edition (1948).

A book treating also classical scattering theory as well as quantum scattering theory is:

R.G. Newton - Scattering Theory of Waves and Particles - McGraw-Hill Book Co., New York (1966).

This book contains many interesting worked examples as well as a of useful information.

Chapter 17 Problems

17.1 A proton beam producing a current of 5×10^{-9} amps is incident on a target of Cu. Assume the target thickness is such that the areal density is 0.2 mgm/cm^2. The detector has an area of 0.5 cm^2 and is 20 cm from the target. If 10 protons are counted by the detector every second at a particular angle, calculate the cross-section for protons scattering off Cu at that angle.

17.2 Use the expressions (17.40) and (17.41) and evaluate the matrix elements

$$\langle\vec{p}|G_o^{\pm}(t)|\vec{k}\rangle$$

where $|\vec{p}\rangle$, $|\vec{k}\rangle$ are free particle states of momentum $\vec{p}$ and $\vec{k}$ respectively.

17.3 Calculate the Fourier transform of the distribution $\theta(t)$. Hint:

$$\theta(\omega) = \lim_{\varepsilon\to 0+} \int_0^{\infty} e^{-i(\omega-i\varepsilon)t} dt .$$

The limit here is to be understood in the sense of distributions.

17.4 Assume that V is independent of time and use (17.53) to obtain an equation for the Fourier transform $G^{\pm}(\omega)$ of $G^{\pm}(t)$ in terms of the Fourier transforms of $G_o^{\pm}(t)$. Write a formal solution for $G^{\pm}(\omega)$.

17.5 In equation (17.84) approximate $|\Psi^{+}(E_i)\rangle$ by a free particle state. If V is a screened Coulomb potential

$$V = -V_o \frac{e^{-\mu r}}{r}$$

calculate the scattering amplitude. This is known as the first Born approximation.

Chapter 18

Scattering Theory: Time Independent Formulation

In the previous chapter we gave a time-dependent formulation of scattering. Eventually by Fourier transforming this led to a time-independent set of equations known as the Lippmann-Schwinger equations. We now start from the time-independent Schrödinger equation and rederive these results along the way.

This chapter is independent of chapter 17 and may be studied before chapter 17. The results obtained here provide an efficient means for computing differential cross- sections. To recall the definition of cross-section it is worthwhile to read or re-read section 17.6.

18.1 The Scattering Amplitude

We start with the time-independent Schrödinger equation for a one-body problem where

$$H = \frac{p^2}{2m} + V(\vec{r}) \qquad 18.1$$

and m may be the reduced mass of a two-particle system in which case $V(\vec{r})$ is the potential between the two particles. It is usual in potential scattering to introduce

$$k^2 = \frac{2mE}{\hbar^2}\,, \quad U(\vec{r}) = \frac{2mV(\vec{r})}{\hbar^2}\,. \qquad 18.2$$

In that case the time-independent Schrödinger equation corresponding to the hamiltonian H becomes:

$$(\nabla^2+k^2)\psi(\vec{r}) = U(\vec{r})\psi(\vec{r})\,. \qquad 18.3$$

In order to have a mathematically simple scattering problem it is desirable that $U(\vec{r}) \to 0$ sufficiently rapidly as $|\vec{r}| \to \infty$. We shall assume this to be the case*. In that case we can look for solutions for (18.3) that consist of a given incoming wave say $e^{i\vec{k}\cdot\vec{r}}$ plus a scattered wave $\psi_{scat}(\vec{r})$ so that

$$\psi(\vec{r}) = e^{i\vec{k}\cdot\vec{r}} + \psi_{scat}(\vec{r})\,. \qquad 18.4$$

The current due to the incident beam is

$$\vec{j}_{inc} = \frac{\hbar\vec{k}}{m} \qquad 18.5$$

and corresponds to a flux of particles moving in the $\vec{k}$ direction with uniform momentum $\hbar\vec{k}$. If we consider a very large sphere of radius R with center at the scattering center (potential) namely $\vec{r} = 0$, then the flux through the surface of that sphere due to the scattered wave is

$$4\pi R^2\vec{j}_{scat}\cdot\hat{r}$$

---

* $r^2V(r) \xrightarrow[r\to\infty]{} 0$ is sufficient.

where $\vec{j}_{scat}\cdot\hat{r}$ is the radial component of the current due to the scattered beam. This is given by

$$\vec{j}_{scat}\cdot\hat{r} = \frac{\hbar}{2im}\left(\psi^*_{scat}\frac{\partial\psi_{scat}}{\partial r} - \frac{\partial\psi^*_{scat}}{\partial r}\psi_{scat}\right). \qquad 18.6$$

If the scattered flux through the surface of the sphere is to tend to a constant value independent of R as $R \to \infty$, we must have

$$\psi^*_{scat}\frac{\partial\psi_{scat}}{\partial r} \underset{R\to\infty}{\longrightarrow} \text{const}/R^2.$$

This means that

$$\lim_{R\to\infty}\psi_{scat}(R) = -\frac{1}{4\pi}T(\vec{k},\vec{k}')\frac{e^{ikR}}{R} = f(k,\theta)\frac{e^{ikR}}{R} \qquad 18.7$$

where $\vec{k}$ is the incident momentum, $\vec{k}'$ is the scattered or outgoing momentum. The quantity $f(k,\theta)$ is called the scattering amplitude.

Thus the flux of scattered particles through an element of solid angle $d\Omega$ is given by

$$R^2 d\Omega.\ \frac{\hbar k}{m}|f(k,\theta)|^2\frac{1}{R^2} = \frac{\hbar k}{m}|f(k,\theta)|^2 d\Omega.$$

Hence using (17.55) we see that the differential cross-section is given by

$$\frac{d\sigma(\theta,\phi)}{d\Omega} = |f(k,\theta)|^2 \qquad 18.8$$

where θ,ϕ are the angles between $\vec{k}$ and $\vec{k}'$. Since the experimentally accessible quantity we are interested in is $\frac{d\sigma(\theta,\phi)}{d\Omega}$ our task is reduced to computing the scattering amplitude $f(k,\theta)$.

18.2 Green's Functions and the Lippmann-Schwinger Equations

In order to find a solution of the form (18.4) we convert the Schrödinger equation (18.3) to an integral equation. To do this we must determine the Green's function satisfying the relation

$$(\nabla^2 + k^2)G(\vec{r},\vec{r}\,') = \delta(\vec{r}-\vec{r}\,') \,. \tag{18.9}$$

This is analogous to solving for an operator $G(E)$ using

$$(-H_o+E)G(E) = 1. \tag{18.10}$$

Now for E complex, say z, the operator-valued function

$$G(z) = (z - H_o)^{-1} \tag{18.11}$$

is well defined* since the eigenfunctions $\Psi^{(o)}(E)$ of H_o form a complete set

$$H_o\Psi^{(o)}(E) = E\Psi^{(o)}(E). \tag{18.12}$$

Thus,

$$G(z)\Psi^{(o)}(E) = \frac{1}{z-E}\,\Psi^{(o)}(E) \,. \tag{18.13}$$

When z approaches real values this operator develops poles in z at the eigenvalues and must be defined in the sense of distributions. We examine this point more explicitly starting from (18.9). If we Fourier transform (18.9) using

$$G(\vec{r},\vec{r}\,') = \frac{1}{(2\pi)^3}\int e^{i\vec{q}\cdot(\vec{r}-\vec{r}\,')}\tilde{G}(q)d^3q \tag{18.14}$$

---

* It is called the resolvent of H_o. See also problem (6.6).

and use the integral representation of the δ function

$$\delta(\vec{r}-\vec{r}\,') = \frac{1}{(2\pi)^3}\int e^{i\vec{q}\cdot(\vec{r}-\vec{r}\,')}\, d^3q$$

we obtain the result

$$\tilde{G}(\vec{q}) = \frac{1}{k^2 - q^2} .$$

Consequently

$$G(\vec{r},\vec{r}\,') = -\frac{1}{(2\pi)^3}\int \frac{e^{i\vec{q}\cdot(\vec{r}-\vec{r}\,')}}{q^2 - k^2}\, d^3q . \qquad 18.15$$

Since the integrand has poles on the path of integration, we must specify how this integral is to be defined. Thus, it can be defined as a principal value integral by omitting the poles or differently by including a portion of the residue due to each pole. Each of these integrations gives rise to a different Green's functions. We are particularly interested in two of these Green's functions, namely

$$G^{\pm}(\vec{r},\vec{r}\,') = \lim_{\varepsilon\to 0+} \frac{-1}{(2\pi)^3}\int \frac{e^{i\vec{q}\cdot(\vec{r}-\vec{r}\,')}}{q^2-(k^2\pm i\varepsilon)}\, d^3q . \qquad 18.16$$

We can immediately integrate over the azimuthal angle if we choose the q_z axis aligned with $\vec{r}-\vec{r}\,'$ so that $\vec{q}\cdot(\vec{r}-\vec{r}\,') = q|\vec{r}-\vec{r}\,'|\cos\theta$. This choice for the z direction of the integration variable is a trick that can be frequently used to evalute three-dimensional Fourier integrals. The integral in (18.16) now reduces to

$$G^{\pm}(\vec{r},\vec{r}\,') = \lim_{\varepsilon\to 0+} \frac{-2\pi}{(2\pi)^3}\int_0^{\infty} \frac{q^2\,dq}{q^2-k^2\mp i\varepsilon}\int_{-1}^{1} e^{iq|\vec{r}-\vec{r}\,'|u}\, du$$

where we have integrated over ϕ and made the substitution $u = \cos\theta$. The integration over u is now easily performed to yield:

$$G^{\pm}(\vec{r},\vec{r}\,') = \lim_{\varepsilon\to 0+} \frac{1}{(2\pi)^2} \frac{1}{i|\vec{r}-\vec{r}\,'|} \int_0^{\infty} \frac{q\,dq}{q^2-k^2\mp i\varepsilon} \left(e^{iq|\vec{r}-\vec{r}\,'|} - e^{-iq|\vec{r}-\vec{r}\,'|}\right)$$

$$= \lim_{\varepsilon\to 0+} \frac{-i}{(2\pi)^2} \frac{1}{|\vec{r}-\vec{r}\,'|} \int_{-\infty}^{\infty} \frac{e^{iq|\vec{r}-\vec{r}\,'|}}{q^2-k^2\mp i\varepsilon} q\,dq$$

To evaluate these integrals we now close the contour with a half circle in the upper half plane so that the exponential is damped. Then the poles are located either at $\pm(k+i\varepsilon)$ or at $\pm(k-i\varepsilon)$. In either case we obtain only the residue from the pole in the upper half plane. Thus we get

$$G^{\pm}(\vec{r},\vec{r}\,') = -\frac{1}{4\pi} \frac{e^{\pm ik|\vec{r}-\vec{r}\,'|}}{|\vec{r}-\vec{r}\,'|} . \qquad 18.17$$

This is simply the configuration space expression for the operator

$$G^{\pm}(E) = \frac{1}{E-H_o \pm i\varepsilon} \qquad 18.18$$

that we considered before. It is now a simple matter to rewrite the time-independent Schrödinger equation as an integral equation. The boundary conditions are incorporated in the $\pm i\varepsilon$.

In fact the equation is given by

$$\Psi^{(\pm)}(\vec{r}) = \Psi^{(o)}(\vec{r}) + \int G^{\pm}(\vec{r},\vec{r}\,')U(\vec{r}\,')\Psi^{(\pm)}(\vec{r}\,')d^3r' \qquad 18.19$$

where $\Psi^{(o)}(\vec{r})$ is a solution of the free Schrödinger equation. The fact that $\Psi^{(\pm)}$ satisfy the correct boundary conditions is clear from the short range nature of $U(\vec{r})$ which allows us to take the limit $|\vec{r}| \to \infty$

under the integral sign. To see that $\Psi^{(\pm)}(\vec{r})$ satisfy equation (18.3) we need only apply the operator ∇^2+k^2 to both sides and use equation (18.9). We then find

$$(\nabla^2+k^2)\Psi^{(\pm)}(\vec{r}) = o + \int \delta(\vec{r}-\vec{r}\,')U(\vec{r}\,')\Psi^{(\pm)}(\vec{r}\,')dr'$$

$$= U(\vec{r})\Psi^{(\pm)}(\vec{r})$$

as required.

Thus, the solutions $\Psi^{(\pm)}$ corresponding to $G^{\pm}$ are indeed

$$\Psi^{(\pm)}(\vec{r}) = \Psi^{(o)}(\vec{r}) + \int G^{\pm}(\vec{r},\vec{r}\,')U(\vec{r}\,')\Psi^{(\pm)}(r')d^3r' \;. \qquad 18.19$$

Going back to the Schrödinger equation in the form

$$(E-H_o)|\Psi(E)\rangle = V|\Psi(E)\rangle$$

we get

$$|\Psi^{\pm}(E)\rangle = |\Psi^{(o)}(E)\rangle + \frac{1}{E-H_o\pm i\varepsilon} V|\Psi^{\pm}(E)\rangle \;. \qquad 18.20$$

Thus (18.19) is simply (18.20) written out in configuration space. In either case the equations are known as the Lippmann-Schwinger equations.

To show the equivalence of (18.19) and (18.20) we simply transcribe (18.20) into the configuration space representation using the relations

$$\Psi^{(\pm)}(\vec{r}) = \langle\vec{r}|\Psi^{\pm}(E)\rangle \qquad 18.21$$

and

$$\Psi^{(o)}(\vec{r}) = \langle\vec{r}|\Psi^{(o)}(E)\rangle = e^{i\vec{p}\cdot\vec{r}} \equiv \langle\vec{r}|\vec{p}\rangle \qquad 18.22$$

where

$$E = \frac{\hbar^2 p^2}{2m} .$$

Then (18.20) becomes

$$\Psi^{(\pm)}(\vec{r}) = \Psi^{(o)}(\vec{r}) + \langle\vec{r}|\,\frac{1}{E-H_o \pm i\varepsilon}\,V|\Psi^{\pm}(E)\rangle$$

$$= \Psi^{(o)}(\vec{r}) + \int d^3k\; d^3q\; d^3r'\; d^3r''\; \langle\vec{r}|\vec{k}\rangle\langle\vec{k}|\frac{1}{E-H_o \pm i\varepsilon}|\vec{q}\rangle \times$$

$$\times\; \langle\vec{q}|\vec{r}\,'\rangle\langle\vec{r}\,'|V|\vec{r}\,''\rangle\langle\vec{r}\,''|\Psi^{\pm}(E)\rangle$$

$$= \Psi^{(o)}(\vec{r}) + \frac{1}{(2\pi)^3}\int \frac{d^3r'd^3k}{E - \frac{\hbar^2 k^2}{2m} \pm i\varepsilon}\, e^{i\vec{k}.(\vec{r}-\vec{r}\,')}V(\vec{r}\,')\Psi^{\pm}(\vec{r}\,') \tag{18.23}$$

where we have used the fact that V is local so that

$$\langle\vec{r}\,'|V|\vec{r}\,''\rangle = V(\vec{r}\,')\delta(\vec{r}\,'-\vec{r}\,'') . \tag{18.24}$$

Now writing $E = \frac{\hbar^2 \vec{p}^{\,2}}{2m}$ the integral over k becomes

$$\frac{2m}{\hbar^2}\,\frac{1}{(2\pi)^3}\int \frac{d^3k}{p^2-k^2+i\varepsilon}\, e^{i\vec{k}\cdot(\vec{r}-\vec{r}\,')} = \frac{2m}{\hbar^2}\, G^{\pm}(\vec{r},\vec{r}\,') .$$

Putting this back in (18.23) and recalling that

$$U(\vec{r}) = \frac{2m}{\hbar^2}\, V(\vec{r})$$

we immediately obtain (18.19).

18.3 The Born Approximation

The Lippmann-Schwinger equations provide an immediate approximation technique when the interaction V is small compared to the energy E. Thus it becomes more exact as the energy increases. To get the Born series* one simply iterates equation 18.20 to get

$$|\Psi^{\pm}(E)\rangle = \sum_{n=0}^{\infty} \left(\frac{1}{E-H_o \pm i\varepsilon} V\right)^n |\Psi^{(o)}(E)\rangle \,. \qquad 18.25$$

The usefulness of this series is that for small V or large E it may be truncated after only a few terms. In fact, in practice one frequently keeps only the first non-trivial term. This is known as the first Born or simply Born approximation. Written out it reads

$$|\Psi^{+}(E)\rangle = |\Psi^{(o)}(E)\rangle + \frac{1}{E-H_o+i\varepsilon} V|\Psi^{(o)}(E)\rangle. \qquad 18.26$$

Going over to configuration space this becomes

$$\Psi^{+}(\vec{r}) = e^{i\vec{k}.\vec{r}} + \frac{2m}{\hbar^2}\int G^{+}(\vec{r}-\vec{r}\,')V(r')\Psi^{(o)}(r')d^3r' \,. \qquad 18.27$$

We could have obtained this directly, of course, by just iterating equation 18.19 once and dropping all other terms.

Substituting for $\phi^{(o)}(r')$ as well as $G^{+}(\vec{r}-\vec{r}\,')$ this becomes

$$\Psi^{(+)}(\vec{r}) = e^{i\vec{k}.\vec{r}} - \frac{2m}{\hbar^2}\,\frac{1}{4\pi}\int \frac{e^{ik|\vec{r}-\vec{r}\,'|}}{|r-r'|} V(\vec{r}\,')e^{i\vec{k}.\vec{r}\,'}d^3r' \,. \qquad 18.28$$

---

* Mathematicians call this the Neumann series - see reference 18.1 chapter 2.

To get the scattering amplitude we require the asymptotic solution for large r. Now

$$|\vec{r}-\vec{r}\,'|^2 = r^2 - 2rr'\cos\alpha + r'^2$$

$$\xrightarrow[r\to\infty]{} r^2\left(1 - \frac{2r'}{r}\cos\alpha\right)$$

Therefore,

$$|\vec{r}-\vec{r}\,'| \xrightarrow[r\to\infty]{} r - r'\cos\alpha \qquad 18.29$$

where $\cos\alpha = \hat{r}\cdot\hat{r}'$.

Similarly,

$$\frac{1}{|\vec{r}-\vec{r}\,'|} \xrightarrow[r\to\infty]{} \frac{1}{r} + \frac{r'\cos\alpha}{r^2} \; .$$

Since we are only interested in the solution which behaves asymptotically as e^{ikr}/r (see the discussion of section 1) we can drop the higher order term in the denominator at this stage. We must, however retain it in the argument of the exponential since the exponential varies more rapidly. Thus, we have

$$\frac{e^{ik|\vec{r}-\vec{r}\,'|}}{|\vec{r}-\vec{r}\,'|} \xrightarrow[r\to\infty]{} \frac{e^{ik(r-r'\cos\alpha)}}{r} \; . \qquad 18.30$$

Since V(r) is assumed short range, the integral in equation (18.28) converges uniformly and we are justified in using the asymptotic form in the integrand. Thus,

$$\Psi^+_{Born}(\vec{r}) \xrightarrow[r\to\infty]{} e^{i\vec{k}\cdot\vec{r}} - \frac{2m}{\hbar^2}\frac{1}{4\pi}\frac{e^{ikr}}{r}\int e^{i\vec{k}\cdot\vec{r}\,'-ikr'\cos\alpha}\, V(r')d^3r' \; . \qquad 18.31$$

If we now call the scattered momentum $\vec{k}'$ such that

$$\vec{k}'.\vec{r}' = k\, r'\cos\alpha \qquad 18.32$$

then the argument of the exponential becomes

$$i(\vec{k} - \vec{k}').\vec{r}' \ .$$

We further define the <u>change in momentum</u> or <u>momentum transfer</u> $\vec{q}$ as

$$\vec{q} = \vec{k} - \vec{k}' \ . \qquad 18.33$$

Then calling θ the scattering angle (angle between $\vec{k}$ and $\vec{k}'$) we have since $|\vec{k}| = |\vec{k}'|$ that

$$q^2 = k^2 - 2k^2\cos\theta + k^2 = 2k^2(1 - \cos\theta) = 4k^2\sin^2\frac{\theta}{2}$$

so that

$$q = 2k\left|\sin\frac{\theta}{2}\right| \ . \qquad 18.34$$

The Born term now yields

$$\Psi^+(\vec{r}) \xrightarrow[r\to\infty]{} e^{i\vec{k}.\vec{r}} - \frac{1}{4\pi}\frac{e^{ikr}}{r}\int e^{i\vec{q}.\vec{r}'}U(r')d^3r' \ . \qquad 18.35$$

Or defining the Fourier transform $\tilde{U}(\vec{q})$ of $U(\vec{r}')$ by

$$\tilde{U}(\vec{q}) = \int e^{i\vec{q}.\vec{r}'}\, U(\vec{r}')d^3r' \qquad 18.36$$

we obtain

$$\Psi^+(\vec{r}) \xrightarrow[r\to\infty]{} e^{i\vec{k}.\vec{r}} - \frac{1}{4\pi}\frac{e^{ikr}}{r}\tilde{U}(\vec{q}) \ . \qquad 18.37$$

Comparing this with (18.7) we see that in the first order Born approximation the scattering amplitude $f(k,\theta)$ is given by

$$f(k,\theta) = -\frac{1}{4\pi}\tilde{U}(\vec{k}-\vec{k}') . \qquad 18.38$$

18.4 Example - The Yukawa Potential

As an example of the application of the Born approximation we consider the Yukawa or screened Coulomb potential

$$V(r) = -V_o \frac{e^{-\alpha r}}{r} . \qquad 18.39$$

In this case

$$\tilde{U}(\vec{q}) = -\frac{2mV_o}{\hbar^2}\int e^{i\vec{q}\cdot\vec{r}}\,\frac{e^{-\alpha r}}{r}\,d^3r .$$

Performing the angular integrations, as before, we now get

$$\tilde{U}(\vec{q}) = -\frac{2mV_o}{\hbar^2}\cdot\frac{2\pi}{iq}\int_0^\infty dr\left[e^{-r(\alpha-iq)}-e^{-r(\alpha+iq)}\right] .$$

Hence

$$\vec{U}(\vec{q}) = -\frac{2mV_o}{\hbar^2}\,\frac{4\pi}{q^2+\alpha^2} . \qquad 18.40$$

Thus,

$$f(k,\theta) = \frac{2mV_o}{\hbar^2}\,\frac{1}{(\vec{k}-\vec{k}')^2+\alpha^2} . \qquad 18.41$$

Substituting this into (18.8) and using (18.34) we obtain the differential cross-section $\frac{d\sigma}{d\Omega}$

$$\frac{1}{2\pi}\frac{d\sigma(\theta)}{d\cos\theta} = \frac{4m^2V_o^2}{\hbar^4}\frac{1}{(4k^2\sin^2\frac{\theta}{2}+\alpha^2)^2} . \qquad 18.42$$

If $V_o = Ze^2$ and $\alpha = 0$ this becomes

$$\frac{1}{2\pi}\frac{d\sigma(\theta)}{d\cos\theta} = \frac{4m^2Z^2e^4}{16\hbar^4k^4\sin^4\frac{\theta}{2}} .$$

Recalling that $\hbar^2k^2/2m = E$ we get

$$\frac{1}{2\pi}\frac{d\sigma(\theta)}{d\cos\theta} = \frac{Z^2e^4}{16E^2\sin^4\frac{\theta}{2}} \qquad 18.43$$

which is just the classical Rutherford cross-section for the scattering of an electron of charge e by a Coulomb potential of charge Ze. Although the formula (18.43) is correct its derivation is not entirely correct since for $\alpha = 0$ the potential (18.39) is no longer short-ranged. This trick of replacing the Coulomb potential $1/r$ by $e^{-\alpha r}/r$ and letting $\alpha \to 0$ at the end permits us to use our standard scattering theory without worrying about the long-range nature of the Coulomb potential. The pure Coulomb potential leads to logarithmically oscillating phase contributions (reference (17.2) section 14.6) in the wave function ψ^+. In fact,

$$\psi^+ \sim e^{i[kz+n\ln k(r-z)]} + f(\theta)\frac{1}{r}e^{i[kr-n\ln 2kr]}$$
$$+ \text{ higher order terms.}$$

where $n = \frac{Ze^2m}{\hbar^2}$. In this case one treats $f(\theta)$ as the scattering amplitude. These complications can be avoided by using the "adiabatic switching off" of the Coulomb potential for long range with the factor $e^{-\alpha r}$ and then letting $\alpha \to 0$.

It is rather fortunate that for the case of Coulomb scattering the quantum mechanical and classical cross-sections agree, for otherwise Rutherford's famous experiments would not have yielded the simple interpretation of a tiny nuclear core inside an atom that they did yield. The results of these experiments were of paramount importance in the development of quantum mechanics since they were the basis of Bohr's model as well as Schrödinger's computation of the spectrum of the hydrogen atom.

The fact that the scattering amplitude and hence the differential cross-section are independent of the azimuthal angle ϕ is not simply a coincidence but is a consequence of the spherical symmetry of the potential. Thus, just as for classical scattering, a spherically symmetric potential gives rise to a differential cross-section independent of the azimuthal angle. We shall begin to exploit this in the next sections. As a first step we shall develop some more mathematical machinery.

18.5 The Free Schrödinger Equation in Spherical Coordinates

In section 10.1 we had occasion to consider the Schrödinger equation in spherical coordinates. The solutions involved spherical Bessel functions. We now reexamine this problem systematically and develop some properties of the Bessel functions.

We start with the free Schrödinger equation

$$(\nabla^2 + k^2)\psi(r) = 0 \quad . \tag{18.44}$$

The solution can be expanded in spherical harmonics

$$\psi(\vec{r}) = \sum_{\ell,m} R_\ell(r) Y^*_{\ell m}(\hat{k}) Y_{\ell m}(\hat{r}) \tag{18.45}$$

where $\hat{k}$, $\hat{r}$ represent the θ,ϕ directions of $\vec{k}$ and $\vec{r}$ respectively.

Substituting (18.45) into (18.44) we are led to the radial equation

$$\left[\frac{d^2}{dr^2} + \frac{2}{r}\frac{d}{dr} - \frac{\ell(\ell+1)}{r^2} + k^2\right] R_\ell(r) = 0 \ . \tag{18.46}$$

As stated in section 10.1, the solutions of this equation are the spherical Bessel functions $j_\ell(kr)$ and $n_\ell(kr)$. In addition to these it is convenient to introduce two additional solutions corresponding to Hankel functions of the first and second kind namely

$$h_\ell^{(1)}(kr) = j_\ell(kr) + i\, n_\ell(kr) \tag{18.47a}$$

$$h_\ell^{(2)}(kr) = j_\ell(kr) - i\, n_\ell(kr) \tag{18.47b}$$

To explore the properties of these functions we derive integral representations for them. We also replace kr by x and $R_\ell(kr)$ by $Z_\ell(x)$ in order to simplify the notation. Thus (18.46) becomes

$$\left[\frac{d^2}{dx^2} + \frac{2}{x}\frac{d}{dx} + 1 - \frac{\ell(\ell+1)}{x^2}\right] Z_\ell(x) = 0 \ . \tag{18.48}$$

To derive the integral representations we look for a solution of the form

$$Z_\ell(x) = x^\lambda \int_a^b e^{xu} f(u)\, du \tag{18.49}$$

where we shall specify the limits a,b later. Substituting this in (18.48) we obtain

$$x^{\lambda-2} \int_a^b f(u)\left[\lambda(\lambda+1) - \ell(\ell+1) + 2(\lambda+1)xu + (u^2+1)x^2\right] e^{xu} du = 0 \ . \tag{18.50}$$

we now choose $\lambda(\lambda+1) = \ell(\ell+1)$ so that

$$\lambda = \ell \quad \text{or} \quad -(\ell+1) \; .$$

We then obtain from (18.50)

$$\int_a^b f(u)\left[2(\lambda+1)u + (u^2+1)\frac{d}{du}\right]e^{xu}du = 0 \; . \tag{18.51}$$

We next rewrite the second term as a total differential by using the identity

$$\frac{d}{du}\left[f(u)(u^2+1)e^{xu}\right] = e^{xu}\,\frac{d}{du}\left[f(u)(u^2+1)\right]+f(u)(u^2+1)\frac{d}{du}\,e^{xu}$$

This yields

$$\int_a^b e^{xu}\left\{2(\lambda+1)uf(u)-\frac{d}{du}\left[f(u)(u^2+1)\right]\right\}du + \int_a^b \frac{d}{du}\left[f(u)(u^2+1)\,e^{xu}\right]du = 0. \tag{18.52}$$

We now make both integrals vanish separately: the first by causing the integrand to vanish and the second by an appropriate choice of the limits of integration. The first condition gives

$$\frac{d}{du}\left[f(u)(u^2+1)\right] = 2(\lambda+1)uf(u) \; . \tag{18.53}$$

The solution of this equation is

$$f(u) = f(0)(u^2+1)^{\lambda} \; . \tag{18.54}$$

To get the second condition we integrate the last term in (18.52) to get

$$f(b)(b^2+1)e^{xb} - f(a)(a^2+1)e^{xa} = 0 .$$

Since x is arbitrary this requires both $f(a)(a^2+1)e^{xa}$ and $f(b)(b^2+1)e^{xb}$ to vanish. After substituting for f from (18.54) this reduces to

$$e^{xa}(a^2+1)^{\lambda+1} = 0 \qquad 18.55a$$

and

$$e^{xb}(b^2+1)^{\lambda+1} = 0 . \qquad 18.55b$$

For $\lambda = \ell \geq 0$ this is easily accomplished by choosing $a = -i$, $b = i$. Another possibility for $0 \leq \arg x \leq \pi/2$ is to take one of the limits as $\infty\, e^{i\alpha}$ where $\pi/2 < \alpha < \pi$. Each of these limits gives rise to a different function $Z_\ell(x)$, corresponding to $j_\ell(x)$, $n_\ell(x)$, $h_\ell^{(1)}(x)$, $h_\ell^{(2)}(x)$. The value of f(x) at the origin f(0) gives the appropriate normalization.

We work out one case in detail and leave the rest as exercises. Thus consider the case

$$Z_\ell(x) = f(0)\, x^\ell \int_{-i}^{i} e^{xu}(u^2+1)^\ell du . \qquad 18.56$$

Since this function is finite at the origin and is a solution of (18.48) it must be proportional to $j_\ell(x)$. To make it equal to $j_\ell(x)$ we choose f(0) appropriately. Now for $x \to 0$ we have

$$j_\ell(x) \xrightarrow[x\to 0]{} \frac{x^\ell}{(2\ell+1)!!} . \qquad 18.57$$

But we also have that

$$Z_\ell(x) \xrightarrow[x\to 0]{} f(0)\, x^\ell \int_{-i}^{i} (u^2+1)^\ell\, du = f(0) x^\ell I_\ell \qquad 18.58$$

where we have made the identification

$$I_\ell = \int_{-i}^{i} (u^2+1)^\ell \, du \; . \tag{18.59}$$

To evaluate this integral put $u = it$ then

$$I_\ell = i \int_{-1}^{1} (1-t^2)^\ell \, dt \; . \tag{18.60}$$

Integrating by parts one can now show (problem 18.1) that

$$I_\ell = \frac{2\ell}{2\ell+1} I_{\ell-1} \qquad \ell \geq 1 \; . \tag{18.61}$$

Since $I_0 = 2i$ we immediately get

$$I_\ell = \frac{2^{\ell+1}\ell!}{(2\ell+1)!!} i \quad . \tag{18.62}$$

Thus we have

$$Z_\ell(x) \xrightarrow[x\to 0]{} i\, f(0) \frac{2^{\ell+1}\ell!}{(2\ell+1)!!} \quad . \tag{18.63}$$

But as stated

$$j_\ell(x) \xrightarrow[x\to 0]{} \frac{x^\ell}{(2\ell+1)!!} \quad . \tag{18.64}$$

Thus if we choose

$$i\, f(0) = \frac{1}{2^{\ell+1}\ell!} \tag{18.65}$$

we get

$$j_\ell(x) = \frac{x^\ell}{i\, 2^{\ell+1} \ell!} \int_{-i}^{i} e^{xu}(u^2+1)^\ell \, du \tag{18.66}$$

Letting $u = it$ again this can be rewritten as

$$j_\ell(x) = \frac{x^\ell}{2^{\ell+1} \ell!} \int_{-1}^{1} (1-t^2)^\ell \cos xt \, dt \, . \tag{18.67}$$

The various properties of the $j_\ell(x)$ can now be easily derived from this integral representation (problem 18.2).

In a similar manner we define

$$h_\ell^{(1)}(x) = \frac{x^\ell}{i\, 2^\ell \ell!} \int_{\infty e^{i\alpha}}^{i} e^{xu}(1+u^2)^\ell \, du \, . \tag{18.68}$$

$$h_\ell^{(2)}(x) = \frac{x^\ell}{i\, 2^\ell \ell!} \int_{-i}^{\infty e^{i\alpha}} e^{xu}(1+u^2)^\ell \, du \, . \tag{18.69}$$

where $\pi/2 < \alpha < \pi$ and $0 \leq \arg x \leq \pi/2$.

We are now in a position to derive the partial wave decomposition of a plane wave, a result that we will need in the immediate future. Thus we write

$$e^{i\vec{k}\cdot\vec{r}} = \sum_{\ell,m} a_\ell(kr) Y^*_{\ell m}(\hat{k}) Y_{\ell m}(\hat{r}) \tag{18.70}$$

where as before $\hat{k},\hat{r}$ indicate the directions of the variables indicated. If we let θ be the angle between $\vec{k}$ and $\vec{r}$ so that $\cos\theta = \hat{k}\cdot\hat{r}$ and then use the addition theorem for the spherical harmonics, namely

$$\sum_{m=-\ell}^{\ell} Y^*_{\ell m}(\hat{k}) Y_{\ell m}(\hat{r}) = \frac{2\ell+1}{4\pi} P_\ell(\cos\theta) \tag{18.71}$$

the expansion (18.70) can be written

$$e^{i\vec{k}\cdot\vec{r}} = \sum_{\ell} \frac{2\ell+1}{4\pi} a_{\ell}(kr)P_{\ell}(\cos\theta). \qquad 18.72$$

Now using the orthogonality relation for the Legendre polynomials and calling $\cos\theta = u$ we get

$$a_{\ell}(kr) = 2\pi \int_{-1}^{1} e^{ikru} P_{\ell}(u)\, du \qquad 18.73$$

But from the Rodrigues formula for the Legendre polynomials (equation 9.103) we have

$$P_{\ell}(u) = \frac{(-1)^{\ell}}{2^{\ell}\ell!} \frac{d^{\ell}}{du^{\ell}} (1-u^2)^{\ell} \,. \qquad 18.74$$

Substituting this into (18.73) and integrating by parts ℓ times we get

$$a_{\ell}(kr) = (4\pi)\, i^{\ell} \frac{(kr)^{\ell}}{2^{\ell+1}\ell!} \int_{-1}^{1} e^{ikru}(1-u^2)^{\ell} du. \qquad 18.75$$

Comparing this with (18.67) gives us the desired formula

$$e^{ikz} = \sum_{\ell=0}^{\infty} (2\ell+1) i^{\ell} j_{\ell}(kr) P_{\ell}(\cos\theta) \,. \qquad 18.76$$

For completeness we now list some of the most useful properties of the Bessel functions. If one takes the integral representations as definitions of the Bessel functions then all their properties can be derived from them.

Asymptotic Behaviour

$$j_\ell(x) \xrightarrow[x\to\infty]{} \frac{1}{x}\cos[x - \tfrac{1}{2}(\ell+1)\pi] = \frac{1}{x}\sin(x - \tfrac{1}{2}\ell\pi) \qquad 18.77$$

$$n_\ell(x) \xrightarrow[x\to\infty]{} \frac{1}{x}\sin[x - \tfrac{1}{2}(\ell+1)\pi] = -\frac{1}{x}\cos(x - \tfrac{1}{2}\ell\pi) \qquad 18.78$$

$$h_\ell^{(1)}(x) \xrightarrow[x\to\infty]{} \frac{1}{x}\, e^{i[x - \frac{1}{2}(\ell+1)\pi]} \qquad 18.79$$

$$h_\ell^{(2)}(x) \xrightarrow[x\to\infty]{} \frac{1}{x}\, e^{-i[x - \frac{1}{2}(\ell+1)\pi]} \qquad 18.80$$

This shows that $h_\ell^{(1)}(x)$ corresponds asymptotically to an outgoing spherical wave and $h_\ell^{(2)}(x)$ corresponds asymptotically to an incoming wave.

Behaviour for Small Argument

$$j_\ell(x) \xrightarrow[x\to 0]{} \frac{x^\ell}{(2\ell+1)!!} \qquad 18.81$$

$$n_\ell(x) \xrightarrow[x\to 0]{} -\frac{(2\ell-1)!!}{x^{\ell+1}} \qquad 18.82$$

Also since $h_\ell^{(1)}(x) = j_\ell(x) + in_\ell(x)$ and $h_\ell^{(2)}(x) = j_\ell(x) - in_\ell(x)$ their behaviour for small values of x is given by

$$h_\ell^{(1)}(x) \xrightarrow[x\to 0]{} -i\,\frac{(2\ell-1)!!}{x^{\ell+1}} \qquad 18.83$$

$$h_\ell^{(2)}(x) \xrightarrow[x\to 0]{} i\,\frac{(2\ell-1)!!}{x^{\ell+1}} \quad . \qquad 18.84$$

This shows that $j_\ell(x)$ is the only one of these functions that is finite at $x = 0$.

18.6 Partial Wave Analysis

We now apply the tools developed in the previous section to a study of scattering from a spherically symmetric potential $V(r)$. Throughout this section we assume that $V(r)$ is short range in the sense that

$$\int_0^\infty r^2 V(r)dr < \infty .$$

Not only does this ensure that we have a well-defined scattering problem but it also ensures that various expressions that we shall encounter converge.

Consider the Schrödinger equation for a particle in such a spherically symmetric potential written in the form (18.3)

$$[\nabla^2 + k^2 - U(r)]\psi(\vec{r}) = 0 . \qquad 18.85$$

As before we now decompose $\psi(\vec{r})$ into spherical harmonics

$$\psi(\vec{r}) = 4\pi \sum_{\ell,m} i^\ell R_\ell(r) Y^*_{\ell m}(\hat{k}) Y_{\ell m}(\hat{r}) \quad . \qquad 18.86$$

The factor $4\pi i^\ell$ is included for later convenience. In terms of $\cos\theta = k.r$ this expression becomes

$$\psi(\vec{r}) = \sum_{\ell=0}^{\infty} i^\ell (2\ell+1) R_\ell(r) P_\ell(\cos\theta) . \qquad 18.87$$

In either case substituting into (18.85) yields the radial equation for $R_\ell(r)$

$$\left[\frac{d^2}{dr^2} + \frac{2}{r}\frac{d}{dr} - \frac{\ell(\ell+1)}{r^2} + k^2 - U(r)\right]R_\ell(r) = 0 . \qquad 18.88$$

We still need to specify the boundary conditions on $R_\ell(r)$. For $r \to 0$ we want $R_\ell(r)$ finite. Thus if $U(r)$ is bounded near the origin by c/r for some constant c then (18.88) reduces for $r \to 0$ to

$$\left[\frac{d^2}{dr^2} - \frac{\ell(\ell+1)}{r^2}\right]R_\ell(r) = 0 \qquad 18.89$$

and the solution finite at the origin is $r^{\ell+1}$. Thus,

$$R_\ell(r) \xrightarrow[r\to 0]{} A\, r^{\ell+1} . \qquad 18.90$$

The remaining boundary condition is determined by the physics of the situation. We start with an incoming beam $e^{i\vec{k}\cdot\vec{r}}$ which can be decomposed in each partial wave into an incoming spherical wave and an outgoing spherical wave as follows:

$$\begin{aligned} e^{i\vec{k}\cdot\vec{r}} &= \sum_{\ell=0}^{\infty} i^\ell(2\ell+1)j_\ell(kr)P_\ell(\cos\theta) \\ &= \frac{1}{2}\sum_{\ell=0}^{\infty} i^\ell(2\ell+1)[h_\ell^{(1)}(kr) + h_\ell^{(2)}(kr)]P_\ell(\cos\theta) \end{aligned} \qquad 18.91$$

where we have used equations 18.76 and 18.47.

The effect of the potential is to cause scattering and hence modify the amplitude of each of the <u>outgoing spherical waves</u>. Thus asymptotically the full solution must be of the form

$$\psi(\vec{r}) \xrightarrow[r\to\infty]{} \sum_{\ell=0}^{\infty} i^\ell(2\ell+1)\frac{1}{2}[h_\ell^{(2)}(kr) + S_\ell(k)h_\ell^{(1)}(kr)]P_\ell(\cos\theta) \qquad 18.92$$

where S_ℓ contains the total effect of the scattering. We have incorporated here the fact that for $r \to \infty$, $\psi(\vec{r})$ must satisfy the free Schrödinger equation due to the short- range nature of the potential as well as the fact that the scattering process affects only the outgoing waves. (18.92) can now be rewritten as

$$\psi(\vec{r}) \xrightarrow[r\to\infty]{} \frac{1}{2} \sum_{\ell=0}^{\infty} i^{\ell}(2\ell+1)\{h_{\ell}^{(1)}(kr) + h_{\ell}^{(2)}(kr)$$

$$+[S_{\ell}(k)-1]h_{\ell}^{(1)}(kr)\}P_{\ell}(\cos\theta)$$

$$= \sum_{\ell=0}^{\infty} i^{\ell}(2\ell+1)\{j_{\ell}(kr) + \frac{1}{2}[S_{\ell}(k)-1]h_{\ell}^{(1)}(kr)\}P_{\ell}(\cos\theta) \qquad 18.93$$

$$= e^{i\vec{k}\cdot\vec{r}} + \sum_{\ell=0}^{\infty} i^{\ell}(2\ell+1)\frac{1}{2}[S_{\ell}(k)-1]h_{\ell}^{(1)}(kr)P_{\ell}(\cos\theta) . \qquad 18.94$$

Since this is the solution for large r we can replace $h_{\ell}^{(1)}(kr)$ by its asymptotic form (18.79) to get the final asymptotic form for $\psi(\vec{r})$.

$$\psi(\vec{r}) \xrightarrow[r\to\infty]{} e^{i\vec{k}\cdot\vec{r}} + f(k,\theta)\frac{e^{ikr}}{r} \qquad 18.95$$

where

$$f(k,\theta) = \frac{i}{2k} \sum_{\ell=0}^{\infty} (2\ell+1)[1-S_{\ell}(k)]P_{\ell}(\cos\theta) . \qquad 18.96$$

The asymptotic expression (18.95) is precisely of the form we previously derived quite generally (equations 18.4, 18.7) and thus shows that $f(k,\theta)\frac{e^{ikr}}{r}$ is the scattered wave. Thus, $f(k,\theta)$ can again be identified as the scattering amplitude. Recalling (18.8) we see that the differential cross-section $d\sigma/d\Omega$ is given by

$$\frac{d\sigma}{d\Omega} = |f(k,\theta)|^2 . \qquad 18.97$$

Thus our scattering problem is solved if we find $f(k,\theta)$ or alternatively $S_{\ell}(k)$. The remaining boundary condition for $R_{\ell}(r)$ can now be extracted from these results. Thus

$$R_\ell(r) \xrightarrow[r\to\infty]{} \frac{1}{2}\left[h_\ell^{(2)}(kr) + S_\ell(k)h_\ell^{(1)}(kr)\right] \qquad 18.98$$

or, using the asymptotic forms of the Hankel functions,

$$R_\ell(r) \xrightarrow[r\to\infty]{} \frac{1}{2kr}\{e^{-i[kr-1/2(\ell+1)\pi]} + S_\ell(k)e^{i[kr-1/2(\ell+1)\pi]}\} \qquad 18.99$$

18.7 Phase Shifts

For elastic scattering, from a spherical potential, probability must be conserved for each partial wave since different ℓ values do not couple. Thus the radial flux for each incoming and outgoing partial wave must be the same. The incoming partial waves obtained from the partial wave decomposition of $e^{i\vec{k}.\vec{r}}$ are given by $i^\ell(2\ell+1)j_\ell(kr)$ and their radial flux is zero. For the outgoing partial waves we can compute the flux using their asymptotic form (18.99). Equating the two fluxes yields:

$$0 = \lim_{r\to\infty} r^2 \frac{\hbar}{2im}\left(R_\ell^* \frac{\partial R_\ell}{\partial r} - R_\ell \frac{\partial R_\ell^*}{\partial r}\right) .$$

Writing this out we find

$$|S_\ell(k)|^2 = 1 . \qquad 18.100$$

Thus, for elastic scattering, the complex numbers $S_\ell(k)$ may be expressed in terms of real numbers $\delta_\ell(k)$ called the phase shifts and defined by

$$S_\ell(k) = e^{2i\delta_\ell(k)} , \qquad 18.101$$

where the factor of two is conventional.

These phase shifts have an intuitive interpretation which we now describe. For this purpose we consider a potential of finite range so that

$$U(r) = 0 \qquad r \geq a \, . \tag{18.102}$$

Then for $r > a$ the most general solution of the radial equation is

$$R_\ell(r) = A_\ell\left[\cos\delta_\ell \, j_\ell(kr) - \sin\delta_\ell \, n_\ell(kr)\right] \qquad k > a. \tag{18.103}$$

The asymptotic behaviour of this solution is for large r

$$R_\ell(r) \to \frac{A_\ell}{kr}\left[\cos\delta_\ell \, \sin(kr - \frac{\ell\pi}{2}) + \sin\delta_\ell \, \cos(kr - \frac{\ell\pi}{2})\right]$$

or

$$R_\ell(r) \xrightarrow[r\to\infty]{} \frac{A_\ell}{kr} \sin(kr - \frac{\ell\pi}{2} + \delta_\ell) \, . \tag{18.104}$$

On the other hand the free incoming wave is

$$j_\ell(kr) \xrightarrow[r\to\infty]{} \frac{1}{kr} \sin(kr - \frac{\ell\pi}{2}) \, . \tag{18.105}$$

Thus, the effect of the potential is to shift the phase of the solution in the region of no interaction. In the next chapter we demonstrate this result explicitly for a square well.

The phase shift here defined coincides with the one in equation 18.101. This also agrees with the definition of phase shifts given for the one-dimensional scattering problems in section 5.7 .

If we now use the definitions of phase shifts to compute the differential cross-section we obtain

$$\frac{d\sigma}{d\Omega} = |f(k,\theta)|^2 = \frac{1}{k^2} \sum_{\ell,\ell'=0}^{\infty} (2\ell+1)(2\ell'+1) \sin\delta_\ell \, \sin\delta_{\ell'} \cos(\delta_\ell - \ell_{\ell'}) \times$$

$$\times P_\ell(\cos\theta) P_{\ell'}(\cos\theta) \, . \tag{18.106}$$

In general these are infinite sums and difficult to evaluate. However, as we shall now indicate, for potentials of short or finite range and for low energies, these sums may be truncated after just a few terms. In other words, only a few partial waves contribute.

Suppose that for $r \geq a$ the potential may be neglected. Then for this region the total energy of a particle is kinetic $\hbar^2 k^2/2m$ and exceeds the energy due to orbital motion alone $\hbar^2 \ell(\ell+1)/2mr^2$. Setting $r = a$ we obtain

$$\frac{\hbar^2 k^2}{2m} \geq \frac{\hbar^2 \ell(\ell+1)}{2ma^2} \quad \text{or} \quad k^2 a^2 \geq \ell(\ell+1)$$

whence we get the inequality

$$\ell < ka \quad . \tag{18.107}$$

Thus, if ℓ violates this inequality the corresponding partial wave cannot participate in the scattering classically and we may truncate the sums in (18.106) by the largest $\ell < ka$.

Clearly for potentials of finite range the length a is well defined. For short range potentials, however, it is not as clear how to perform the approximation. In chapter 19 we shall replace the above heuristic argument with a rigorous derivation and prove that those higher partial waves that violate (18.107) do not contribute appreciably to the differential cross-section.

18.8 The Optical Theorem - Unitarity Bound

If we integrate (18.106) to get the total cross-section we find using

$$\int P_\ell(\cos\theta) P_{\ell'}(\cos\theta)\, d\Omega = \frac{4\pi}{2\ell+1}\, \delta_{\ell,\ell'} \tag{18.108}$$

that

$$\sigma = \frac{4\pi}{k^2} \sum_{\ell=0}^{\infty} (2\ell+1)\sin^2\delta_\ell \ . \qquad 18.109$$

On the other hand

$$\text{Im } f(k,\theta) = \frac{1}{2k} \sum_{\ell=0}^{\infty} (2\ell+1)\text{Re}(1-S_\ell)P_\ell(\cos\theta)$$

$$= \frac{1}{2k} \sum_{\ell=0}^{\infty} (2\ell+1)2\sin^2\delta_\ell \, P_\ell(\cos\theta) \ .$$

If we now set $\theta = 0$ and use $P_\ell(1) = 1$ we get

$$\text{Im } f(k,0) = \frac{1}{k} \sum_{\ell=0}^{\infty} (2\ell+1)\sin^2\delta_\ell \ . \qquad 18.110$$

Comparing this with (18.109) we obtain the optical theorem

$$\sigma = \frac{4\pi}{k} \text{Im } f(k,0) \qquad 18.111$$

The total cross-section (18.109) can also be written

$$\sigma = \sum_{\ell=0}^{\infty} \sigma_\ell \qquad 18.112$$

where the ℓ'th partial cross-section σ_ℓ is given by

$$\sigma_\ell = \frac{4\pi}{k^2}(2\ell+1)\sin^2\delta_\ell \ . \qquad 18.113$$

For elastic scattering, the phase shifts are real and it therefore follows that for elastic scattering

$$\sigma_\ell \leq \frac{4\pi}{k^2}(2\ell+1) \ . \qquad 18.114$$

The value $\frac{4\pi}{k^2}(2\ell+1)$ is known as the unitarity bound and is reached only for $\delta_\ell = (n+1/2)\pi$.

This is the condition for resonance and shows up as a local maximum in the cross-section for the corresponding partial wave.

Notes, References and Bibliography

18.1 A very readable little book on integral equations is:

F. Smithies - Integral Equations - Cambridge University Press (1962).

All the references of chapter 17 are applicable to this chapter as well.

A very enjoyable book treating the more mathematical aspects of potential scattering is:

V. de Alfaro and T. Regge, Potential Scattering, North-Holland Pub. Co., Amsterdam (1965).

Chapter 18 Problems

18.1 Derive equation 18.61.

18.2 Use the integral representation for the spherical Bessel functions $Z_\ell(x)$ to show that

a) $Z_{\ell-1}(x) + Z_{\ell+1}(x) = \frac{2\ell+1}{x} Z_\ell(x) \quad \ell \geq 1$

b) $\frac{d}{dx} Z_\ell(x) = Z_{\ell-1}(x) - \frac{\ell+1}{x} Z_\ell(x) \quad \ell \geq 1$

c) $\frac{d}{dx}[x^{-\ell} Z_\ell(x)] = -x^{-\ell} Z_{\ell+1}(x)$.

Here $Z_\ell(x)$ is any one of the four spherical Bessel functions.

18.3 A useful formula for generating any one of the spherical Bessel functions is

$$Z_\ell = x^\ell \left(-\frac{1}{x}\frac{d}{dx}\right)^\ell Z_o \ .$$

Combine the results of problem 18.2 to get

$$Z_\ell = \left(-\frac{d}{dx} + \frac{\ell-1}{x}\right) Z_{\ell-1} = -x^{\ell-1}\frac{d}{dx}\left(\frac{Z_{\ell-1}}{x^{\ell-1}}\right)$$

and hence derive the generating formula above.

18.4 Show that the Wronskian

$$W = j_\ell(x)\ n_\ell'(x) - n_\ell(x)\ j_\ell'(x)$$

satifies the differential equation

$$W' = -\frac{2}{x} W.$$

Solve this equation and use the behaviour of j_ℓ, n_ℓ for small x to fix the constant of integration to get

$$W = \frac{1}{x^2} \ .$$

Hint: Start with the differential equations for $j_\ell(x)$ and $n_\ell(x)$.

18.5 a) Verify equation (18.68).

b) Verify equation (18.69).

18.6 Find the equation for the phase shifts for scattering off a hard sphere potential.

$$V(r) = \begin{cases} \infty & r = a \\ o & r > a \end{cases} .$$

Solve this for $\ell = 0,1$.

18.7 Calculate the first Born approximation for the scattering amplitude for the potential

$$V = V_o \, e^{-\mu r^2} .$$

18.8 Repeat 18.7 for the potential

$$V = \begin{cases} -V_o & r < a \\ 0 & r \geq a \end{cases}$$

Chapter 19

Further Topics in Potential Scattering

In this chapter we apply some of the technqiues developed in the previous chapter to a few specific problems. Generally the application of scattering theory involves finding the scattering amplitude for a given potential. Since most potentials do not admit closed solutions it is necessary to have recourse to approximate techniques. With this in mind we shall develop another approximation method useful for high energy scattering, known as the Glauber or Eikonal approximation. However, to apply approximation techniques intelligently requires a feeling for the kind of results that may be obtained. The example we solve, namely the square well, is the archetype of many problems and will therefore serve to illustrate the concepts and sharpen our intuition.

There is another use of potential scattering, in practice, and that is to parametrize data obtained from scattering experiments. This is of particular importance in low-energy nuclear physics since the interactions between nucleons are only approximately understood. It is essential here to have a knowledge of which results of an experiment depend on various dynamical properties. Another way of phrasing this is to ask on what properties of a potential the phase shifts depend the most. Historically a very important tool for answering this question was effective range theory. We therefore also devote a section to this topic.

19.1 Example: The Square Well

One of the simplest potentials for which the scattering problem can be solved in closed form is the square well. Furthermore this potential illustrates many of the results to be expected in potential scattering. Also there are many instances in which more realistic potentials can be approximated by an appropriate square well potential.

For a square well we have

$$V(r) = \begin{matrix} -V_o & V_o > 0 \; , \; r \leq a \\ 0 & r > a \; . \end{matrix} \qquad 19.1$$

If we now define

$$K_o^2 = \frac{2m}{\hbar^2} V_o \quad , \quad k^2 = \frac{2m}{\hbar^2} E \qquad 19.2$$

and

$$K^2 = k^2 + K_o^2 \qquad 19.3$$

the radial equation becomes

$$\left[\frac{d^2}{dr^2} + \frac{2}{r}\frac{d}{dr} - \frac{\ell(\ell+1)}{r^2} + K^2\right]R_\ell(r) = 0 \quad r \leq a$$

$$\left[\frac{d^2}{dr^2} + \frac{2}{r}\frac{d}{dr} - \frac{\ell(\ell+1)}{r^2} + k^2\right]R_\ell(r) = 0 \quad r > a. \qquad 19.4$$

The solutions of these equations satisfying the conditions that $R_\ell(0)$ is finite and $R_\ell(r)$ corresponds to a fixed incoming flux for large values of r are

$$R_\ell(r) = A_\ell j_\ell(Kr) \qquad r \leq a \tag{19.5}$$

$$R_\ell(r) = j_\ell(kr) + \frac{1}{2}[S_\ell(k)-1]h_\ell^{(1)}(kr) \qquad r > a \tag{19.6}$$

Replacing the $S_\ell(k)$ by the phase shifts we obtain an alternate expression for $r > a$

$$R_\ell(r) = e^{i\delta_\ell}[\cos\delta_\ell j_\ell(kr) - \sin\delta_\ell n_\ell(kr)] \tag{19.7}$$

where we have also used the relation $h_\ell^{(1)} = j_\ell + in_\ell$. Since R_ℓ and its first derivatives are continuous we can match the solutions at $r = a$. To eliminate the constant A_ℓ it is convenient to match logrithmic derivatives. We thus get

$$\frac{Kj_\ell'(Ka)}{j_\ell(Ka)} = \frac{k[j_\ell'(ka)\cos\delta_\ell - n_\ell'(ka)\sin\delta_\ell]}{j_\ell(ka)\cos\delta_\ell - n_\ell(ka)\sin\delta_\ell} \tag{19.8}$$

where the primes indicate derivatives with respect to the argument. A little algebra then yields

$$\tan\delta_\ell = \frac{kj_\ell(Ka)j_\ell'(ka) - Kj_\ell'(Ka)j_\ell(ka)}{kj_\ell(Ka)n_\ell'(ka) - Kj_\ell'(Ka)n_\ell(ka)} . \tag{19.9}$$

We now restrict ourselves to very low energies. In that case the most important contribution is from the s-wave or $\ell = 0$ phase shift. However for $\ell = 0$ we can replace the Bessel functions by trigonometric functions since

$$j_o(kr) = \frac{\sin kr}{kr} \tag{19.10}$$

and

$$n_o(kr) = -\frac{\cos kr}{kr} . \tag{19.11}$$

After some further algebra we then obtain

$$\tan \delta_o = \frac{k \tan Ka - K \tan ka}{K + k \tan Ka \tan ka} . \tag{19.12}$$

This can be rewritten as

$$\tan \delta_o = \tan[\tan^{-1}(\frac{k}{K} \tan Ka) - ka]$$

so that

$$\delta_o = \tan^{-1}(\frac{k}{K} \tan Ka) - ka . \tag{19.13}$$

Suppose the potential is weak and the energy is low so that both $ka \ll 1$ as well as $Ka \ll 1$. In that case

$$\delta_o \simeq ka(\frac{\tan Ka}{Ka} - 1) \underset{k \to 0}{\longrightarrow} 0. \tag{19.14}$$

Suppose now we still have very low energy $ka \ll 1$ but the potential is allowed to be large. Then for $Ka < \pi/2$

$$\delta_o \simeq ka(\frac{\tan Ka}{Ka} -1). \tag{19.15}$$

This approximation breaks down as k increases so that Ka approaches $\pi/2$. At this point ($Ka = \pi/2$), δ_o changes sign and decreases as shown in figure 19.1. This can only occur for $K_o a < \pi/2$.

If $K_o a$ is somewhat larger than $\pi/2$ then δ_o will pass through $\pi/2$ and, as $k \to 0$, keep on increasing to π. In this case we must replace (19.15) by

$$\delta_o \simeq \pi + ka(\frac{\tan Ka}{Ka} -1) . \tag{19.16}$$

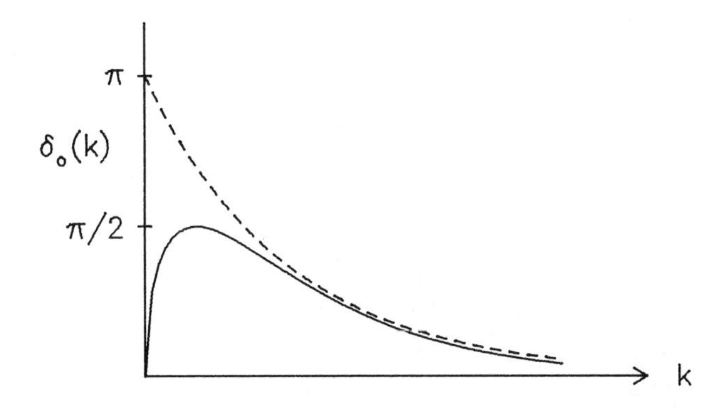

Fig. 19–1 S – Wave Phase Shift

The value $K_o a = \pi/2$ has a simple physical interpretation. It is precisely the minimum strength of the potential necessary for a bound state to exist as is easily seen from the results of problem 10.5.

Thus we can conclude that the s-wave phase shift approaches zero for zero energy if there is no bound state and approaches π if there is one bound state. This is also illustrated in figure 19.1. The result is applicable to other potentials and can be generalized to several bound states. In that case it is known as Levinson's Theorem.

The s-wave phase shift at zero energy is given by

$$\delta_o(0) = n\pi \,, \tag{19.17}$$

where n is the number of s-wave bound states, unless the potential has a transitional strength; in that case

$$\delta_o(0) = (n+ \frac{1}{2})\pi \,. \tag{19.18}$$

To summarize, we have found the phase shifts for a square well potential. In particular we have examined the s-wave phase shift and found the condition for an s-wave resonance. Also we have seen that the value of the s-wave phase shift at zero energy is related to the number of s-wave bound states via Levinson's Theorem. This last result is useful in removing the ambiguity (modulo π) that can result in solving for phase shifts.

19.2 Partial Wave Analysis of the Lippmann-Schwinger Equation

The Lippmann-Schwinger equation in configuration space reads

$$\psi^{(\pm)}(\vec{r}) = \psi^{(o)}(\vec{r}) + \int G^{\pm}(\vec{r},\vec{r}\,')V(r')\psi^{(\pm)}(\vec{r}\,')d^3r' \qquad 19.19$$

where $G^{\pm}(\vec{r},\vec{r}\,')$ are the free Green's functions satisfying

$$(\nabla^2 + k^2)G^{\pm}(\vec{r},\vec{r}\,') = \delta(\vec{r}-\vec{r}\,') \qquad 19.20$$

with outgoing and incoming wave boundary conditions.

To obtain the partial wave decompositions of these equations, we set

$$\psi^{(\pm)}(\vec{r}) = \sum_{\ell,m} \psi_\ell^{(\pm)}(r)Y^*_{\ell m}(\hat{k})Y_{\ell m}(\hat{r}) \qquad 19.21$$

$$\psi^{(o)}(\vec{r}) = \sum_{\ell,m} \psi_\ell^{(o)}(r)Y^*_{\ell m}(\hat{k})Y_{\ell m}(\hat{r}) \qquad 19.22$$

where as always $\hat{k}$, $\hat{r}$ indicate the θ,ϕ directions of $\vec{k}$ and $\vec{r}$ respectively. We also expand the Green's functions

$$G^{\pm}(\vec{r},\vec{r}\,') = \sum_{\ell,m} G_\ell^{(\pm)}(\vec{r},\vec{r}\,')Y^*_{\ell m}(\hat{r}\,')Y_{\ell m}(\hat{r}). \qquad 19.23$$

We then substitute all of these expressions into (19.19) and integrate out explicitly the angular variables corresponding to $\vec{r}\,'$. This yields the expression

$$\psi_\ell^{(\pm)}(r) = \psi_\ell^{(o)}(r) + \int G_\ell^{(\pm)}(\vec{r},\vec{r}\,')V(r')\psi_\ell^{(\pm)}(r')r'^2dr' \qquad 19.24$$

where we have also used the fact that the potential is spherically symmetric and the spherical harmonics form a complete orthonormal set. The explicit form for the partial wave Green's functions $G_\ell^{(\pm)}(\vec{r},\vec{r}\,')$ is obtained by substituting the expansion (19.23) into the defining equation 19.20 for the full Green's functions $G^{(\pm)}(\vec{r},\vec{r}\,')$. This yields

$$\left(\frac{d^2}{dr^2} + \frac{2}{r}\frac{d}{dr} - \frac{\ell(\ell+1)}{r^2} + k^2\right)G_\ell^{(\pm)}(r,r') = \frac{1}{rr'}\,\delta(r-r') \qquad 19.25$$

where the partial wave Green's functions are finite at $r = 0$ and have asymptotic behaviour specified by the superscripts $(\pm)$. These Green's functions are furthermore continuous at $r = r'$ and have a discontinuity in their first derivative. This discontinuity is obtained by integrating both sides of equation 19.25 about $r = r'$ to get

$$\frac{d}{dr}\,G_\ell^{(\pm)}(r,r')\Big|_{r=r'+0} - \frac{d}{dr}\,G_\ell^{(\pm)}(r,r')\Big|_{r=r'-0} = \frac{1}{r^2} \qquad 19.26$$

We therefore have, to begin with, as general solutions of (19.25)

$$G_\ell^{(\pm)}(r,r') = A_\ell^{(\pm)}(r')j_\ell(kr) \qquad r < r' \qquad 19.27a$$

$$G_\ell^{(+)}(r,r') = B_\ell^{(+)}(r')h_\ell^{(1)}(kr) \qquad r > r' \qquad 19.27b$$

$$G_\ell^{(-)}(r,r') = B_\ell^{(-)}(r')h_\ell^{(2)}(kr) \qquad r > r' \qquad 19.27c$$

We now impose the condition of continuity in these functions as well as the given discontinuity in the first derivatives to get the "constants" $A_\ell^{(\pm)}(r')$, $B_\ell^{(\pm)}(r')$

$$A_\ell^{(+)}(r') = -ikh_\ell^{(1)}(kr')$$

$$A_\ell^{(-)}(r') = ikh_\ell^{(2)}(kr')$$

$$B_\ell^{(\pm)}(r') = \mp ikj_\ell(kr')\ .$$

Thus, the partial wave Green's functions are given by

$$G_\ell^{(+)}(r,r') = -ik\ j_\ell(kr_<)h_\ell^{(1)}(kr_>) \qquad 19.28a$$

$$G_\ell^{(-)}(r,r') = \ ik\ j_\ell(kr_<)h_\ell^{(2)}(kr_>) \qquad 19.28b$$

where

$$r_< = \begin{matrix} r & \text{for} & r < r' \\ r' & \text{for} & r > r' \end{matrix}$$

$$r_> = \begin{matrix} r & \text{for} & r > r' \\ r' & \text{for} & r' > r \end{matrix}$$

In obtaining these results we have also used the Wronskian (see problem 18.5)

$$j_\ell(x)\ n_\ell'(x) - n_\ell(x)j_\ell'(x) = x^{-2}\ . \qquad 19.29$$

This completes the partial wave decomposition of the Lippmann-Schwinger equations. They are given by (19.24) with the explicit forms (19.28a) and (19.28b) for $G^\pm(\vec{r},\vec{r}\,')$ respectively. We shall have occasion to use these results repeatedly in the next sections, since these equations are completely equivalent to the radial Schrödinger equation plus boundary conditions. It is the fact that the integral equations already incorporate the boundary conditions that makes them so useful.

19.3 Effective Range Approximation

It is frequently desirable to have a parmetrization of scattering data in terms of variables more readily interpreted than phase shifts. For low energy scattering a set of such variables is provided by the scattering length and effective range. In this section we develop expressions for these two parameters. We furthermore supplant the heuristic arugment of section (18.7), that only a few partial waves contribute for low energy scattering, by rigorous bounds.

We therefore consider the Schrödinger equation for the ℓ'th partial wave in a spherically symmetric potential

$$\left[\frac{d^2}{dr^2} + \frac{2}{r}\frac{d}{dr} + k^2 - \frac{\ell(\ell+1)}{r^2} - U(r)\right]R_\ell(r) = 0. \qquad 19.30$$

We want to compare the solution to this equation with the solution for the ℓ'th partial wave of the free Schrödinger equation that is finite at the origin

$$\left[\frac{d^2}{dr^2} + \frac{2}{r}\frac{d}{dr} + k^2 - \frac{\ell(\ell+1)}{r^2}\right]j_\ell(kr) = 0\ . \qquad 19.31$$

To compare these solutions we multiply (19.30) by $j_\ell(kr)$, (19.31) by $R_\ell(r)$ and subtract to get

$$j_\ell(kr)\frac{d^2}{dr^2}R_\ell(r) - R_\ell(r)\frac{d^2}{dr^2}j_\ell(kr) + \frac{2}{r}\left[j_\ell(kr)\frac{d\,R_\ell(r)}{dr} - R_\ell(r)\frac{d\,j_\ell(kr)}{dr}\right] = j_\ell(kr)U(r)R_\ell(r). \qquad 19.32$$

We next integrate from 0 to a and use integration by parts on the first term to get

$$\int_0^a r^2 dr\,\frac{d}{dr}\left[j_\ell\frac{dR_\ell}{dr} - R_\ell\frac{dj_\ell}{dr}\right] + 2\int_0^a r\,dr\left[j_\ell\frac{dR_\ell}{dr} - R_\ell\frac{dj_\ell}{dr}\right]$$

$$= a^2\left[j_\ell(kr)\frac{dR_\ell(r)}{dr} - R_\ell(r)\frac{dj_\ell(kr)}{dr}\right]_{r=a} \qquad 19.33$$

$$= \int_0^a r^2 dr\; j_\ell(kr)U(r)R_\ell(r).$$

Now choose a sufficiently large a so that for $j_\ell(ka)$ as well as for

$R_\ell(a)$ we can use their asymptotic forms

$$j_\ell(ka) \simeq \frac{\sin(ka - \frac{\ell\pi}{2})}{ka}$$

$$R_\ell(a) \simeq e^{i\delta_\ell}[\cos\delta_\ell\, j_\ell(ka) - \sin\delta_\ell\, n_\ell(ka)]$$
$$\simeq e^{i\delta_\ell}[\cos\delta_\ell \frac{\sin(ka - \frac{\ell\pi}{2})}{ka} + \sin\delta_\ell \frac{\cos(ka - \frac{\ell\pi}{2})}{ka}]$$

or

$$R_\ell(a) \simeq e^{i\delta_\ell} \frac{\sin(ka - \frac{\ell\pi}{2} + \delta_\ell)}{ka}. \qquad 19.34$$

Substituting these results into (19.33) we obtain the following integral representation for the phase shift

$$e^{i\delta_\ell} \sin\delta_\ell \simeq k \int_0^a r^2 dr\, j_\ell(kr) U(r) R_\ell(r). \qquad 19.35$$

For a sufficiently large a this equation becomes exact. To obtain an estimate for the phase shift, we replace $R_\ell(r)$ by $e^{i\delta_\ell} j_\ell(kr)$. This extremely crude approximation assumes δ_ℓ is small and yields

$$\sin\delta_\ell \simeq k \int_0^a dr\, r^2 j_\ell^2(kr) U(r). \qquad 19.36$$

A much more accurate approach using a variational technique was introduced by Schwinger (ref 19.3) and used by Blatt and Jackson (ref. 19.4). Now suppose the potential has a range ρ and the energy is sufficiently low so that $k\rho \ll 1$. In this case we can (firstly) replace the upper limit in the integral by ρ and (secondly) use the asymptotic behaviour for small argument of the Bessel functions. This yields the relation

$$\sin\delta_\ell \simeq \frac{(k\rho)^{2\ell+1}}{[(2\ell+1)!!]^2} \rho \int_0^\rho U(r) (\frac{r}{\rho})^{2(\ell+1)} dr. \qquad 19.37$$

From this relation we can make the following deductions:

1) The phase shifts are odd functions of k.

2) For low energies $(k\rho \ll 1)$ the phase shifts decrease rapidly with ℓ.

In order to prove this second observation we maximize the integral by setting $r = \rho$. Also we replace $\sin \delta_\ell$ by δ_ℓ. Then we get

$$\frac{\delta_1}{\delta_o} = \left(\frac{k\rho}{3}\right)^2 \quad , \qquad \frac{\delta_2}{\delta_o} = \frac{(k\rho)^4}{(15)^2} \text{ , etc.}$$

This shows that for low energy the scattering occurs predominantly in the lower partial waves, particularly the $\ell = 0$ or s-waves. This also justifies our previous heuristic arguments that partial wave analysis is useful for low energy scattering since only a few partial waves contribute. With this in mind we now examine in more detail the s-wave scattering for low energy.

Restricting the discussion to s-wave solutions, we can repeat the previous computation starting from the radial equations

$$\left[\frac{d^2}{dr^2} + \frac{2}{r}\frac{d}{dr} + k^2 - U(r)\right]R_o(k,r) = 0 \tag{19.38}$$

$$\left[\frac{d^2}{dr^2} + \frac{2}{r}\frac{d}{dr} - U(r)\right]R_o(0,r) = 0 \ . \tag{19.39}$$

Again, multiplying (19.38) by $R_o(0,r)$, (19.39) by $R_o(k,r)$, and substracting we obtain

$$\begin{aligned} R_o(0,r)R_o''(k,r) &- R_o(k,r)R_o''(0,r) \\ &+ \frac{2}{r}\left[R_o(0,r)R_o'(k,r)-R_o(k,r)R_o'(0,r)\right] \\ &= -k^2 R_o(0,r)R_o(k,r). \end{aligned} \tag{19.40}$$

For large r we have from (18.104)

$$R_o(k,r) \simeq c \frac{\sin(kr+\delta_o)}{r} \equiv f(k,r) \; . \qquad 19.41$$

This function $f(k,r)$ clearly satisfies the free $\ell = 0$ radial Schrödinger equation as does $f(0,r)$. Hence we obtain

$$\begin{aligned} f(0,r)f''(k,r) &- f(k,r)f''(0,r) \\ &+ \frac{2}{r}[f(0,r)f'(k,r)-f(k,r)f''(0,r] \\ &= -k^2 f(0,r)f(k,r). \end{aligned} \qquad 19.42$$

If we further choose $c = 1/\sin\delta_o$ then

$$f(0,r) = \frac{1}{r}\,(1-\frac{r}{c_o}) \qquad 19.43$$

where by taking the limit $k \to 0$ in equation (19.41) we find

$$\frac{1}{c_o} = -\lim_{k\to 0} k \cot \delta_o \; . \qquad 19.44$$

We now subtract (19.42) from (19.40) and integrate the resultant equation from 0 to ∞. If we further use the easily computed result that

$$r^2[f(k,r)f'(0,r) - f(0,r)f'(k,r)]_{r=0} = \frac{1}{c_o} + k\cot\delta_o$$

we get

$$k\cot\delta_o = -\frac{1}{c_o} + k^2\int_0^\infty r^2 dr[f(k,r)f(0,r)-R_o(k,r)R_o(0,r)]. \qquad 19.45$$

Our final low energy approximation is now to set $k = 0$ in the integral and define

$$r_o = 2 \int_0^\infty r^2 dr[f^2(0,r) - R_o^2(0,r)] \ . \qquad 19.46$$

The final formula for the low energy s-wave phase shift in the effective range approximation is then

$$k \cot \delta_o \simeq -\frac{1}{c_o} + \frac{1}{2} k^2 r_o \ . \qquad 19.47$$

The constant c_o is known as the scattering length and r_o is called the effective range.

Actually equation 19.47 says only that $k \cot \delta_o$ is an even function of k for small k. The usefulness of this formula derives from the interpretation that can be given to it for potentials strong compared to the energy. For smooth potentials that do not change sign, r_o is proportional to the range of the potentials. However for more complicated potentials r_o may even be negative and the above simple interpretation does not follow.

Since for "simple" potentials r_o depends only on the range and depth of the potential, equation 19.47 is also known as the shape-independent approximation.

The scattering length is related to the existence of bound states. From (19.43) we see that c_o gives the location of the zero in $f(0,r)$, the zero-energy form of the asymptotic solution (19.41). If the potential has a range ρ and a depth V_o then for $V_o\rho$ so small that there are no s-wave bound states we have that $\frac{\pi}{2} > \delta_o > 0$ and, as we see form (19.44), $c_o < 0$. If $V_o\rho$ is at the transitional strength for producing the first bound state, $\delta_o = \frac{\pi}{2}$ and $c_o = \pm\infty$. For $V_o\rho$ sufficiently strong to produce a bound state, $\frac{\pi}{2} < \delta_o \leq \pi$ so that $c_o > 0$. The three cases are depicted in figs. 19.2 with a corresponding square well potential.

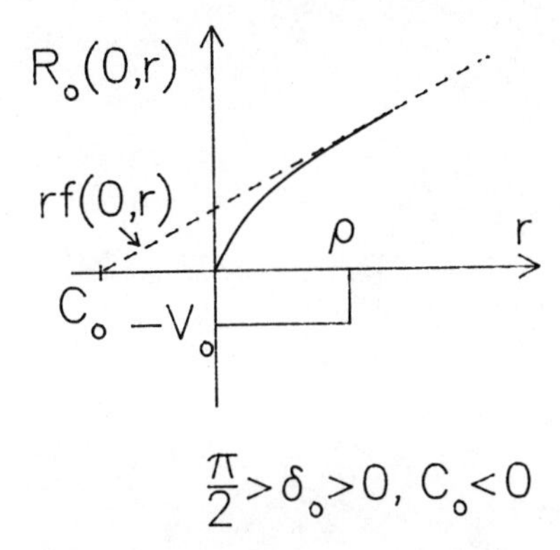

$\frac{\pi}{2} > \delta_o > 0,\ C_o < 0$

Fig. 19–2a No S–Wave Bound States

19.4 The Glauber or Eikonal Approximation

We have previously studied the W.K.B. approximation. This approximation turned out to be useful for potentials varying slowly so that the change $\Delta V = \frac{dV}{dx} \lambda$ occurring in the potential over one wavelength was small. We shall now consider another semi-classical method based on the approximation that λ is very small, or alternatively, that the wave vector $\vec{k}$ is very large. This is clearly a <u>high energy approximation</u> and as such has found many applications in recent years.

We start with the Lippmann-Schwinger equation 19.19 with the explicit form 19.17 for the free Green's functions. Thus

$$\psi_{\vec{k}}^{(+)}(r) = e^{i\vec{k}\cdot\vec{r}} - \frac{1}{4\pi} \int d^3r' \, \frac{e^{ik|\vec{r}-\vec{r}'|}}{|\vec{r}-\vec{r}'|} U(\vec{r}')\psi_{\vec{k}}^{(+)}(\vec{r}') \qquad 19.48$$

As a first step we now set

$$\psi_{\vec{k}}^{(+)}(\vec{r}) = e^{i\vec{k}\cdot\vec{r}} \phi_{\vec{k}}^{(+)}(\vec{r}) \ . \qquad 19.49$$

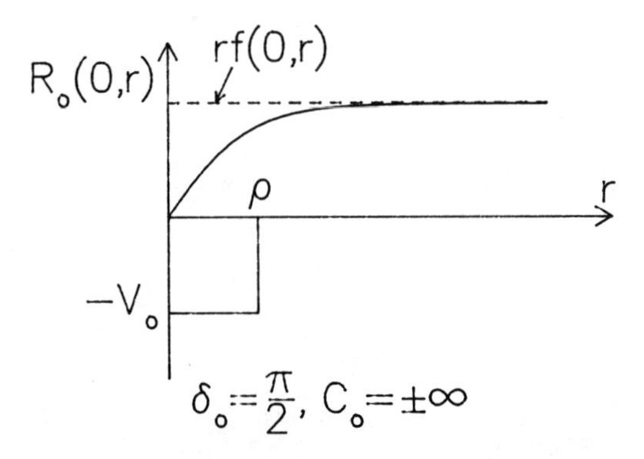

Fig. 19–2b Transitional Strength

Then 19.48 becomes:

$$\phi^{(+)}_{\vec{k}}(\vec{r}) = 1 - \frac{1}{4\pi}\int d^3r' \, \frac{e^{ik|\vec{r}-\vec{r}\,'|-i\vec{k}\cdot(\vec{r}-\vec{r}\,')}}{|\vec{r} - \vec{r}\,'|} \, U(\vec{r}\,')\phi^{(+)}_{k}(\vec{r}\,') \qquad 19.50$$

After putting $\vec{r}-\vec{r}\,' = \vec{\rho}$ this becomes

$$\phi^{(+)}_{\vec{k}}(\vec{r}) = 1 - \frac{1}{4\pi}\int \frac{d^3\rho}{\rho} \, e^{i(k\rho-\vec{k}\cdot\vec{\rho})} U(\vec{r}-\vec{\rho})\phi^{(+)}_{k}(\vec{r}-\vec{\rho}) \qquad 19.51$$

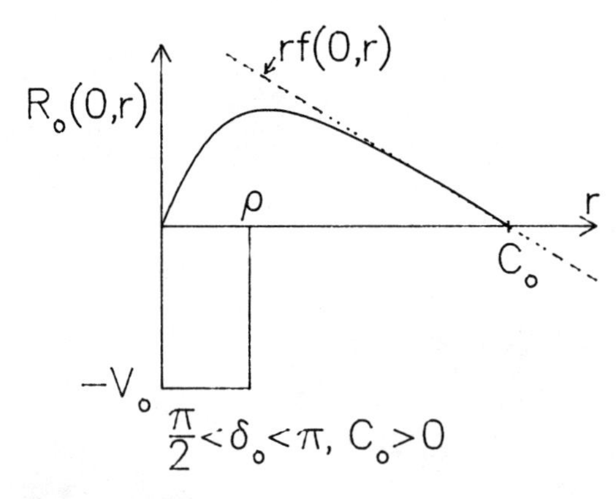

Fig. 19–2c One S–Wave Bound State

So far the equation is exact. Our approximation now consists in realizing that for k large the exponential in 19.51 oscillates rapidly except where $\vec{k}\cdot\vec{\rho} = k\rho$ which is the forward direction. Thus we replace $\vec{\rho}$ everywhere except in the exponential by $\vec{\rho}\,' = \hat{k}\rho$, where $\hat{k}$ is a unit vector in the $\vec{k}$ direction. This yields the approximate formula

$$\phi_{\vec{k}}^{(+)}(\vec{r}) \simeq 1 - \frac{2\pi}{4\pi}\int_0^\infty \rho' d\rho' U(\vec{r}-\vec{\rho}\,')\phi_{\vec{k}}^{(+)}(\vec{r}-\vec{\rho}\,')\int_{-1}^{1} e^{ik\rho(1-u)}du \tag{19.52}$$

where $k\rho u = \vec{k}\cdot\vec{\rho}$.

Evaluating the second integral

$$\int_{-1}^{1} e^{ik\rho(1-u)}du = -\frac{1}{ik\rho}(1-e^{2ik\rho}) \simeq -\frac{1}{ik\rho}$$

and using the fact that $\rho = \rho'$ we obtain the relation

$$\phi_{\vec{k}}^{(+)}(\vec{r}) \simeq 1 + \frac{1}{2ik}\int_0^\infty d\rho' U(\vec{r}-\vec{\rho}\,')\phi_{\vec{k}}^{(+)}(\vec{r}-\vec{\rho}\,'). \tag{19.53}$$

We next decompose $\vec{r}$ into a vector $\vec{b}$ normal to $\vec{k}$ and a vector parallel to $\vec{k}$. Thus

$$\vec{r} = \vec{b} + \hat{k}z. \tag{19.54}$$

The vector $\vec{b}$ clearly corresponds to the classical impact parameter. We then get

$$\begin{aligned}\phi_{\vec{k}}^{(+)}(\vec{b}+kz) &\simeq 1 - \frac{i}{2k}\int_0^\infty d\rho' U(\vec{b}+\hat{k}z-\vec{\rho}\,')\phi_{\vec{k}}^{(+)}(\vec{b}+\hat{k}z-\vec{\rho}\,') \\ &= 1 - \frac{i}{2k}\int_{-\infty}^{z} dz' U(\vec{b}+\hat{k}z')\phi_{\vec{k}}^{(+)}(\vec{b}+\hat{k}z')\end{aligned} \tag{19.55}$$

where

$$\hat{k}z' = \hat{k}z - \vec{\rho}' . \qquad 19.56$$

Equation 19.55 is of the Volterra type which is always solvable by iteration. The final result can be summed in this case and yields

$$\phi^{(+)}_{\vec{k}}(\vec{b}+\hat{k}z) \simeq \exp[-\frac{i}{2k}\int_{-\infty}^{z} dz' U(\vec{b}+\hat{k}z')] . \qquad 19.57$$

Therefore the scattering solution is given by

$$\phi^{(+)}_{\vec{k}}(\vec{b}+\hat{k}z) \simeq \exp ik[z- \int_{-\infty}^{z} dz' \frac{U(\vec{b}+\hat{k}z')}{2k^2}] . \qquad 19.58$$

But

$$U = \frac{2mV}{\hbar^2}$$

so that

$$\frac{U}{2k^2} = \frac{2mV}{2\hbar^2 k^2} = \frac{V}{2E} . \qquad 19.59$$

With these results we can obtain a very straightforward physical interpretation for (19.58). The integral in (19.58) is

$$\int_{-\infty}^{z} dz' \frac{V(\vec{b}+\hat{k}z')}{2E} \qquad 19.60$$

and gives the shift in phase from the undeflected value kz. On the other hand, (19.60) represents a linear interpolation for the path of the undeflected particle through the potential as shown in fig. (19.3a).

So far we were able to compute only scattering in the forward

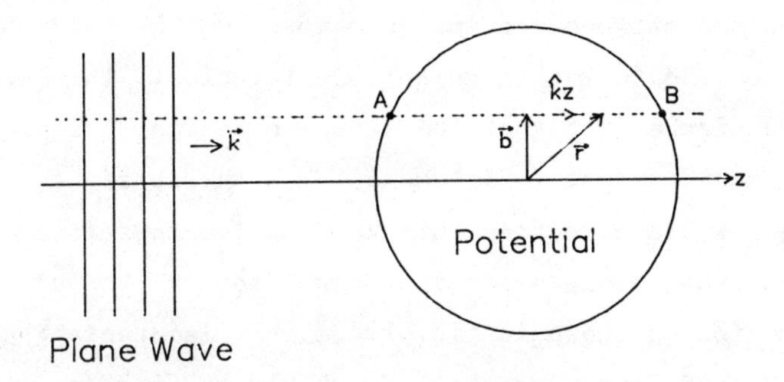

Fig. 19–3a Eikonal Approximation

direction. Since the scattering angle θ will be small for these high energies we can consider scattering from $\vec{k}$ to $\vec{k}'$, where $\cos\theta = \hat{k}\cdot\hat{k}'$, by interpolating through the interaction region using the classical trajectory as shown in fig. (19.3b). To accomplish this we simply recall that the local wave number at any point is given by

$$k(\vec{r}) = \frac{1}{\hbar}\sqrt{2m(E-V)}\ , \tag{19.61}$$

or for $V/E \ll 1$

$$k(\vec{r}) \simeq k\left(1-\frac{V}{2E}\right)\ , \tag{19.62}$$

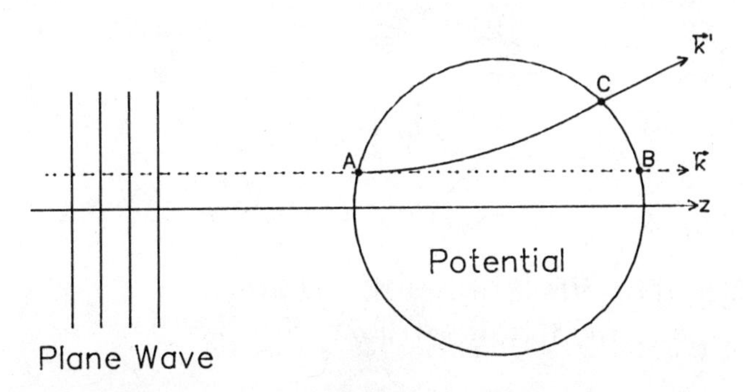

Fig. 19–3b Integration Along Classical Trajectory

which was the form used for scattering in the forward direction. To use (19.61) to interpolate between the incident momentum $\hbar\vec{k}$ and the final momentum $\hbar\vec{k}'$ would lead to rather cumbersome formulae. This would also not take advantage of the fact that the actual classical trajectory differs by only a small amount (θ is small) from the straight-line trajectory in the forward direction for $V \ll E$. We therefore compromise and integrate along a straight line from $\vec{k}$ to $\vec{k}'$ through the interaction region as shown in fig. 19.3c. It is crucial here that not only should $V \ll E$ but also that V should be of short range.

This interpolation is most easily accomplished by replacing $\hat{k}$ by a unit vector

$$\hat{K} = \frac{\vec{k} + \vec{k}'}{|\vec{k} + \vec{k}'|} \tag{19.63}$$

everywhere in (19.58). This then yields for the approximate full wave function for scattering from $\vec{k}$ to $\vec{k}'$

$$\psi^{(+)}_{\vec{k}}(\vec{b}+\hat{K}z) \simeq \exp i[\vec{k}(\vec{b}+\hat{K}z) - k \int_{-\infty}^{z} dz' \, \frac{V(\vec{b}+\hat{K}z')}{2E(k)}]. \tag{19.64}$$

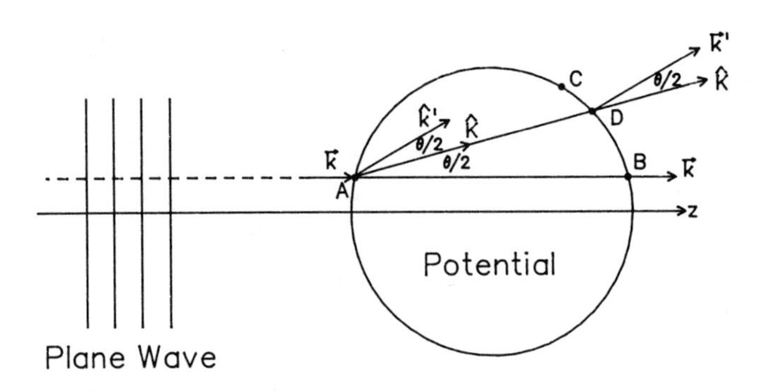

Fig. 19–3c Integration Along Interpolated Trajectory

Having obtained $\psi^{(+)}_{\vec{k}}$ we are now in a position to compute the scattering amplitude. Rather than evaluate the asymptotic behaviour of $\psi^{(+)}_{\vec{k}}$ we shall use the expression (17.84) for the scattering amplitude

$$T(\vec{k}',\vec{k}) = \langle\psi^{o}_{\vec{k}'}|V\psi^{(+)}_{\vec{k}}\rangle = -\frac{m}{2\pi\hbar^2}\int d^3r e^{-i\vec{k}'\cdot\vec{r}}V(\vec{r})\psi^{(+)}_{\vec{k}}(\vec{r}).$$

Now recalling that $\vec{r} = \vec{b} + \hat{k}z$ and that

$$(\vec{k}-\vec{k}').\hat{K} = 0 \tag{19.65}$$

we obtain upon writing the volume element as $d^3r = d^2b\,dz$ that

$$T(\vec{k}',\vec{k}) = -\frac{m}{2\pi\hbar^2}\int d^2b\,dz\,V(\vec{b}+\hat{K}z)e^{i(\vec{k}-\vec{k}')\cdot(\vec{b}+\hat{K}z)}$$

$$\times \exp - i[\frac{k}{2E(k)}\int_{-\infty}^{z} dz'V(\vec{b}+\hat{K}z') \tag{19.66}$$

$$= -\frac{m}{2\pi\hbar^2}\int d^2b\,dz\,V(\vec{b}+\hat{K}z)e^{i(\vec{k}-\vec{k}')\cdot\vec{b}}$$

$$\times \exp - i[\frac{k}{2E(k)}\int_{-\infty}^{z} dz'\,V(\vec{b}+\hat{K}z')]$$

We can carry out the z integration by writing

$$T(\vec{k}',\vec{k}) = \frac{-m}{2\pi\hbar^2}\int d^2b\,\frac{2iE}{k}\,e^{i(\vec{k}-\vec{k}')\cdot\vec{b}}\int_{-\infty}^{\infty} dz$$

$$\times \frac{d}{dz}\exp - i[\frac{k}{2E(k)}\int_{-\infty}^{z} dz'\,V(\vec{b}+\hat{K}z')]$$

Then,

$$T(\vec{k}',\vec{k}) = -\frac{ik}{2\pi}\int d^2b\; e^{i(\vec{k}-\vec{k}')\cdot\vec{b}}\left\{\exp-\frac{ik}{2E(k)}\int_{-\infty}^{\infty} dz\; V(\vec{b}+\hat{K}z)-1\right\}.$$

Now define the momentum transfer

$$\vec{q} = \vec{k}' - \vec{k}\ , \qquad q = 2k \sin\theta/2 \tag{19.67}$$

where θ is shown in fig. 19.3c. This vector $\vec{q}$ lies in the plane swept out by the vector $\vec{b}$ during the integration over d^2b.

If we now assume V to be a spherically symmetric potential, then we can write

$$d^2b = bdb\; d\phi$$

and carry out the ϕ integration

$$T(\vec{k}',\vec{k}) = -\frac{ik}{2\pi}\int_0^{\infty} bdb\;\left[\exp-\frac{ik}{2E(k)}\int_{-\infty}^{\infty} dz\; V(\sqrt{b^2+z^2})-1\right]\times$$

$$\times\int_0^{2\pi} e^{iqb\cos\phi}d\phi\ .$$

But the last integral is simply a representation for a Bessel function

$$\int_0^{2\pi} e^{iqb\cos\phi}d\phi = 2\pi\; J_o(qb).$$

Hence we can write

$$T(\vec{k},\vec{k}') = -ik\int_0^{\infty} db\; bJ_o(qb)[e^{2i\chi(b)} - 1] \tag{19.68}$$

where we have defined the phase angle

$$\chi(b) = -\frac{k}{4E}\int_{-\infty}^{\infty} dz\ V(\sqrt{b^2+z^2}) = -\frac{k}{2E}\int_{0}^{\infty} dz\ V(\sqrt{b^2+z^2})$$

or

$$\chi(b) = -\frac{k}{2E}\int_{b}^{\infty} \frac{rdr}{\sqrt{r^2-b^2}}\, V(r). \tag{19.69}$$

The interpretation of the scattering amplitude 19.68 becomes somewhat more transparent if we carry the semiclassical approximation somewhat further and replace kb by $(\ell + \frac{1}{2})$. Furthermore for large ℓ we have that

$$P_\ell(\cos\theta) \simeq (\cos\theta)^\ell J_o[(\ell+\frac{1}{2})\sin\theta] \tag{19.70}$$

and hence

$$P_\ell(\cos\theta) \underset{\theta\to 0}{\longrightarrow} J_o[(\ell+\frac{1}{2})\theta] . \tag{19.71}$$

Then recalling that $q = 2k\sin\frac{\theta}{2} \simeq \theta k$ we can replace the integral $\int_0^\infty d(kb)\ldots$ by a sum $\sum_{\ell=0}^{\infty} \ldots$ and get

$$T(\vec{k},\vec{k}') \simeq \frac{1}{2ik}\sum_\ell (2\ell+1)[e^{2i\chi_\ell}-1]P_\ell(\cos\theta).$$

This shows that the phase angle $\chi(b)$ is just an approximation for the phase shift corresponding to the ℓ'th partial wave where $\ell \simeq kb - \frac{1}{2}$.

Since the Glauber method is a semi-classical approximation it should be related to the WKB method. This is indeed the case. In fact the Glauber method can be derived from the WKB method if a small angle approximation is made at the appropriate stage in the WKB approach.

This completes our treatement of scattering problems. There are many more specialized approximation techniques. Also much physical insight has been gained from an analysis of the scattering amplitude as an analytic function of the energy and angular momentum. The singularities in this function (poles and branch points) are directly related to the occurrence of bound states, resonances and thresholds for reactions. The reader is strongly urged to consult some of the references listed.

So far we have also only considered problems that are reducible to one-body problems. In the next chapter we shall consider many particle systems and in particular the effects arising from the indistinguishability of particles.

Notes, References and Bibliography

The treatment of the eikonal approximation is based on the 1958 Boulder lectures by R.J. Glauber, Lectures in Theoretical Physics Vol. I; edited by W.E. Britten and L.G. Durham, Interscience Publishers, Inc., New York (1959).

All the references of chapter 17 and 18 are useful here. The book by de Alfaro and Regge is especially nice for a treatement of complex energies and complex angular momentum.

A variational principle for the s-wave phase shift was introduced by J. Schwinger, Phys. Rev. 72 742 (1947).

Schwinger's approach was used by J.M. Blatt and J.D. Jackson Phys. Rev. 76 18(1949).

See also:

H. Bethe Phys. Rev. 76 38(1949).

Chapter 19 Problems

19.1 Find the effective range and scattering length for the ($\ell = 0$) s-wave, given the potential

$$V = \begin{matrix} -V_0 & r < a \\ 0 & r \geq a . \end{matrix}$$

19.2 Repeat 19.1 for the Yukawa potential

$$V = -V_o \frac{e^{-\mu r}}{r} \; .$$

19.3 Both the Born and eikonal approximation apply for very high energies.
Calculate the scattering amplitude using the eikonal approximation for the Gaussian potential

$$V = V_o e^{-\mu r^2} \; .$$

Compare this with the Born amplitude (problem 18.8) for small angle scattering.

19.4 Repeat 19.3 for the attractive square well

$$V = \begin{matrix} -V_o & r < a \\ 0 & r \geq a \end{matrix}$$

and compare with the result of problem 18.9.

19.5 Use the results of problems 19.1 and 19.2 and fix the parameters of the Yukawa potential in terms of those of the square well so that both yield the same s-wave scattering length and effective range. The fact that this is possible is what is meant by calling this a shape-independent approximation.

19.6 a) What effect does a repulsive δ-function located at the origin have on the various partial waves?
b) Solve the partial-wave Lippmann-Schwinger equation for a potential

$$V = V_o \delta(r-a) \; .$$

19.7 a) Find the phase shifts for scattering by a hard sphere

$$V(r) = \begin{matrix} \infty & r < a \\ 0 & r \geq a \, . \end{matrix}$$

b) Find the total cross-section for an incoming energy $E = \frac{\hbar^2 k^2}{2m}$ in the two limits $k \to 0$, $k \to \infty$. Explain the factor of 2 difference.

Chapter 20

Systems of Identical Particles

One of the most profound and far-reaching consequences of quantum mechanics results from the indistinguishability of two identical particles. That two identical particles are indistinguishable seems a tautology. Nevertheless classically it is possible to follow (at least in principle) the trajectory of any particle. Thus in classical mechanics, if two identical particles interact we can, in principle, follow each particle even throughout the region of interaction until they are separated. In this sense the particles retain their individuality and are in fact distinguishable, in principle. Thus, in classical mechanics, no difference occurs in the treatment of a system of distinguishable or indistinguishable particles.

In quantum mechanics the situation is very different and that is why the discussion has been delayed until now. If we consider two identical particles that come together, interact and then separate, their individuality is lost. This occurs because during the interaction their wave functions must overlap (occupy the same portion of space at the same time). When the particles move apart it is impossible to tell which was which. This is a simple consequence of the fact that we cannot follow their individual trajectories and since they are identical we cannot say, even when they are again separated, where each one came from.

Of course this is not built into the theory a priori and we must now do so. To simplify the discussion we begin by considering a system of only two identical particles. Later we shall generalize this to an arbitrary number.

20.1 Two Identical Particles

Consider a system of two identical particles interacting via a two-body potential $V_{int}(\vec{x}_1,\vec{x}_2)$. Since action and reaction are equal and opposite we must have

$$V_{int}(\vec{x}_1,\vec{x}_2) = V_{int}(\vec{x}_2,\vec{x}_1) \ . \tag{20.1}$$

In addition to the two-body potential we assume the existence of an external force described by a potential $V(\vec{x})$. The hamiltonian for the two particles is then given by

$$H = T_1 + T_2 + V(\vec{x}_1) + V(\vec{x}_2) + V_{int}(\vec{x}_1,\vec{x}_2) \tag{20.2}$$

where T_i is the kinetic energy for the particle labelled i. Thus

$$T_i = \frac{\vec{p}_i^{\,2}}{2m} \ . \tag{20.3}$$

We write this hamiltonian as $H(1,2)$ to indicate the dependence on the particle labels. The indistinguishability of the particles is reflected in the fact that

$$H(1,2) = H(2,1) \tag{20.4}$$

If we let $\psi(1,2)$ be the wave function for the two particles then the

Schrödinger equation reads

$$H(1,2)\psi(1,2) = E\psi(1,2). \tag{20.5}$$

We now define a particle-exchange operator P_{12} with the property that

$$P_{12}\psi(1,2) = \psi(2,1). \tag{20.6}$$

Then as a consequence of 20.4 we have that

$$P_{12}H(1,2)P_{12}^{-1} = H(2,1) = H(1,2)$$

so that

$$[P_{12},H(1,2)] = 0. \tag{20.7}$$

It therefore follows that the eigenstates of $H(1,2)$ can be labelled with the eigenvalues E of $H(1,2)$ and the eigenvalues α of P_{12} as well as whatever other labels are necessary. The eigenvalues of P_{12} are easily found since as 20.6 shows

$$P_{12}^2 = 1. \tag{20.8}$$

This implies that

$$\alpha^2 = 1 \quad , \quad \alpha = \pm 1 . \tag{20.9}$$

The two eigenvalues correspond to two physically very different types of particles, known as bosons for $\alpha = 1$ and fermions for $\alpha = -1$.

For bosons the wave function is symmetric under the interchange of particle labels

$$\psi(1,2) = \psi(2,1).$$

For fermions the wave function is antisymmetric under the interchange of a pair of particle labels

$$\psi(1,2) = -\psi(2,1).$$

These simple rules when generalized to a system of N particles have very far-reaching consequences. Before carrying out this generalization we consider in more detail one specific problem involving two electrons.

20.2 The Hydrogen Molecule

The hydrogen molecule consists of two hydrogen atoms bound together. Thus we have two protons and two electrons as shown in fig. 20.1. Since the protons are much more massive than the electrons we neglect their motion and treat them as fixed centers of force. With this approximation the hamiltonian for a hydrogen molecule becomes

$$H = \frac{p_1^2}{2m} - \frac{e^2}{r_1} + \frac{p_2^2}{2m} - \frac{e^2}{r_2} + \frac{e^2}{R} + \frac{e^2}{r_{12}} - \frac{e^2}{r_{1B}} - \frac{e^2}{r_{2A}} \tag{20.10}$$

where the various quantities are labelled in fig. 20.1. We write this hamiltonian as

$$H = H_o + H' \tag{20.11}$$

where H_o is the "hydrogen atom" part of the hamiltonian

$$H_o = \frac{p_1^2}{2m} - \frac{e^2}{r_1} + \frac{p_2^2}{2m} - \frac{e^2}{r_2} . \tag{20.12}$$

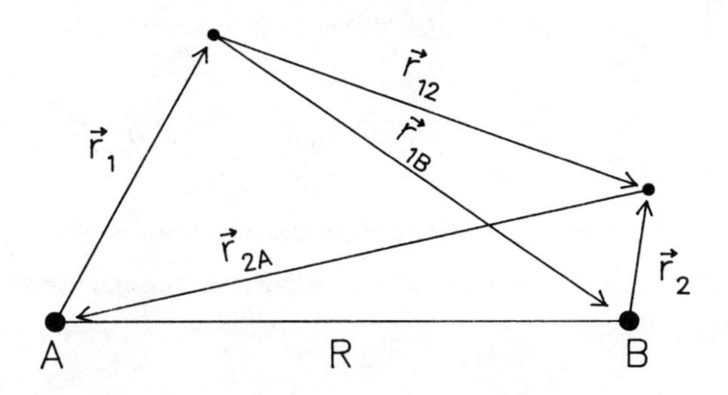

Fig. 20–1 Hydrogen Molecule

and

$$H' = \frac{e^2}{r_{12}} - \frac{e^2}{r_{1B}} - \frac{e^2}{r_{2A}} + \frac{e^2}{R} . \qquad 20.13$$

The hamiltonian 20.10 is of precisely the same form as the hamiltonian 20.2. Here the two-body potential is e^2/r_{12} , whereas $-(\frac{e^2}{r_1} + \frac{e^2}{r_{1B}})$ is the "external" potential for particle 1 and $-(\frac{e^2}{r_2} + \frac{e^2}{r_{2A}})$ is the "external" potential for particle 2 . The repulsive potential e^2/R between the two protons is just a constant.

To find the ground state energy of the hydrogen molecule we treat H' as a perturbation. In that case we only need to find a good approximation for the ground state wave function. The energy is then approximated by the expectation value of H in this approximate ground state. Now for R sufficiently large we have two unperturbed hydrogen atoms with the hamiltonian H_o. We use this for our approximate ground state.

Electrons are fermions. Thus the total wavefunction must be antisymmetric in the particle labels, and hence we must include the spin. There are two states of good total spin: the triplet state $\chi^{(t)}$, with $s = 1$, which is even in the two particle labels, and the singlet state $\chi^{(s)}$, with $s = 0$, which is odd in the two particle labels. The corresponding spatial wave functions are therefore,

$$\psi_s(r_1,r_2) = A_s[\psi_A(1)\psi_B(2) + \psi_A(2)\psi_B(1)]$$

and

$$\psi_t(r_1,r_2) = A_t[\psi_A(1)\psi_B(2) - \psi_A(2)\psi_B(1)]$$

where A_s and A_t are normalization constants, and $\psi_A(j)$, $\psi_B(k)$ represent hydrogen atom ground state wave functions centered at A and B respectively. Thus,

$$\psi_A(1) = (\pi a^3)^{-1/2} e^{-r_1/a} \qquad \psi_B(2) = (\pi a^3)^{-1/2} e^{-r_2/a}$$
$$\psi_A(2) = (\pi a^3)^{-1/2} e^{-r_{2A}/a} \qquad \psi_B(1) = (\pi a^3)^{-1/2} e^{-r_{1B}/a} \tag{20.14}$$

where a is the Bohr radius $a = \hbar^2/2me^2$.

The normalizations A_t and A_s are given by

$$1 = |A_s|^2\{1+2|(\psi_A,\psi_B)|^2+1\}$$

$$1 = |A_t|^2\{1-2|(\psi_A,\psi_B)|^2+1\}$$

so

$$A_s = [2(1+A^2)]^{-1/2} \tag{20.15}$$

$$A_t = [2(1-A^2)]^{-1/2} \tag{20.16}$$

where the overlap integral A is given by

$$A = \frac{1}{\pi a^3}\int e^{-(r_1+r_{1B})/a} d^3r_1 = (\psi_A,\psi_B) \tag{20.17}$$

To evaluate this integral we transform to elliptical coordinates

$$\xi = \frac{r_1+r_{1B}}{R}, \quad \eta = \frac{r_1-r_{1B}}{R}, \quad \phi \tag{20.18}$$

where ϕ is the angle of rotation about the line joining the two protons. The volume element in these coordinates is (ref. 20.2)

$$d^3r = \frac{R^3}{8}(\xi^2 - \eta^2)d\xi d\eta d\phi \tag{20.19}$$

with the range of integration

$$1 \leq \xi < \infty, \quad -1 \leq \eta \leq 1, \quad 0 \leq \phi \leq 2\pi. \tag{20.20}$$

Thus,

$$\begin{aligned} A &= \frac{1}{8\pi}\left(\frac{R}{a}\right)^3 \int_1^\infty e^{-\frac{R}{a}\xi} d\xi \int_{-1}^{1} (\xi^2 - \eta^2) d\eta \int_0^{2\pi} d\phi \\ &= \left[1 + \frac{R}{a} + \frac{1}{3}\left(\frac{R}{a}\right)^2\right] e^{-R/a} . \end{aligned} \tag{20.21}$$

To the extent that ψ_s and ψ_t are good wave functions, the ground state energy of the hydrogen molecule can be approximated in the singlet and triplet states by

$$\begin{aligned} E_s(R) &= \int \psi_s(1,2) H \psi_s(1,2) d^3r_1 d^3r_2 \\ E_t(R) &= \int \psi_t(1,2) H \psi_t(1,2) d^3r_1 d^3r_2 . \end{aligned} \tag{20.22}$$

These integrals can be rewritten with a little algebra in the form

$$E_s(R) = \frac{J+K}{1+A^2} \qquad E_t(R) = \frac{J-K}{1-A^2} \tag{20.23}$$

where

$$\begin{aligned} J &= \int |\psi_A(1)\psi_B(2)|^2 \left(\frac{e^2}{r_{12}} + \frac{e^2}{R} - \frac{e^2}{r_{2A}} - \frac{e^2}{r_{1B}}\right) d^3r_1 d^3r_2 \\ &= \frac{e^2}{R} - \int |\psi_A(1)|^2 \frac{e^2}{r_{1B}} d^3r_1 - \int |\psi_B(2)|^2 \frac{e^2}{r_{2A}} d^3r_2 \\ &\quad + \int |\psi_A(1)|^2 \frac{e^2}{r_{12}} |\psi_B(2)|^2 d^3r_1 d^3r_2 \end{aligned} \tag{20.24}$$

and

$$K = \int \phi_A(1)\phi_B(2)\left(\frac{e^2}{r_{12}} + \frac{e^2}{R} - \frac{e^2}{r_{2A}} - \frac{e^2}{r_{1B}}\right)\phi_A(2)\phi_B(1)\; d^3r_1 d^3r_2$$

$$= \frac{e^2}{R} A^2 - A\int\phi_B(2)\,\frac{e^2}{r_{2A}}\,\phi_A(2)d^3r_2 - A\int\phi_A(1)\,\frac{e^2}{r_{1B}}\,\phi_B(1)d^3r_1$$

$$+ \int\phi_A(1)\phi_B(2)\,\frac{e^2}{r_{12}}\,\phi_A(2)\phi_B(1)d^3r_1 d^3r_2 \;. \qquad 20.25$$

The integral J is called the Coulomb Interaction Integral and contains the following terms. The first term gives the Coulomb repulsion of the two protons. The second term gives the energy due to the interaction of the proton at B with the "charge density" $-e|\phi_A(1)|^2$ due to electron 1 at A, while the third term gives the interaction energy between the proton at A and the "charge density" $-e|\phi_B(2)|^2$ of electron 2 at B and is therefore equal to the second term. The last term gives the interaction between the two "charge densities" $-e|\phi_A(1)|^2$ and $-e|\phi_B(2)|^2$ of the two electrons centered at A and B.

The integral K is called the exchange energy and results strictly from the indistinguishability of the two electrons. It is this type of term that gives rise to covalent bonding.

The various integrals are evaluated as follows:

$$\int|\phi_A(1)|^2\,\frac{e^2}{r_{1B}}\,d^3r_1 = \int|\phi_B(2)|^2\,\frac{e^2}{r_{2A}}\,d^3r_2$$

are evaluated by the use of elliptical coordinates (eqns. 20.18,

20.19, 20.20) and yield

$$\int |\psi_A(1)|^2 \frac{e^2}{r_{1B}} d^3r_1 = \frac{\pi R^2}{2} \left[\int_1^\infty \xi\, e^{-\frac{R}{a}\xi} d\xi \int_{-1}^1 e^{-\frac{R}{a}\eta} d\eta + \int_1^\infty e^{-\frac{R}{a}\xi} d\xi \int_{-1}^1 \eta e^{-\frac{R}{a}\eta} d\eta \right]$$

$$= \frac{\pi a^3}{R} [1 - e^{-\frac{2R}{a}}(1 + \frac{R}{a})]. \qquad 20.26$$

Also we can write

$$\int |\psi_A(1)|^2 \frac{e^2}{r_{12}} |\psi_B(2)|^2 d^3r_1 d^3r_2 = \int \rho_A(1)\phi_B(1) d^3r_1 \qquad 20.27$$

where

$$\rho_A(1) = -e|\psi_A(1)|^2 \qquad 20.28$$

is the "charge density" due to electron 1 centered at A and $\phi_B(1)$ is the "potential" at r_1 due to electron 2 with charge density $-e|\psi_B(2)|^2$ centered at B. Thus $\phi_B(1)$ satifies the Poisson equation

$$\nabla^2 \phi_B(\vec{r}_1) = 4\pi e|\psi_B(2)|^2 . \qquad 20.29$$

This "potential" ϕ_B is calculated in several books on electrostatics (ref. 20.1). Hence we can combine all these results and get

$$J = \frac{e^2}{R} e^{-2\frac{R}{a}} [1 + \frac{5}{8}\frac{R}{a} - \frac{3}{4}(\frac{R}{a})^2 - \frac{1}{6}(\frac{R}{a})^3]. \qquad 20.30$$

For the exchange integral K the terms involving only one integration are again obtained by going to elliptical coordinates. The last term cannot be expressed in terms of elementary functions. It can however be expressed in terms of exponential integrals

$$E_i(x) = -\int_{-x}^{\infty} \frac{e^{-t}}{t}\, dt \ . \qquad 20.31$$

This was done by Sugiura (ref. 20.1). The final result for K is

$$\begin{aligned} K = {} & \frac{e^2}{R} A^2 [1+ \frac{6}{5}(C + \ln \frac{R}{a})] \\ & + \frac{e^2}{a} e^{-2\frac{R}{a}} [\frac{11}{8} + \frac{103}{20}\frac{R}{a} + \frac{49}{15}(\frac{R}{a})^2 + \frac{11}{15}(\frac{R}{a})^3] \\ & + \frac{e^2}{R}\frac{6}{5} e^{2\frac{R}{a}} [1- \frac{R}{a} + \frac{1}{3}(\frac{R}{a})^2]^2 E_i(-4\frac{R}{a}) \\ & - \frac{12}{5} A\, e^{\frac{R}{a}} [1- \frac{R}{a} + \frac{1}{3}(\frac{R}{a})^2] E_i(-2\frac{R}{a}) \end{aligned} \qquad 20.32$$

where $C = 0.577215...$ is Euler's constant.

If all these results are combined we obtain the results sketched in fig. 20.2 for the energies $E_s(R)$, $E_t(R)$.

Although the quantitative agreement with experiment is rather poor for the above computation, the qualitative features are as shown in fig. 20.2. The computation can again be improved by replacing ψ_A , ψ_B by

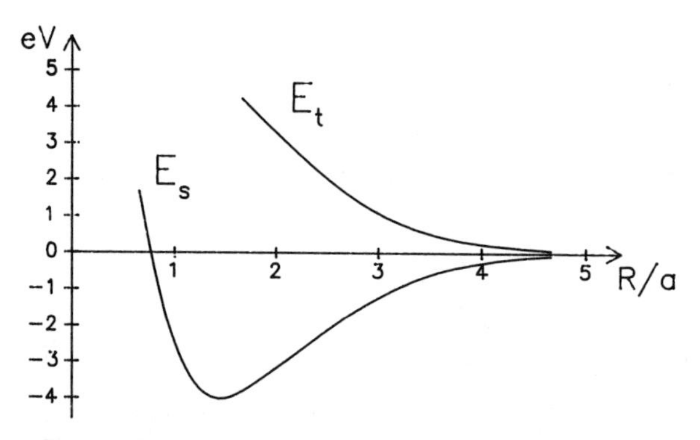

Fig. 20–2 Energy as a Function of Separation

wave functions with an effective proton charge Ze and treating Z as a variational parameter. This improves the quantitative agreement with experiment considerably.

The main point of this calculation, however, was to bring out the effect of the exchange energy. This energy is entirely a consequence of the antisymmetry of the electron wave function under the interchange of the two electrons. We shall now consider what happens in the case of several particles.

20.3 N Identical Particles

We now take up again the discussion started in section 1, except that we do not limit ourselves to two particles. The argument follows very closely the argument of section 1. The hamiltonian for a system of N identical particles is

$$H(1,2,\ldots,N) = \sum_{i=1}^{N} \left[\frac{p_i^2}{2m} + V(i)\right] + V_{int}(1,2,\ldots,N) \qquad 20.33$$

where $V(i)$ is an external potential acting on the i-th particle and $V_{int}(1,2,\ldots,N)$ represents the mutual interaction between the particles. If, as is usually the case, the interaction between the particles is due only to two- body interactions then

$$V_{int}(1,2,\ldots,N) = \frac{1}{2} \sum_{i \neq j}^{N} V_{int}(i,j) \qquad 20.34$$

where due to Newton's third law

$$V_{int}(i,j) = V_{int}(j,i) \; . \qquad 20.35$$

The labels i,j etc. represent all of the particle coordinates, internal as well as external. Combining equations 20.33, 20.34 and 20.35 we see that the N-particle hamiltonian is invariant under the interchange of particles. Thus it is again possible to introduce a particle exchange operator P_{ij} that commutes with the total hamiltonian. The action of P_{ij} is

$$P_{ij}\psi(1,2,\ldots,i,\ldots,j\ldots N) = \psi(1,2,\ldots j,\ldots,i\ldots N) \tag{20.36}$$

As for the case of two particles the eigenvalues of P_{ij} are ± 1 and can be used together with the energy E to label the eigenstates of the hamiltonian.

It is a remarkable fact, that greatly simplifies all considerations, that for a given type of particle all the eigenvalues of P_{ij} are either $+1$ or -1 and are never mixed. Particles with the eigenvalue $+1$ are called bosons and particles with the eigenvalue -1 are called fermions.

To illustrate these considerations we consider a wave function for three identical particles.

1) Three identical bosons:

$$\psi(1,2,3) = \psi(1,3,2) = \psi(2,1,3) = \psi(2,3,1)$$

$$= \psi(3,1,2) = \psi(3,2,1) \ .$$

2. Three identical fermions:

$$\psi(1,2,3) = -\psi(1,3,2) = -\psi(2,1,3) = \psi(2,3,1)$$

$$= \psi(3,1,2) = -\psi(3,2,1) \ .$$

The nature of a given type of particle is furthermore intimately connected with its spin. This result is summarized in the following theorem.

Spin and Statistics Theorem

All particles with integer spin (including spin zero) are bosons and all particles with half-odd integer spins are fermions.

The proof of this theorem is beyond the scope of this book since it requires the machinery of relativistic quantum field theory. We, therefore, content ourselves with the statement of this theorem. Experimentally it has been verified to a very high degree for electrons and also for photons.

There is one result, however, which follows quite straightforwardly from the requirement that a fermion wave function is totally antisymmetric. Thus, for a system of N fermions no two particles may occupy the same state.

This result follows quite trivially. Let $\psi_{k_1 k_2 \ldots k_N}(x_1, x_2, \ldots, x_N)$ be the totally antisymmetric wave function for a system of N non-interacting fermions. Here k_i is a complete labelling of the state occupied by particle i. If such a labelling is possible also for interacting fermions then the rest of our argument also applies to them. This antisymmetry means that

$$\begin{aligned} &\psi_{k_1 \ldots k_i \ldots k_j \ldots k_N}(x_1 \ldots x_i \ldots x_j \ldots x_N) \\ &= -\psi_{k_1 \ldots k_i \ldots k_j \ldots k_N}(x_1 \ldots x_j \ldots x_i \ldots x_N) \\ &= -\psi_{k_1 \ldots k_j \ldots k_i \ldots k_N}(x_1 \ldots x_i \ldots x_j \ldots x_N). \end{aligned} \tag{20.37}$$

If we now assume that two particles are in the same state, or what is equivalent that $k_i = k_j$ then we have the further result that

$$\begin{aligned} &\psi_{k_1 \ldots k_i \ldots k_j \ldots k_N}(x_1 \ldots x_i \ldots x_j \ldots x_N) \\ &= \psi_{k_1 \ldots k_j \ldots k_i \ldots k_N}(x_1 \ldots x_i \ldots x_j \ldots x_N) \\ &\text{for } k_i = k_j \,. \end{aligned} \tag{20.38}$$

Combining this result with eqn. 20.37 we obtain that for these fermions

$$\psi_{k_1..k_i..k_i..k_N}(x_1,x_2,..x_i,..,x_j..,x_N) = 0. \qquad 20.39$$

This result is known as the Pauli Exclusion Principle. It states that no two fermions (both interacting as well as non-interacting) may simultaneously occupy the same state. Under state we must understand here all the quantum numbers of the particles involved. Thus, particles with different mass or charge automatically have different states. This principle was originally postulated by Pauli to provide an explanation for the periodic table. Its consequence are, however, of much greater generality. Thus, for example, many of the properties of solids are a consequence of this principle.

20.4 Non-Interacting Fermions

Consider a system of N fermions that do not interact with each other but are subject to an external potential $V(x)$. The hamiltonian for this system is

$$H = \sum_{j=1}^{N} [\frac{p_j^2}{2m} + V(x_j)] = \sum_{j=1}^{N} H(\vec{x}_j) = \sum_{j=1}^{N} H_j . \qquad 20.40$$

Since the index j simply labels the various particles, their corresponding single particle hamiltonians H_j commute

$$[H_j,H_k] = 0 \qquad 20.41$$

and hence

$$[H_j,H] = 0 . \qquad 20.42$$

Thus, to diagonalize the N-particle hamiltonian 20.40 we need only diagonalize the single-particle hamiltonians. This is, of course, a consequence of the unrealistic assumption of no interaction between the particles.

We now let $\{|k\rangle, k = 0,1,2,\ldots\}$ be a complete set of one-particle states satisfying the one-particle Schrödinger equation

$$H_j|k\rangle = E_k|k\rangle \ . \tag{20.43}$$

An eigenstate of the total hamiltonian is then given by $\prod_{j=1}^{N}|k_j\rangle$. This state must of course be totally antisymmetrized. If A is an antisymmetrization operator then the N particle fermion eigenfunction of the hamiltonian 20.40 is

$$|k_1,k_2,\ldots,k_N\rangle = A|k_1\rangle|k_2\rangle \ldots |k_N\rangle \ . \tag{20.44}$$

Here k_j is the value of k for the j-th particle. Clearly, we also have

$$H|k_1,\ldots,k_N\rangle = (E_{k_1} + \ldots + E_{k_N})|k_1,\ldots,k_N\rangle \ . \tag{20.45}$$

We shall now repeat the procedure above for a fixed representation. This will make it easy to carry out the appropriate antisymmetrization.

The one-particle states in configuration space are given by

$$\phi_k(x_j) = \langle x_j|k\rangle. \tag{20.46}$$

We again emphasize that the labels k are a <u>complete set of labels for the one-particle states</u> and that x_j denotes the configuration space variable for the j-th particle. In this fixed coordinate representation, a totally antisymmetric eigenstate of the full hamiltonian is given by a so- called Slater determinant.

Thus, if the single-particle wave functions are normalized then the normalized and anti-symmetrized N-particle wave function is given by the following determinant known as a Slater determinant.

$$\psi_{k_1\ldots k_N}(x_1,\ldots,x_N)= \frac{1}{\sqrt{N!}} \det\begin{pmatrix} \phi_{k_1}(x_1) & \phi_{k_1}(x_2) & \cdots & \phi_{k_1}(x_N) \\ \phi_{k_2}(x_1) & \phi_{k_2}(x_2) & \cdots & \phi_{k_2}(x_N) \\ \vdots & \vdots & & \\ \phi_{k_N}(x_1) & \phi_{k_N}(x_2) & \cdots & \phi_{k_N}(x_N) \end{pmatrix} \qquad 20.47$$

The antisymmetry of a determinant under the interchange of rows or columns make it obvious that this wave function has the required antisymmetry. The fact that the wave function can be written as a sum of products of single-particle wave functions is again a consequence of the unphysical assumption of no interaction between the individual particles.

Nevertheless a wave-function of the form of a Slater determinant is frequently the starting point for a computation involving an interacting fermion system. The main justification for this is the simplicity of the approach.

20.5 Non-Interacting Bosons

We now repeat the discussion of the previous section for a system of N bosons that do not interact with each other but are subject to an external potential $V(x)$. Again we have an hamiltonian consisting of a sum of single- particle hamiltonians

$$H = \sum_{j=1}^{N} [\frac{p_j^2}{2m} + V(x_j)] \equiv \sum_{j=1}^{N} H_j \,. \qquad 20.48$$

We again start with a complete set of one-particles states $\{|k\rangle, k = 0,1,2,\ldots\}$ that are eigenstates of the single particle hamiltonian

$$H_j|k\rangle = E_k|k\rangle \,. \qquad 20.49$$

An eigenstate of the hamiltonian is then given by a symmetrized product of such single particle states. Thus, we have the N boson state

$$|k_1,\ldots,k_N\rangle = \frac{1}{\sqrt{N!}}\sum_P |k_1\rangle|k_2\rangle\ldots|k_N\rangle \qquad 20.50$$

where the sum is over all $N!$ permutations of the N particle labels.

It is again straighforward to go to a fixed representation and construct the analogue of a Slater determinant. We start with a normalized set of one-particle wave-functions in configuration space

$$\phi_k(x_j) = \langle x_j|k\rangle \ . \qquad 20.51$$

In terms of these an N particle boson state is given by the following permanent

$$\phi_{k_1\ldots k_N}(x_1,\ldots,x_N) = \frac{1}{\sqrt{N!}}\,\mathrm{Perm}\begin{pmatrix} \phi_{k_1}(x_1) & \phi_{k_1}(x_2) & \cdots & \phi_{k_1}(x_N) \\ \phi_{k_2}(x_1) & \phi_{k_2}(x_2) & \cdots & \phi_{k_2}(x_N) \\ \vdots & \vdots & & \\ \phi_{k_N}(x_1) & \phi_{k_N}(x_2) & \cdots & \phi_{k_N}(x_N) \end{pmatrix} \qquad 20.52$$

where "Perm" means a permanent, which is the same thing as a determinant without changes of sign.

Thus, again we have the N-particle state expressed as a sum of products of one-particle states. As for the case of fermions this is again a consequence of the lack of interaction between the particles.

In the next section we shall refine this method of description even further. It is to be remembered, however, that the labels are single-particle labels. Thus, we do not have a convenient machinery for handling collective phenomena such as particle correlations or phase transitions, starting from a basis of one-particle states.

20.6 Occupation Number Space and Second Quantization for Bosons

In this section we develop an elegant method for handling the symmetrization of boson states. In order to do this we first generalize from a system with a definite number N of particles to a system with an arbitrary number of particles.

We again start with a complete basis of one particle states $\{|k\rangle, k = 0,1,2,\ldots\}$, and wavefunctions $\phi_k(x_j)$. An N-boson state can now be described by stating that particle 1 is in the state $|k_1\rangle$, particle 2 is in the state $|k_2\rangle$ etc. and then symmetrizing to get the state $|k_1,\ldots,k_N\rangle$.

Another but completely equivalent way of stating this is to say that there are n_o particles in the state $|0\rangle$, n_1 particles in the state $|1\rangle$ etc. up to n_∞. The reason this specification of an N boson state by the sequence of integers $|n_1,n_2,\ldots,n_\infty\rangle$ is equivalent to giving the state $|k_1,\ldots,k_N\rangle$ explicitly is due to the fact that the state $|k_1,\ldots k_N\rangle$ is totally symmetric. Thus we can define a state of N bosons in occupation number space by an infinite sequence of non-negative integers $|n_1,n_2,\ldots n_\infty\rangle$ such that

$$\sum_{k=0}^{\infty} n_k = N. \qquad 20.53$$

In this way of specifying states, the four particle states $|0,2,5,5\rangle$ corresponding to a symmetric product of the states $|0\rangle$, $|2\rangle$, $|5\rangle$, $|5\rangle$ is given by $|1,0,1,0,0,2,0, 0,0,\ldots,0,\ldots\rangle$. This says that there is one particle in the ground state $k = 0$, one particle in the state $k = 2$ and two particles in the state $k = 5$.

If the one-particle states are the eigenstates of a one-particle hamiltonian $H(x_j)$ such that the N-particle hamiltonian is

$$H = \sum_{j=1}^{N} H(x_j) \qquad 20.54$$

then the state $|n_o,n_1,n_2\ldots\rangle$ is an eigenstate of H. The eigenvalue of H is given in terms of the single particle energies E_k

$$H(x_j)|k\rangle = E_k|k\rangle \ . \tag{20.55}$$

In fact,

$$H|n_o,n_1,\ldots\rangle = (\sum_{k=0}^{\infty} n_k E_k)|n_o,n_1,\ldots\rangle \ . \tag{20.56}$$

It therefore becomes convenient to introduce operators N_k such that

$$N_k|n_o,n_1,\ldots,n_k,\ldots\rangle = n_k|n_o,n_1,\ldots,n_k\ldots\rangle. \tag{20.57}$$

Thus the operators N_k "count" the number of particles in the state $|k\rangle$ and are appropriately named number operators. The hamiltonian may now be written

$$H = \sum_{k=0}^{\infty} E_k N_k \ . \tag{20.58}$$

We now take the further step and write N_k in terms of annihilation and creation operators

$$N_k = a_k^\dagger a_k \tag{20.59}$$

where we assume the commutation relation

$$[a_k,a_\ell^\dagger] = \delta_{k\ell} \tag{20.60}$$

with all other operators commuting

$$[a_k,a_\ell] = [a_k^\dagger,a_\ell^\dagger] = 0 \ . \tag{20.61}$$

With this notation we now have

$$H = \sum_{k=0}^{\infty} E_k a_k^{\dagger} a_k \ . \qquad 20.62$$

The ground state (no-particle state or vacuum) is $|0,0, 0,\ldots,0,\ldots\rangle$ and is annihilated by all the annihilation operators. A general state can now be written as

$$|n_o,n_1,\ldots\rangle = \frac{(a_o^{\dagger})^{n_o}}{\sqrt{n_o!}} \frac{(a_1^{\dagger})^{n_1}}{\sqrt{n_1!}} \ldots |0,0,0,\ldots\rangle \qquad 20.63$$

An N-particle state in configuration space is furthermore given by

$$\Psi^{(N)}_{\{n_o,n_1,\ldots\}}(x_1,x_2,\ldots,x_N) = \langle x_1,x_2,\ldots,x_N|n_o,n_1,\ldots\rangle. \qquad 20.64$$

Thus, the completeness relation for states in occupation number space is

$$\sum_{\{n\}} |n_o,n_1,\ldots,n_k,\ldots\rangle\langle n_o,n_1,\ldots,n_k,\ldots| = 1 \qquad 20.65$$

where the sum extends over all sequences $\{n_o,n_1,\ldots\}$ with a fixed number N of particles $\sum_{k=0}^{\infty} n_k = N$.

Similarly the orthonormality of these states is expressed by

$$\langle n_o,n_1,\ldots,n_k,\ldots|n_o'n_1',\ldots,n_k'\ldots\rangle = \prod_{k=0}^{\infty} \delta_{n_k n_k'} \ . \qquad 20.66$$

It is therefore straightforward to write an arbitrary state in terms of the occupation number space basis

$$|\Psi\rangle = \sum_{\{n\}} |n_o,n_1,\ldots\rangle\langle n_o,n_1,\ldots|\Psi\rangle \ . \qquad 20.67$$

The coefficients $\langle n_o, n_1, \ldots | \Psi \rangle$ are the probability amplitudes for finding n_o particles in the state $|0\rangle$, n_1 particles in the state $|1\rangle$, etc. if the state of the system is $|\Psi\rangle$. These coefficients are, in fact, nothing else but the permanent given by eqn. 20.52.

In arriving at eqn. 20.63 we have in fact rewritten a system of N non-ineracting bosons like an infinite system of harmonic oscillators. This procedure, at present, is nothing other than an elegant formalism for incorporating the symmetry of bose states. It turns out, however, to have far reaching consequences in solid state physics where many computations are facilitated, and much physical insight is gained from this machinery.

20.7 Occupation Number Space and Second Quantization for Fermions

We shall now carry out a discussion parallel to that of section 6 except that we shall be dealing with a system of fermions. It will again be convenient to generalize from a system with a fixed number N of particles to a system with an arbitrary number of particles.

As before let $\{|k\rangle, k = 0,1,2,\ldots\}$ be a complete basis of one-particle states. An N-fermion state is then given if we state that particle 1 is in the state $|k_1\rangle$, particle 2 is in the state $|k_2\rangle$ etc. We then simply antisymmetrize and obtain the N-fermion state $|k_1, k_2, \ldots, k_N\rangle$. Due to the antisymmetry none of the k_j may coincide.

The next step now is to introduce the occupation number space representation for these states. Thus we define a state $|n_o, n_1, n_2, \ldots\rangle$ where each n_k is either 0 or 1. If $n_k = 0$ it means there is no particle in the state $|k\rangle$, whereas if $n_k = 1$ there is exactly one particle in the state $|k\rangle$. The antisymmetry of fermion states restricts the occupation numbers n_k to the two values (0,1). On the other hand the antisymmetry further gives a one to one relation between an N-fermion state $|k_1, k_2, \ldots, k_N\rangle$ and a state $|n_o, n_1, \ldots\rangle$ where

$$\sum_{k=0}^{\infty} n_k = N \, . \tag{20.68}$$

For example, the 2-particle state $|0,3\rangle$ corresponding to

$$|0,3\rangle = \frac{1}{\sqrt{2}} (|0\rangle|3\rangle - |3\rangle|0\rangle \qquad 20.69$$

where the first state on the right of 20.69 refers to particle 1 and the second state to particle 2 is now written as $|1,0,0,1,0,0,\ldots\rangle$. This states that there is one particle in the ground state $k = 0$ and one particle in the excited state $k = 3$. Notice again that due to the Pauli exclusion principle we can have at most one particle in a given state.

We now assume that the one-particle states $\{|k\rangle,\ k = 0,1,2,\ldots\}$ are eigenstates of a one-particle hamiltonian $H(x_j)$ such that the N-particle hamiltonian is

$$H = \sum_{j=1}^{N} H(x_j) \ . \qquad 20.70$$

Then we have that

$$H(x_j)|k\rangle = E_k|k\rangle \qquad 20.71$$

and the state $|n_o,n_1,\ldots,n\rangle$ is an eigenstate of H

$$H|n_o,n_1,\ldots\rangle = (\sum_{k=0}^{\infty} n_k E_k)|n_o,n_1,\ldots\rangle \ . \qquad 20.72$$

We now again introduce number operators N_k that state whether the state $|k\rangle$ is occupied or not. Thus they must have the eigenvalues 0,1. This requires that

$$N_k^2 = N_k \ . \qquad 20.73$$

We next try writing the number operators in terms of annihilation and creation operators

$$N_k = a_k^\dagger a_k. \tag{20.74}$$

This time, however, a_k and $a_k^\dagger$ cannot satisfy the commutation relations 20.60, 20.61 for otherwise we would have boson operators as before and the number operators would have all non-negative integers as eigenvalues and would therefore not satisfy 20.73. Furthermore the states produced would be symmetric.

To ensure that a state $|n_o, n_1, \ldots\rangle$ can still be written in the form

$$|n_o, n_1, n_2, \ldots\rangle = (a_o^\dagger)^{n_o} (a_1^\dagger)^{n_1} \ldots |0,0,\ldots\rangle \tag{20.75}$$

with not more than one particle in a given state requires that

$$(a_k^\dagger)^2 = 0. \tag{20.76}$$

This immediately implies that

$$(a_k)^2 = 0 \,. \tag{20.77}$$

Furthermore the two-particle state

$$|0,\ldots 0,\underset{k}{1},0\ldots,0,\underset{\ell}{1},0,\ldots\rangle = a_k^\dagger a_\ell^\dagger |0,\ldots,0,0,\ldots,0,\ldots\rangle \tag{20.78}$$

is antisymmetric in k and ℓ. This requires that

$$a_k^\dagger a_\ell^\dagger = -a_\ell^\dagger a_k^\dagger$$

or

$$a_k^\dagger a_\ell^\dagger + a_\ell^\dagger a_k^\dagger = 0. \tag{20.79}$$

Again the hermitian adjoint of this equation yields

$$a_k a_\ell + a_\ell a_k = 0. \qquad 20.80$$

Equations 20.79 and 20.80 imply 20.76 and 20.77. If we now further postulate that

$$a_k^\dagger a_\ell + a_\ell a_k^\dagger = \delta_{k\ell} \qquad 20.81$$

it is a simple matter to verify 20.73. Thus we need $N_k^2 = N_k$, but this reads

$$a_k^\dagger a_k a_k^\dagger a_k = a_k^\dagger a_k (a_k^\dagger a_k + a_k a_k^\dagger) = a_k^\dagger a_k \qquad 20.82$$

as required. The first equality follows from $a_k^2 = 0$ and the second from 20.81.

Thus fermion creation and annihilation operators satisfy the anti-commutation relations given by 20.79, 20.80 and 20.81. We shall henceforth write these equations as

$$[a_k, a_\ell]_+ = [a_k^\dagger, a_\ell^\dagger]_+ = 0 \qquad 20.83$$

and

$$[a_k^\dagger, a_\ell]_+ = \delta_{k\ell} \qquad 20.84$$

where

$$[A,B]_+ \equiv AB + BA \qquad 20.85$$

is called the anti-commutator of A and B.

Computations for fermions may now be carried out in the same manner as for bosons except that the creation and annihilation operators satisfy anti-commutation instead of commutation relations. Thus the hamiltonian 20.70 may now be written

$$H = \sum_{k=0}^{\infty} E_k a_k^{\dagger} a_k . \qquad 20.86$$

An N-particle state is now picked out by restricting ourselves to that subspace of occupation number space for which

$$\sum_{k=0}^{\infty} n_k = N. \qquad 20.87$$

Corresponding to this it is convenient to define a total particle number operator $\mathcal{N}$ such that

$$\mathcal{N} = \sum_{k=0}^{\infty} N_k = \sum_{k=0}^{\infty} a_k^{\dagger} a_k . \qquad 20.88$$

An N particle state then belongs to the subspace for which the total number operator $\mathcal{N}$ has the eigenvalue N.

Again an arbitrary state $|\Psi\rangle$ may be expanded in a basis of occupation number states. Thus,

$$|\Psi\rangle = \sum_{\{n\}} |n_o, n_1, \ldots\rangle\langle n_o, n_1, \ldots|\Psi\rangle \qquad 20.89$$

just as for the case of bosons. The completeness and orthonormality relations are also the same as for bosons namely 20.65 and 20.66. The expansion coefficients $\langle n_o, n_1, \ldots|\Psi\rangle$ are just the Slater determinant 20.47 and give the probability amplitude for finding n_o particles in the state $|0\rangle$, n_1 particles in the state $|1\rangle$, etc. Throughout, of course, all the n_k are either 0 or 1.

20.8 Field Operators

In this section we shall treat the fermi and bose case at the same time. Where necessary we shall indicate the differences. We begin by defining the field operators $\psi(\vec{x})$, $\psi^\dagger(\vec{x})$

$$\psi(\vec{x}) = \sum_{k=0}^{\infty} \langle\vec{x}|k\rangle a_k \qquad 20.90$$

$$\psi^\dagger(\vec{x}) = \sum_{k=0}^{\infty} \langle k|\vec{x}\rangle a_k^\dagger \ . \qquad 20.91$$

We then have the following (anti) commutation relations

$$\begin{aligned}[\psi(\vec{x}),\psi^\dagger(\vec{y})]_\pm &= \sum_{j,k=0}^{\infty} \langle\vec{x}|j\rangle\langle k|\vec{y}\rangle[a_j,a_k^\dagger]_\pm \\ &= \sum_{k=0}^{\infty} \langle\vec{x}|k\rangle\langle k|\vec{y}\rangle = \langle\vec{x}|\vec{y}\rangle = \delta(\vec{x}-\vec{y}).\end{aligned} \qquad 20.92$$

Also

$$[\psi(\vec{x}),\psi(\vec{y})]_\pm = [\psi^\dagger(\vec{x}),\psi^\dagger(\vec{y})]_\pm = 0 \ . \qquad 20.93$$

Throughout, the upper sign applies to fermions and the lower sign to bosons.

The inverse of eqns. 20.90, 20.91 is

$$a_k = \int \langle\vec{k}|\vec{x}\rangle\psi(\vec{x})d^3x \qquad 20.94$$

$$a_k^\dagger = \int \langle\vec{x}|k\rangle\psi^\dagger(\vec{x})d^3x \qquad 20.95$$

where we have used the closure or completeness condition

$$\int |\vec{x}\rangle\langle\vec{x}|d^3x = 1 \ . \qquad 20.96$$

The total number operator N may now be expressed in terms of the field operators. Thus we have

$$N = \sum_{k=0}^{\infty} a_k^\dagger a_k = \sum_{k=0}^{\infty} \int d^3x d^3y \langle \vec{x}|k\rangle\langle \vec{k}|\vec{y}\rangle \psi^\dagger(\vec{x})\psi(\vec{y})$$

$$= \int d^3x\, \psi^\dagger(\vec{x})\psi(\vec{x}). \tag{20.97}$$

The operator $\psi^\dagger(\vec{x})\psi(\vec{x})$ may therefore be interpreted as a particle density operator. To further obtain an interpretation of the field operators we calculate their <u>commutator</u> with N for both fermion and boson operators.

$$[N,\psi^\dagger(\vec{x})] = \int d^3y\{\psi^\dagger(\vec{y})\psi(\vec{y})\psi^\dagger(\vec{x}) - \psi^\dagger(\vec{x})\psi^\dagger(\vec{y})\psi(\vec{y})\}$$

$$= \int d^3y\, \psi^\dagger(\vec{y})\delta(\vec{x}-\vec{y}) = \psi^\dagger(\vec{x}) . \tag{20.98}$$

This identifies $\psi^\dagger(\vec{x})$ as a creation operator for those particles for which N is a number operator. Similarly we get that

$$[N,\psi(\vec{x})] = -\psi(\vec{x}) \tag{20.99}$$

and that $\psi(\vec{x})$ is an annihilation operator.

We now define a <u>vacuum</u> or <u>no-particle</u> state $|\Omega\rangle$ such that

$$\psi(\vec{x})|\Omega\rangle = 0. \tag{20.100}$$

This implies that $a_k|\Omega\rangle = 0$ for all k and hence this vacuum is the same as the state $|0,0,0,\ldots\rangle$ of occupation number space. That this is a no-particle state is further demonstrated by the fact that

$$N|\Omega\rangle = 0. \tag{20.101}$$

It then follows that $\psi^\dagger(\vec{x})|\Omega\rangle$ is a one-particle state, since using

20.98 we obtain

$$N\psi^\dagger(\vec{x})|\Omega\rangle = \{\psi^\dagger(\vec{x})N + [\psi^\dagger(\vec{x}),N]\}|\Omega\rangle$$
$$= 1\cdot\psi^\dagger(\vec{x})|\Omega\rangle. \tag{20.102}$$

The probability amplitude for finding the particle, which is in the state $\psi^\dagger(\vec{x})|\Omega\rangle$, in the state $|\alpha\rangle$ is $\langle\alpha|\psi^\dagger(\vec{x})|\Omega\rangle$. Thus if $|\alpha\rangle = |\vec{y}\rangle$ we get

$$\langle\vec{y}|\psi^\dagger(\vec{x})|\Omega\rangle = \langle\Omega|\psi(\vec{y})\psi^\dagger(\vec{x})|\Omega\rangle$$
$$= \delta(\vec{y}-\vec{x})\langle\Omega|\Omega\rangle = \delta(\vec{y}-\vec{x}) \tag{20.103}$$

showing that $\psi^\dagger(\vec{x})|\Omega\rangle$ is the state for a particle located at the point $\vec{x}$. Thus $\psi^\dagger(\vec{x})$ creates a particle localized at the point $\vec{x}$. Such states are not normalized and must be turned into wave-packets (smeared) before they belong to Hilbert space. That is, we find $\langle\Omega|\psi(\vec{x})\psi^\dagger(\vec{x})|\Omega\rangle = \delta(0)$.

To form wave-packets we can either smear the states $\psi^\dagger(\vec{x})|\Omega\rangle$ to get the states

$$|f\rangle = \int d^3x\ f(\vec{x})\psi^\dagger(\vec{x})|\Omega\rangle \tag{20.104}$$

or else we smear the operators to form operators

$$\psi^\dagger(f) = \int d^3x f(\vec{x})\psi^\dagger(\vec{x}) \tag{20.105}$$

from which we form states

$$|f\rangle = \psi^\dagger(f)|\Omega\rangle\ .$$

In either case the resultant states have a finite norm if $f(\vec{x})$ is square integrable since

$$\langle f|f\rangle = \int d^3x\, d^3y\, f^*(\vec{x})f(y)\langle\Omega|\psi(\vec{x})\psi^\dagger(\vec{y})|\Omega\rangle$$
$$= \int d^3x|f(\vec{x})|^2 \ . \qquad 20.106$$

The smeared operators may be viewed as creation and annihilation operators for particles localized according to the wave-packets $f(\vec{x})$. They satisfy the following (anti) commutation relations:

$$[\psi(f),\psi^\dagger(g)]_\pm = \int f^*(x)g(x)dx \qquad 20.107$$

$$[\psi(f),\psi(g)]_\pm = [\psi^\dagger(f),\psi^\dagger(g)]_\pm = 0 \ . \qquad 20.108$$

It is, therefore, quite natural to treat the probability amplitude $\langle\alpha|\psi^\dagger(x)|\Omega\rangle$ as a wavefunction $F^*_\alpha(\vec{x})$, the corresponding amplitude $\langle\Omega|\psi(\vec{x})|\alpha\rangle$ is the wavefunction $F_\alpha(\vec{x})$. The reason for our choice of complex conjugates will become clear a little later. Similarly a two particle wave function in configuration space is then given by $\langle\Omega|\psi(\vec{x})\psi(\vec{y})|\alpha_1\alpha_2\rangle$ and an N-particle wavefunction in configuration space is given by $\langle\Omega|\psi(\vec{x}_1)\psi(\vec{x}_2)\ldots\psi(\vec{x}_N)|\alpha_1,\alpha_2,\ldots,\alpha_N\rangle$. We shall use these results very soon to show how the second quantized formalism in terms of field operators may be used to describe N-particle Schrödinger equations.

20.9 Representation of Operators

In practice we know the representation of operators in configuration space. We shall now first rewrite them in the occupation number space representation. This will allow us to express them in terms of the field operators.

Thus, suppose we are given a local operator F in configuration space. If this operator does not change the number of particles then its configuration space representation is

$$\langle \vec{x}_1, \vec{x}_2, \ldots, \vec{x}_N | F | \vec{y}_1, \vec{y}_2, \ldots, \vec{y}_M \rangle$$
$$= \delta_{NM} F^{(N)}(\vec{x}_1, \vec{x}_2, \ldots, \vec{x}_N) \prod_{j=1}^{N} \delta(\vec{x}_j - \vec{y}_j) \ . \qquad 20.109$$

We now use the completeness of the states $|\vec{x}_1, \vec{x}_2, \ldots, \vec{x}_N\rangle$ to write

$$\langle n_o, n_1, \ldots | F | m_o, m_1, \ldots \rangle = \int d^3x_1 \ldots d^3x_N d^3y_1 \ldots d^3y_N$$
$$\langle n_o, n_1, \ldots | \vec{x}_1, \ldots \vec{x}_N \rangle \langle \vec{x}_1, \ldots \vec{x}_N | F | \vec{y}_1, \ldots \vec{y}_N \rangle \times$$
$$\langle \vec{y}_1 \ldots \vec{y}_N | m_o, m_1, \ldots \rangle = \int d^3x_1 \ldots d^3x_N F^{(N)}(\vec{x}_1, \ldots \vec{x}_N) \times$$
$$\times \langle n_o, n_1, \ldots | \vec{x}_1, \ldots, \vec{x}_N \rangle \langle \vec{x}_1, \ldots, \vec{x}_N | m_o, m_1, \ldots \rangle$$
$$= \frac{1}{N!} \int d^3x_1 \ldots d^3x_N F^N(\vec{x}_1, \ldots \vec{x}_N) \langle n_o, n_1, \ldots . | \psi^\dagger(\vec{x}_N) \ldots \psi^\dagger(\vec{x}_1) | \Omega \rangle$$
$$\cdot \langle \Omega | \psi(\vec{x}_1) \ldots \psi(\vec{x}_N) | m_o, m_1, \ldots \rangle \ ,$$

where $\sum n_i = \sum m_i = N$. Now $\psi(\vec{x}_1) \ldots \psi(\vec{x}_N) | m_o, m_1, \ldots \rangle$ with $\sum_{k=0}^{\infty} m_k = N$ has a non-vanishing inner product only with the no-particle state. We may therefore replace the intermediate state $|\Omega\rangle\langle\Omega|$ with a sum over all possible intermediate states, or

$$1 = \sum_{\{n\}} | n_o, n_1, \ldots \rangle \langle n_o, n_1, \ldots |$$

to get

$$\langle n_o, n_1, \ldots | F | m_o, m_1, \ldots \rangle = \frac{1}{N!} \langle n_o, n_1, \ldots | \int d^3x_1, \ldots d^3x_N \cdot$$
$$\psi^\dagger(\vec{x}_N) \ldots \psi^\dagger(\vec{x}_1) F^{(N)}(\vec{x}_1 \ldots \vec{x}_N) \psi(\vec{x}_1) \ldots \psi(\vec{x}_N) | m_o, m_1, \ldots \rangle . \qquad 20.110$$

To proceed further we first consider the hamiltonian

$$H = \sum_{j=1}^{N} H(\vec{x}_j) \qquad 20.40$$

consisting of a sum of one-particle hamiltonians. In this case a considerable simplification occurs. This can be seen by taking the terms in the above sum, one at a time. The last term yields

$$\langle n_o, n_1, \ldots | \frac{1}{N!} \int d^3x_1 \ldots d^3x_N H(\vec{x}_N) \psi^\dagger(\vec{x}_N) \ldots \psi^\dagger(\vec{x}_1) \psi(\vec{x}_1) \ldots \psi(\vec{x}_N) | m_o, m_1, \ldots \rangle$$

where as before $\sum n_i = \sum m_i = N$. The integral over d^3x_1 produces the number operator $\int d^3x_1 \psi^\dagger(\vec{x}_1)\psi(\vec{x}_1)$ applied to the states $\psi(\vec{x}_2)\ldots\psi(\vec{x}_N)|m_o, m_1, \ldots\rangle$ and thus yields 1 times the same state. After that, the integral over d^3x_2 produces the number operator $\int d^3x_2 \psi^\dagger(\vec{x}_2)\psi(\vec{x}_2)$ applied to the state $\psi(\vec{x}_3)\ldots\psi(\vec{x}_N)|m_o, m_1, \ldots\rangle$ and yields 2 times this state. Proceeding in this fashion we eventually get

$$\begin{aligned} &\langle n_o, n_1, \ldots | \frac{1}{N!} \int d^3x_1 \ldots d^3x_N H(\vec{x}_N) \psi^\dagger(\vec{x}_N) \ldots \psi^\dagger(\vec{x}_1) \cdot \\ &\quad \cdot \psi(\vec{x}_1) \ldots \psi(\vec{x}_N) | m_o, m_1 \ldots \rangle \\ &= \frac{1}{N} \langle n_o, n_1, \ldots | \int d^3x_N \psi^\dagger(\vec{x}_N) H(\vec{x}_N) \psi(\vec{x}_N) | m_o, m_1, \ldots \rangle . \end{aligned} \qquad 20.111$$

Now every term in the sum can be brought to this form by commuting, or anticommuting all the $\psi^\dagger(\vec{x}_j)H(\vec{x}_j)$ to the extreme left. Hence we get that

$$\begin{aligned} &\langle n_o, n_1, \ldots | H | m_o, m_1, \ldots \rangle \\ &= \frac{1}{N} \langle n_o, n_1, \ldots | \sum_{j=1}^{N} \int d^3x_j \psi^\dagger(\vec{x}_j) H(\vec{x}_j) \psi(\vec{x}_j) | m_o, m_1 \ldots \rangle \\ &= \langle n_o, n_1, \ldots | \int d^3x \, \psi^\dagger(\vec{x}) H(\vec{x}) \psi(\vec{x}) | m_o, m_1, \ldots \rangle . \end{aligned} \qquad 20.112$$

Thus the occupation number space representation of the hamiltonian corresponding to the sum of one-particle hamiltonians 20.4 is given by

$$\begin{aligned} \mathbf{H} &= \int d^3x\, \psi^\dagger(\vec{x}) H(\vec{x}) \psi(\vec{x}) \\ &= \int d^3x\, \psi^\dagger(\vec{x}) [-\frac{\hbar^2}{2m} \nabla^2 + V(\vec{x})] \psi(\vec{x}). \end{aligned} \tag{20.113}$$

To complete our treatment we also need to consider a general N-body interaction. Fortunately, it appears that for systems whose interactions are describable by potentials the most general interaction involves a sum of two-body interactions of the form

$$V(\vec{x}_1, \vec{x}_2, \ldots \vec{x}_N) = \frac{1}{2} \sum_{j \neq k}^{N} V(\vec{x}_j, \vec{x}_k) \tag{20.114}$$

No intrinsic three-body forces or higher have so far been found necessary for a description of actual systems.

Applying the results of equation 20.110 to the interaction 20.114 we obtain in occupation number space the following operator (see problem 20.4)

$$\mathbf{V} = \frac{1}{2} \int d^3x\, d^3y \psi^\dagger(\vec{x}) \psi^\dagger(\vec{y}) V(\vec{x}, \vec{y}) \psi(\vec{y}) \psi(\vec{x}) \ . \tag{20.115}$$

The order of the last two operators is deliberately reversed to remove the <u>self-interaction</u> $V(\vec{x}, \vec{x})$ which does not occur in (20.114).

Thus a hamiltonian containing an external potential $V_o(\vec{r})$ as well as two-body interactions $V(\vec{x}, \vec{y})$ would have the second-quantized form

$$\begin{aligned} \mathbf{H} &= \int d^3x\, \psi^\dagger(\vec{x}) [-\frac{\hbar^2}{2m} \nabla^2 + V_o(\vec{x})] \psi(\vec{x}) \\ &+ \frac{1}{2} \int d^3x\, d^3y\, \psi^\dagger(\vec{x}) \psi^\dagger(\vec{y}) V(\vec{x}, \vec{y}) \psi(\vec{y}) \psi(\vec{x}). \end{aligned} \tag{20.116}$$

20.10 Heisenberg Picture

We next transform to the Heisenberg picture and obtain the equations of motion for the field operators. Many theoretical discussions commence with these field equations, which are known as the Heisenberg equations. They are a sophisticated summary of the Planck frequency condition $E = h\omega$, as we shall see.

The field operators we have used so far have all been in the Schrödinger picture. To distinguish them from the Heisenberg picture operators, we are about to introduce, we label them with a subscript s.Thus the states in the Heisenberg picture are related to the states in the Schrödinger picture through

$$|\Psi\rangle = e^{iHt/\hbar}|\Psi_s(t)\rangle \tag{20.117}$$

and the operators are related by

$$\psi(\vec{x},t) = e^{iHt/\hbar}\psi_s(\vec{x})e^{-iHt/\hbar} \quad . \tag{20.118}$$

Since these transformations are unitary they preserve the <u>equal time</u> (anti) commutation relations. Thus

$$[\psi(\vec{x},t),\psi^\dagger(\vec{y},t)]_\pm = \delta(\vec{x}-\vec{y}) \tag{20.119}$$

$$[\psi(\vec{x},t),\psi(\vec{y},t)]_\pm = [\psi^\dagger(\vec{x},t),\psi^\dagger(\vec{y},t)]_\pm = 0. \tag{20.120}$$

By a straightforward differentiation of equation 20.118 we obtain the <u>Heisenberg equation of motion</u> for the field operators

$$\begin{aligned} i\hbar \frac{\partial}{\partial t}\psi(\vec{x},t) &= [\psi(\vec{x},t),\mathbf{H}] \\ i\hbar \frac{\partial}{\partial t}\psi^\dagger(\vec{x},t) &= [\psi^\dagger(\vec{x},t),\mathbf{H}] \quad . \end{aligned} \tag{20.121}$$

If we consider the hamiltonian 20.116 then since $\mathbf{H}$ is time independent

we get

$$[\psi(x,t),H] = \int d^3y[\psi(x,t),\psi^\dagger(y,t)\{-\frac{\hbar^2}{2m}\nabla^2+V_o(\vec{y})\}$$

$$\psi(\vec{y},t)] + \frac{1}{2}\int d^3yd^3z[\psi(x,t),\psi^\dagger(y,t)\psi^\dagger(z,t)$$

$$\psi(\vec{z},t)\psi(\vec{y},t)]V(\vec{y},\vec{z}). \qquad 20.122$$

This yields the following field equation

$$i\hbar\frac{\partial}{\partial t}\psi(\vec{x},t) = [-\frac{\hbar^2}{2m}\nabla^2 + V_o(\vec{x})]\psi(\vec{x},t)$$

$$+ \int d^3y\ \psi^\dagger(\vec{y},t)V(\vec{y},\vec{x})\psi(\vec{y},t)\psi(\vec{x},t). \qquad 20.123$$

To illustrate how this is obtained we consider the last term in equation 20.122. As before we treat fermions and bosons simultaneously; the upper sign always applies to fermions.

We first note that

$$[\psi(\vec{x},t),\psi(\vec{z},t)\psi(\vec{y},t)] = 0 \qquad 20.124$$

both for fermions and bosons. Therefore the commutator in the last term in 20.122 reduces to

$$[\psi(\vec{x},t),\psi^\dagger(\vec{y},t)\psi^\dagger(\vec{z},t)]\psi(\vec{z},t)\psi(\vec{y},t)\ .$$

But,

$$[\psi(\vec{x},t),\psi^\dagger(\vec{y},t)\psi^\dagger(\vec{z},t)] = \psi(\vec{x},t)\psi^\dagger(\vec{y},t)\psi^\dagger(\vec{z},t) - \psi^\dagger(\vec{y},t)\psi^\dagger(\vec{z},t)\psi(\vec{x},t)$$

$$= [\psi(\vec{x},t)\psi^\dagger(\vec{y},t) \pm \psi^\dagger(\vec{y},t)\psi(\vec{x},t)]\psi^\dagger(\vec{z},t)$$

$$\mp \psi^\dagger(\vec{y},t)[\psi(\vec{x},t)\psi^\dagger(\vec{z},t) \pm \psi^\dagger(\vec{z},t)\psi(\vec{x},t)] \qquad 20.125$$

$$= \delta(\vec{x}-\vec{y})\psi^\dagger(\vec{z},t) \mp \delta(\vec{x}-z)\psi^\dagger(\vec{y},t)$$

Combining all these results and integrating out the delta functions yields

$$\frac{1}{2}\int d^3y d^3z [\psi(\vec{x},t), \psi^\dagger(\vec{y},t)\psi^\dagger(\vec{z},t)\psi(\vec{z},t)\psi(\vec{y},t)]V(\vec{y},\vec{z})$$

$$= \frac{1}{2}\int d^3y\ \psi^\dagger(\vec{y},t)V(\vec{x},\vec{y})[\psi(\vec{y},t)\psi(\vec{x},t) \mp \psi(\vec{x},t)\psi(\vec{y},t)]$$

$$= \int d^3y\ \psi^\dagger(\vec{y},t)V(\vec{x},\vec{y})\psi(\vec{y},t)\psi(\vec{x},t) \ .$$

To obtain the first equality we used Newton's third law in the form $V(\vec{x},\vec{y}) = V(\vec{y},\vec{x})$.

Equation 20.123 occurs whenever we have two-body interactions. Thus in a sense it contains almost all of solid state physics. The problem is that it is not only an equation for operators and thus represents an infinite number of scalar equations, but it is also non-linear. There is therefore no hope of an analytic solution of this equation except in the most trivial cases. In practice it is frequently more convenient to work with the annihilation and creation operators. In that case the hamiltonian (20.116) becomes

$$\mathbf{H} = \sum_{k,\ell} E_{k,\ell} a_k^\dagger a_\ell + \sum_{k,\ell,m,n} V_{k,\ell,m,n} a_k^\dagger a_\ell^\dagger a_m a_n \qquad 20.126$$

where

$$E_{k,\ell} = \int d^3x \langle k|\vec{x}\rangle[-\frac{\hbar^2}{2m}\nabla^2 + V_o(\vec{x})]\langle\vec{x}|\ell\rangle \qquad 20.127$$

and

$$V_{k,\ell,m,n} = \int d^3x d^3y \langle k|\vec{x}\rangle\langle\ell|\vec{y}\rangle V(\vec{x},\vec{y})\langle\vec{x}|m\rangle\langle\vec{y}|n\rangle \ . \qquad 20.128$$

The states $|k\rangle$ here denote one-particle basis states. If they are chosen as eigenstates of the single particle hamiltonian

$$H_o = -\frac{\hbar^2}{2m}\nabla^2 + V_o(\vec{x})$$

with eigenvalues E_k, then equation 20.127 reduces to

$$E_{k\ell} = E_k \delta_{k\ell} \ . \qquad 20.129$$

In that case

$$H = \sum_{k=0}^{\infty} E_k a_k^\dagger a_k + \sum_{k,\ell,m,n} V_{k,\ell,m,n} a_k^\dagger a_\ell^\dagger a_m a_n \ . \qquad 20.130$$

As a final consideration we assume that we have eigenstates of energy E of the hamiltonian 20.126 of our system. Denoting these states by $|E\rangle$ and taking matrix elements on this basis of the Heisenberg equation of motion 20.121) for the field operator we get

$$i\hbar \langle E| \frac{\partial \psi(\vec{x},t)}{\partial t} |E'\rangle = (E' - E)\langle E|\psi(\vec{x},t)|E'\rangle. \qquad 20.131$$

If we Fourier-decompose $\psi(\vec{x},t)$ as

$$\psi(\vec{x},t) = \frac{1}{2\pi} \int_{-\infty}^{\infty} \psi(\vec{x},\omega) e^{-i\omega t} d\omega \qquad 20.132$$

we find that

$$\frac{1}{2\pi} \int_{-\infty}^{\infty} [h\omega-(E'-E)]\langle E|\psi(\vec{x},\omega)|E'\rangle e^{-i\omega t} d\omega = 0 \ ,$$

and we have that

$$\langle E|\psi(\vec{x},\omega)|E'\rangle = 0 \quad \text{unless}$$

$$E' - E = \hbar\omega. \qquad 20.133$$

Thus we have rederived the Planck frequency relation from the Heisenberg equation of motion.

Notes, References and Bibliography

Electrostatics are treated in chapters 1-3 of:

J.D. Jackson - Classical Electrodynamics - John Wiley and Sons, Inc., New York (1962).

20.1 Y. Sugiura - Z. Physik 45 484 (1927).

The reader interested in the further application of quantum mechanics to chemistry may consult:

W. Kauzmann - Quantum Chemistry - Academic Press Inc., New York (1957).

There are many ways to approach second quantization. For an approach different from ours see:

A.L. Fetter and J.D. Walecka - Quantum Theory of Many-Particle Systems - McGraw-Hill Book Co., New York (1971).

20.2 P.M. Morse and H. Feshbach - Methods of Theoretical Physics, Vol. 1 - McGraw-Hill Book Co., Inc., New York (1953).

Chapter 20 Problems

20.1 Using the Pauli exclusion principle, determine the maximum number of electrons in any energy level n of an atom. Neglect the interactions between the electrons.

20.2 Consider a system of non-interacting bosons and write the hamiltonian in the form

$$\mathbf{H} = \sum_{k=0}^{\infty} \hbar\, \omega_k a_k^\dagger a_k \ .$$

Find an explicit expression for

$$\psi(\vec{x},t) = e^{iHt/\hbar}\psi_s(\vec{x})e^{-iHt/\hbar}$$

where

$$\psi_s(\vec{x}) = \sum_{k=0}^{\infty} \langle \vec{x}|k\rangle a_k \ .$$

Hint: Expand the second exponential and commute a_k through showing that

$$a_k\, e^{\lambda a_k^\dagger a_k} = e^{\lambda(a_k^\dagger a_k + 1)} a_k \ .$$

20.3 Repeat problem 20.2 for fermions.

20.4 Obtain equation (20.115) for the occupation number space representation of the interaction specified by equation (20.114).

20.5 Show that the hamiltonian

$$H = \Sigma(E_k a_k^\dagger a_k + \lambda_k a_k^\dagger a_k^\dagger + \lambda_k^* a_k a_k) \quad , \ E_k > 2|\lambda_k|$$

can be diagonalized if a_k , $a_k^\dagger$ are bose operators. Hint: introduce operators

$$b_k = u_k a_k^\dagger + v_k a_k$$

and choose the constants u_k and v_k appropriately. What happens if a_k, $a_k^\dagger$ are fermi operators?

20.6 Let a_k , $a_k^\dagger$ be fermi operators $k = 1,2$

$$b_1^\dagger = ua_1^\dagger - va_2 \qquad b_1 = u^* a_1 - v^* a_2^\dagger$$

$$b_2 = va_1^\dagger + ua_2 \qquad b_2^\dagger = v^* a_1 + u^* a_2^\dagger$$

with

$$|u|^2 + |v|^2 = 1 \qquad uv^* - u^*v = 0.$$

a) Verify that b_k , $b_k^\dagger$ are fermi operators.

b) Show that for an appropriate choice of c

$$b_1^\dagger = Ua_1^\dagger U^\dagger \qquad b_1 = Ua_1U^\dagger$$

$$b_2 = Ua_2U^\dagger \qquad b_2^\dagger = Ua_2^\dagger U^\dagger$$

where

$$U = \exp c(a_1^\dagger a_2^\dagger - a_2 a_1)$$

is a unitary operator. This is known as a Bogoliubov transformation.

Chapter 21

Quantum Statistical Mechanics

21.1 Introduction

In the previous chapter we developed techniques for dealing with systems of identical particles. If the system consists of a very large number of, say 10^{23}, particles then it is neither possible nor desirable to have exact knowledge of the state of the system. In this case, statistical techniques are required to handle this incomplete knowledge. The procedure here is quite analogous to classical statistical mechanics as developed by Boltzmann and Gibbs.

In quantum statistical mechanics, the probability or statistical concepts enter at two levels. There is the statistical distribution of results of a measurement of an observable on identically prepared systems. This has been the subject of our discussion up until now. These quantum mechanical probabilites add coherently and are described by probability amplitudes. There is also the statistical distribution due to an incomplete knowledge of the state of the system. These are the incoherent probabilities used in classical statistical mechanics. Thus, in equilibrium, this second level of probabilities is determined by an ensemble, as developed by Gibbs. We now develop techniques to incorporate both effects.

The approach we take is more heuristic than rigorous, but it brings out the relevant physical input. We start by dividing the universe into two parts:

1) The system of interest to us; this we shall simply call our "system".

2) The external world.

Thus, for example, we could consider a gas in a container as our system and the rest of the universe including the physical container as the external world.

We furthermore assume that our system interacts "weakly" with the external world through the boundaries of the system (walls of the container).

Let $|\psi\rangle$ be a ket decribing both our system and the external world, and let $\{|k\rangle\}$ be a complete orthonormal set of kets for our system. In that case we can write

$$|\psi\rangle = \sum_k |c_k\rangle|k\rangle \qquad 21.1$$

where the $\{|c_k\rangle\}$ are a complete set of kets for the external world.

Now suppose A is an operator corresponding to an observable of our system. Thus, A operates on the space of kets $\{|k\rangle\}$. An instantaneous expectation value of this observable is given by $\langle\psi|A|\psi\rangle/\langle\psi|\psi\rangle$. This quantity represents an average result for a large number of identical measurements performed at the same time. We now rewrite this expression

$$\frac{\langle\psi|A|\psi\rangle}{\langle\psi|\psi\rangle} = \frac{\sum\limits_{n,m} \langle c_n|c_m\rangle\langle n|A|m\rangle}{\sum\limits_n \langle c_n|c_n\rangle} . \qquad 21.2$$

We have used here the fact that A corresponds to an observable of our system and does not operate on the states of the external world. Now $\langle\psi|\psi\rangle$ is time independent and therefore so is also $\sum\limits_n \langle c_n|c_n\rangle$.

In a laboratory, we do not perform instantaneous measurements on a many-particle system, but rather time averaged measurements. Thus, we measure

$$\langle\langle A\rangle\rangle = \overline{\frac{\langle\psi|A|\psi\rangle}{\langle\psi|\psi\rangle}} = \frac{\sum\limits_{n,m} \overline{\langle c_n|c_m\rangle}\langle n|A|m\rangle}{\sum\limits_{n} \langle c_n|c_n\rangle} . \tag{21.3}$$

Here the bar represents a time average. The assumptions of quantum statistical mechanics, when referring to a macroscopic observable of a macroscopic system in thermodynamic equilibrium, are assumptions on the time-averaged coefficients $\overline{\langle c_n|c_m\rangle}$.

21.2 The Density Matrix

We begin with some examples to illustrate the previous ideas. These examples conform as closely as possible to classical statistical mechanics.

21.2.1 The Microcanonical Ensemble.

In this case we assume that our system consists of N particles in a volume V. We further assume that their energy is fixed between E and E + Δ where Δ << E.

If H is the (N-particle) hamiltonian of our system we choose as our basis set $\{|k\rangle\}$ the set of eigenkets of H.

$$H\,|k\rangle = E_k\,|k\rangle . \tag{21.4}$$

Here $|k\rangle$ represents a state for N particles with total energy E_k enclosed in a volume V.

We now make the following statistical assumptions.

M1. $\overline{\langle c_n|c_n\rangle} = \begin{cases} 1 & E < E_n < E + \Delta \\ 0 & \text{otherwise} \end{cases}$ — Assumption of equal a priori probabilities

M2. $\overline{\langle c_n|c_m\rangle} = 0$ for $n \neq m$ — Assumption of random phases.

We thus can write

$$|\psi\rangle = \sum_n b_n \, |n\rangle \tag{21.5}$$

where the b_n are numbers with random phases such that

$$|b_n|^2 = \left\{ \begin{matrix} 1 & E < E_n < E + \Delta \\ 0 & \text{otherwise} \end{matrix} \right\}$$

Specification by outside world or heat bath.

We thus get

$$\langle\langle A \rangle\rangle = \frac{\sum_n |b_n|^2 \langle n|A|n\rangle}{\sum_n |b_n|^2} \, . \tag{21.6}$$

The postulate of random phases implies that the equilibrium state is an <u>incoherent</u> superposition of eigenstates.

The quantities, $\dfrac{|b_n|^2}{\sum |b_m|^2}$ may clearly be considered as classical probabilities p_n . Since this is the case, we can rewrite the above expression so that no reference to phases need occur. Thus

$$\begin{aligned} \langle\langle A \rangle\rangle &= \sum_n p_n \langle n|A|n\rangle \\ &= \sum_{n,m} \langle m|A|n\rangle p_n \langle n|m\rangle \\ &= \mathrm{Tr}\left[\sum_n A|n\rangle p_n \langle n|\right] \end{aligned} \tag{21.7}$$

where Tr means "trace" of the operator following. This expression can also be written as

$$\langle\langle A \rangle\rangle = \mathrm{Tr}\,[A\rho] \tag{21.8}$$

where

$$\rho = \sum_n |n\rangle p_n \langle n| \ . \tag{21.9}$$

The operator ρ is called the density matrix.

Thus in our example of the microcanonical ensemble the density matrix is, up to normalization, given by

$$\rho = \sum_{E<E_n<E+\Delta} |n\rangle\langle n| \ . \tag{21.10}$$

It is now an easy matter to display the density matrix for other ensembles.

21.2.2 The Canonical Ensemble

In this case we have N particles in a volume V. The total energy however is no longer fixed, instead states are weighted with the Boltzmann factor $e^{-\beta E_n}$ where $\beta = 1/(k_B T)$, with k_B Boltzmann's constant. Thus, in analogy to the microcanonical ensemble we make two statistical assumptions.

C1. $\overline{\langle c_n | c_n \rangle} = e^{-\beta E_n}$ Assumption of weighting according to Boltzmann factor

21.11

C2. $\overline{\langle c_n | c_m \rangle} = 0 \quad n \neq m$ Assumption of random phases.

We then obtain for the <u>unnormalized</u> density matrix

$$\rho = \sum_n |n\rangle e^{-\beta E_n} \langle n| \tag{21.12}$$

$$= e^{-\beta H} \sum_n |n\rangle\langle n| \tag{21.13}$$

or

$$\rho = e^{-\beta H} \ . \tag{21.14}$$

One also defines the normalization or partition function for N particles Z_N by

$$Z_N = \mathrm{Tr}\, e^{-\beta H} . \qquad 21.15$$

This quantity yields the interesting macroscopic or thermodynamic observables. We defer a discussion of this until later.

21.2.3 Grand Canonical Ensemble

If we wish to consider a system in which even the particle number N is not specified, but only the average number, we can introduce a further "Boltzmann factor" $e^{\beta\mu N}$ for the particle number. In this case μ is the chemical potential. Thus, in this case, we have for the unnormalized density matrix

$$\rho = e^{-\beta(H-\mu N)} . \qquad 21.16$$

This yields a grand partition function (normalization)

$$Z_G = \mathrm{Tr}\, e^{-\beta(H-\mu N)} . \qquad 21.17$$

The connection between statistical mechanics and thermodynamics is made through the thermodynamic functions. This is the same as in classical statistical mechanics. Thus the internal energy U of a system in the Grand Canonical Ensemble is given by the average value of the hamiltonian. This means

$$U = Z_G^{-1}\, \mathrm{Tr}(\rho H) . \qquad 21.18$$

Since ρ is given by 21.16 it is easily seen that

$$U = -\frac{\partial}{\partial\beta} \ln Z_G . \qquad 21.19$$

Similarly, the average number of particles in our system is given by

$$\langle\langle N \rangle\rangle = Z_G^{-1} \text{ Tr } (\rho N) \ . \tag{21.20}$$

To evaluate this it is convenient to introduce the fugacity

$$z = e^{\beta\mu} \ . \tag{21.21}$$

Then,

$$\rho = e^{\beta\mu N} e^{-\beta H} = z^N e^{-\beta H} \ . \tag{21.22}$$

Hence we get that the grand partition function is

$$Z_G = \sum_{n=0}^{\infty} z^n Z_n \ . \tag{21.23}$$

Combining this result with 21.20 we obtain

$$\langle\langle N \rangle\rangle = z \frac{\partial}{\partial z} \ln Z_G \ . \tag{21.24}$$

Equation 21.24 is usually inverted to express the chemical potential in terms of the average density of particles. This result is then substituted into equation 21.19 to express the internal energy as a function of temperature and density. In fact, by performing Legendre transformations (see section 2.2) it is now possible to compute all other thermodynamic functions. Thus once Z_G is obtained, all thermodynamic quantities are determined in principle.

We simply list the relevant relations. Their derivation is left as an exercise.

$$Z_G = \text{Tr } e^{-\beta(H-\mu N)} = \sum z^n Z_n \ . \tag{21.25}$$

The average number $\langle\langle N \rangle\rangle$ of particles in our system (of volume V) is given by

$$\langle\langle N \rangle\rangle = z \frac{\partial}{\partial z} \ln Z_G \ . \tag{21.26}$$

The internal energy has already been shown to be

$$U = -\frac{\partial}{\partial\beta}\,\ell n\, Z_G \; . \tag{21.19}$$

Just as in classical statistical mechanics we also have

$$\frac{PV}{k_B T} = \ell n\, Z_G \; . \tag{21.27}$$

The three equations 21.19, 21.26, and 21.27 suffice to determine all thermodynamic quantities. Thus, for example, to compute the entropy S one uses 21.26 to eliminate z from 21.19 and to express the internal energy U as a function of $\langle\langle N\rangle\rangle$, V and T. The specific heat at constant volume C_V is now obtained as

$$C_V = \left(\frac{\partial U}{\partial T}\right)_V \; . \tag{21.28}$$

This allows us to express the entropy as

$$dS = C_V\,\frac{dT}{T} \; . \tag{21.29}$$

For more details on thermodynamic relations we direct the reader to the references at the end of this chapter.

Next we consider, by way of illustration, the simplest possible systems, the ideal gases.

21.3 The Ideal Gases

We now combine the discussion of the previous section with the second quantization techniques developed in sections 6. and 7. of chapter 20. This allows us to illustrate the usefulness of these techniques as well as to derive the Bose-Einstein and Fermi-Dirac distributions.

The reason, the second quantized formalism is so useful, is that for the grand canonical ensemble the number of particles is not fixed.

This makes the occupation number space representation ideally suited for computations in the grand canonical ensemble.

The system we consider consists of an indefinite number of non-interacting (free) particles confined to a fixed volume V. This system can therefore be described by a hamiltonian

$$H = \sum_{\vec{k}=0} E_{\vec{k}} a^{\dagger}_{\vec{k}} a_{\vec{k}} = \sum_{\vec{k}=0}^{\infty} E_{\vec{k}} N_{\vec{k}} \qquad 21.30$$

where as discussed in section 6. and 7. of chapter 20, the $E_{\vec{k}}$ are one-particle energies and we are working in the occupation number space representation.

To evaluate

$$Z_G = \mathrm{Tr}\; e^{-\beta(H-\mu N)} \qquad 21.31$$

we simply take matrix elements for a complete set of states and sum over the diagonal elements.

$$Z_G = \sum_{n_1 \ldots n_\infty} \langle n_1 \ldots n_\infty | e^{-\beta(H-\mu N)} | n_1 \ldots n_\infty \rangle \qquad 21.32$$

$$= \sum_{n_1 \ldots n_\infty} \langle n_1 \ldots n_\infty | e^{-\beta(\sum_{\vec{k}} E_{\vec{k}} N_{\vec{k}} - \mu \sum N_{\vec{k}})} | n_1 \ldots n_\infty \rangle \;. \qquad 21.33$$

We have simply used 21.30 and the definition

$$N = \sum_{\vec{k}=0}^{\infty} N_{\vec{k}} \qquad 21.34$$

of the total number operator.

Using,

$$N_{\vec{k}} \, |n_1 \ldots n_\infty\rangle = n_{\vec{k}} |n_1 \ldots n_\infty\rangle \, , \qquad 21.35$$

the expression for the grand partition function becomes

$$Z_G = \sum_{n_1} \langle n_1 | e^{-\beta(E_1 n_1 - \mu n_1)} | n_1 \rangle \ldots \sum_{n_\infty} \langle n_\infty | e^{-\beta(E_\infty n_\infty - \mu n_\infty)} | n_\infty \rangle \qquad 21.36$$

or

$$Z_G = \prod_{\vec{k}}^{\infty} \mathrm{Tr}_{\vec{k}} \; e^{-\beta(E_{\vec{k}} N_{\vec{k}} - \mu N_{\vec{k}})} \qquad 21.37$$

where $\mathrm{Tr}_{\vec{k}}$ means a trace over the subspace corresponding to the one-particle states $|\vec{k}\rangle$.

Up to here, the treatment did not depend on whether we were dealing with bosons or fermions. We must now treat the two cases separately.

1) Bosons:

Since for bosons the occupation numbers $n_{\vec{k}}$ are unrestricted we must sum $n_{\vec{k}}$ over all integers. Thus, we get

$$Z_G = \prod_{\vec{k}}^{\infty} \sum_{n=0}^{\infty} e^{-\beta(E_{\vec{k}} - \mu) n} = \prod_{\vec{k}}^{\infty} \left[1 - e^{-\beta(E_{\vec{k}} - \mu)}\right]^{-1} \; . \qquad 21.38$$

2) Fermions:

For fermions the occuptation numbers $n_{\vec{k}}$ can take on only the values 0 and 1. Thus, we get

$$Z_G = \prod_{\vec{k}}^{\infty} \sum_{n=0,1} e^{-\beta(E_{\vec{k}}-\mu)n} = \prod_{\vec{k}}^{\infty} \left[1+e^{-\beta(E_{\vec{k}}-\mu)}\right] . \qquad 21.39$$

If we take the logarithm of Z_G we can treat both systems (bosons and fermions) simultaneously since we have

$$\ln Z_G = \pm \sum_{\vec{k}} \ln\left[1 \pm e^{-\beta(E_{\vec{k}}-\mu)}\right] . \qquad 21.40$$

From now on the upper sign will always refer to fermions and the lower sign will always refer to bosons.

Introducing the fugacity equation 21.21 we get

$$\ln Z_G = \pm \sum_{\vec{k}} \ln\left[1 \pm ze^{-\beta E_{\vec{k}}}\right] . \qquad 21.41$$

The average number of particles $\langle\langle N\rangle\rangle$ is then, as before, given by

$$\langle\langle N\rangle\rangle = z \frac{\partial}{\partial z} \ln Z_G \qquad 21.24$$

and yields

$$\langle\langle N\rangle\rangle = \sum_{\vec{k}} \frac{ze^{-\beta E_{\vec{k}}}}{1 \pm ze^{-\beta E_{\vec{k}}}} = \sum_{\vec{k}} \left[e^{\beta(E_{\vec{k}}-\mu)} \pm 1\right]^{-1} . \qquad 21.42$$

Similarly, we find that the internal energy which is again given by

$$U = - \frac{\partial}{\partial \beta} \ln Z_G \qquad 21.19$$

yields

$$U = \sum_{\vec{k}} \frac{E_{\vec{k}} ze^{-\beta E_{\vec{k}}}}{1 \pm ze^{-\beta E_{\vec{k}}}} = \sum_{\vec{k}} E_{\vec{k}} \left[e^{\beta(E_{\vec{k}}-\mu)} \pm 1\right]^{-1} . \qquad 21.43$$

To proceed further we must evaluate the sums $\sum_{\vec{k}}$. This is most easily accomplished by taking the so-called thermodynamic limit. This simply means that we count the density of states (modes) in the range k_x and $k_x + dk_x$, k_y and $k_y + dk_y$ as well as k_z and $k_z + dk_z$. This result as already found in chapter 1 yields $\frac{V}{(2\pi)^3} d^3k$. One then takes the limit $V \to \infty$. Thus the sums go over into integrals; in fact

$$\sum_{\vec{k}} \to \frac{V}{(2\pi)^3} \int d^3k = \frac{V}{(2\pi)^3} \cdot 4\pi \int k^2 dk . \qquad 21.44$$

If we now further use that for non-interacting particles

$$E_{\vec{k}} = \frac{\hbar^2 k^2}{2m} \qquad 21.45$$

and write simply E instead of $E_{\vec{k}}$ we get

$$\sum_{\vec{k}} \to \frac{V}{4\pi^2} \left(\frac{2m}{\hbar^2}\right)^{3/2} \int E^{1/2} \, dE . \qquad 21.46$$

We can apply these results immediately to the ideal Fermi gas (later we shall also apply them to the ideal Bose gas) and obtain for the average number of particles $\langle\langle N \rangle\rangle$ and the internal energy U

$$\langle\langle N \rangle\rangle = \frac{V}{4\pi^2} \left(\frac{2m}{\hbar^2}\right)^{3/2} \int_0^\infty \frac{E^{1/2} dE}{z^{-1} e^{\beta E} + 1} \qquad 21.47$$

and

$$U = \frac{V}{4\pi^2}\left(\frac{2m}{\hbar^2}\right)^{3/2}\int_0^\infty \frac{E^{3/2}\,dE}{z^{-1}e^{\beta E}+1} \qquad 21.48$$

The next step is to invert equation 21.47 and solve for the fugacity z in terms of the mean density of particles

$$\frac{1}{v} = \frac{\langle\langle N\rangle\rangle}{V}\,. \qquad 21.49$$

This result is then substituted into equation 21.48 to express the internal energy per unit volume U/V as a function of the mean density v^{-1} and the temperature β^{-1}.

Unfortunately, even for the simple case of an ideal Fermi gas it is not possible to express the integrals 21.47 and 21.48 in terms of elementary functions. It is therefore usual to define certain functions in terms of these integrals. Thus putting

$$\beta E = x^2 \qquad 21.50$$

and introducing the thermal wavelength

$$\lambda = \left(\frac{2\pi\hbar^2}{mk_BT}\right) \qquad 21.51$$

we get

$$\frac{1}{v} = \lambda^{-3}\,\frac{4}{\sqrt{\pi}}\int_0^\infty \frac{x^2\,dx}{z^{-1}e^{x^2}+1} \qquad 21.52$$

and

$$U/V = \beta^{-1}\,\lambda^{-3}\,\frac{4}{\sqrt{\pi}}\int_0^\infty \frac{x^4\,dx}{z^{-1}e^{x^2}+1} \qquad 21.53$$

We now introduce the functions

$$f_{5/2}(z) = \frac{4}{\sqrt{\pi}} \int_0^\infty dx\, x^2 \ln\left(1 + ze^{-x^2}\right) = \sum_{n=1}^\infty \frac{(-1)^{n+1} z^n}{n^{5/2}} \qquad 21.54$$

and

$$f_{3/2}(z) = z \frac{\partial}{\partial z} f_{5/2}(z) = \sum_{n=1}^\infty \frac{(-1)^{n+1} z^n}{n^{3/2}} \quad . \qquad 21.55$$

Differentiating 21.54 under the integral sign shows that

$$f_{3/2}(z) = \frac{4}{\sqrt{\pi}} \int_0^\infty \frac{x^2\, dx}{z^{-1} e^{x^2} + 1} \qquad 21.56$$

Also an integration by parts of $f_{5/2}(z)$ shows that

$$f_{5/2}(z) = \frac{2}{3} \frac{4}{\sqrt{\pi}} \int_0^\infty \frac{x^4\, dx}{z^{-1} e^{x^2} + 1} \quad . \qquad 21.57$$

Thus we have

$$\frac{1}{v} = \lambda^{-3} f_{3/2}(z) \qquad 21.58$$

$$U/V = \frac{3}{2} \frac{k_B T}{\lambda^3} f_{5/2}(z) \quad . \qquad 21.59$$

To obtain the corresponding results for an ideal Bose gas requires some care in replacing the sum $\sum_{\vec{k}}$ by an integral. This is due to the fact that as $z \to 1$ the single term due to $\vec{k} = 0$ diverges and may be as important as the entire sum. We therefore first remove the term corresponding to $\vec{k} = 0$ and replace the rest of the sum by an integral.

Thus we get

$$\langle\langle N\rangle\rangle = \frac{z}{1-z} + \frac{V}{4\pi^2}\left(\frac{2m}{\hbar^2}\right)^{3/2}\int_0^\infty \frac{E^{1/2}dE}{z^{-1}e^{\beta E}-1} \quad . \tag{21.60}$$

and

$$U = \frac{V}{4\pi^2}\left(\frac{2m}{\hbar^2}\right)^{3/2}\int_0^\infty \frac{E^{3/2}dE}{z^{-1}e^{\beta E}-1} \tag{21.61}$$

Again introducing $x^2 = \beta E$, $\frac{1}{v} = \frac{\langle\langle N\rangle\rangle}{V}$, and the thermal wavelength $\lambda = \left(\frac{2\pi\hbar^2}{mk_BT}\right)^{1/2}$ we get

$$\frac{1}{v} = \frac{1}{V}\frac{z}{1-z} + \lambda^{-3}\, g_{3/2}(z) \tag{21.62}$$

and

$$U/V = \frac{3}{2}\frac{k_BT}{\lambda^3}\, g_{5/2}(z) \tag{21.63}$$

where

$$g_{5/2}(z) = -\frac{4}{\sqrt{\pi}}\int_0^\infty dx\, x^2 \ln(1 - ze^{-x^2}) = \sum_{n=1}^\infty \frac{z^n}{n^{5/2}} \tag{21.64}$$

and

$$g_{3/2}(z) = z\frac{\partial}{\partial z} g_{5/2}(z) = \sum_{n=1}^\infty \frac{z^n}{n^{3/2}} \quad . \tag{21.65}$$

To invert the functions $f_{3/2}(z)$ or $g_{3/2}(z)$ to express z in terms of v leads to very complicated expressions. Since our purpose was only to introduce the techniques of statistical mechanics, for further discussions we refer the reader to one of the standard texts listed at the end of this chapter.

21.4 General Properties of the Density Matrix

The use of the density matrix extends beyond the confines of statistical mechanics to all systems for which it is desirable to have a description of the extent to which the states are specified. Thus even pure, or completely specified states can be described by means of the density matrix. It is desirable to have criteria on the density matrix itself such that one can distinguish density matrices for pure and impure or incompletely specified systems. We develop such criteria as well as other formal properties of the density matrix in this section.

As our starting point we take the defining equation 21.9 for the density matrix

$$\rho = \sum_n |n\rangle p_n \langle n| \tag{21.9}$$

where the p_n are classical probabilities so that

$$\sum_n p_n = 1 . \tag{21.66}$$

With this condition we immediately obtain the normalization

$$\mathrm{Tr}\,\rho = \sum_{m,n} \langle m|n\rangle p_n \langle n|m\rangle$$

$$= \sum_{n,m} \delta_{m\,n}\, p_n = \sum_n p_n = 1$$

so that

$$\mathrm{Tr}\,\rho = 1 . \tag{21.67}$$

It is also immediately clear that ρ is self adjoint

$$\rho^\dagger = \rho .$$

The diagonal matrix elements of ρ are

$$\langle m|\rho|m\rangle = p_m \tag{21.68}$$

and are clearly real and satisfy $0 \leq p_m \leq 1$.

Consider a density matrix for a pure state. In this case the probability for the pure state say m is $p_m = 1$. All other states have zero probability. Thus,

$$p_n = \delta_{nm} \tag{21.69}$$

so

$$\rho = \sum_n |n\rangle\delta_{nm}\langle n| = |m\rangle\langle m| \ . \tag{21.70}$$

Clearly

$$\rho^2 = \rho \ . \tag{21.71}$$

Thus, if ρ describes a pure state it is necessarily idempotent i.e. it satisfies equation 21.71. We now show that the converse is also true.

Assume ρ is idempotent, that is, equation 21.71 is valid. We then have

$$\sum_n |n\rangle p_m\langle n| = \sum_{n,m} |m\rangle p_m\langle m|n\rangle p_n\langle n| \ . \tag{21.72}$$

Hence

$$\sum_n |n\rangle p_n\langle n| = \sum_n |n\rangle p_n^2\langle n| \ . \tag{21.73}$$

It therefore follows that

$$p_n^2 = p_n \tag{21.74}$$

so that

$$p_n = 0,1 \quad \text{for all} \quad n.$$

But

$$\sum p_n = 1 \ .$$

It therefore follows that only one of the probabilities $p_n = 1$ and all others are zero. Thus ρ describes a pure state.

We have so far been working in a diagonal representation for ρ. This is a direct consequence of the assumption of random phases for the wave functions of the external world, i.e. $\overline{\langle c_n | c_m \rangle} = 0$ for $n \neq m$. One can, however, can, however, start from a different viewpoint. In that case the density matrix is not automatically diagonal, but nevertheless self-adjoint, so that it can still be brought to diagonal form by a unitary transformation. In the representation in which ρ is diagonal, one can again interpret the wave functions for the external world as satisfying the assumption of random phases.

Now suppose we can write the hamiltonian for our system as well as the external world as

$$H = H_o + H_I + H_{ext} \qquad 21.75$$

where H_o is the hamiltonian for our system, H_{ext} is the hamiltonian for the external world and H_I is the coupling of our system to the external world. If we now consider the external world as providing only a heat bath then, in fact, we perform time averages over the wave functions of the external world. Thus we are led to consider

$$H = H_o + \tilde{H}_I + \langle H_{ext} \rangle \qquad 21.76$$

where now $\langle H_{ext} \rangle$ is a constant and $\tilde{H}_I$ acts as an external interaction (potential) on our system which still has the internal dynamics H_o. We can now drop all reference to the external world and ignore the constant $\langle H_{ext} \rangle$.

Under the action of this hamiltonian, the density matrix will evolve according to the evolution operator.

$$U(t,t_o) = \exp - \frac{i}{\hbar} (H_o + \tilde{H}_I)(t - t_o) \tag{21.77}$$

so that

$$\rho(t) = U(t,t_o)\rho(t_o)U^\dagger(t,t_o) \; . \tag{21.78}$$

Thus the equation of motion for the density operator is

$$i\hbar \frac{\partial \rho}{\partial t} = [H,\rho] \; . \tag{21.79}$$

This equation although similar to the Heisenberg equations of motion clearly has the commutator reversed. Furthermore, we are working in the Schrödinger picture, that is, the wave functions carry all the time dependence and the operators are time independent. If we transform to the Heisenberg picture and H is independent of time then ρ becomes a constant (time independent) operator. Equation 21.79 is the quantum mechanical version of Liouville's theorem.

This concludes our formal treatment of the density operator. In the next section we apply some of these results to a system with finite degrees of freedom.

21.5 The Density Matrix and Polarization

As stated before, the density matrix is particularly useful in a discussion of systems for which the states are not pure or completely specified. This occurs in the case of scattering of particles with spin when the particles are only partially (or not at all) polarized. The density matrix is well suited to a discussion of this case, as well as the case of complete polarization.

To illustate this use of the density matrix, we apply it to a discussion of spin 1/2 particles. For a discussion of arbitrary spins, the reader is referred to the literature cited at the end of this chapter. (ref. 21.1)

If we consider a beam of spin 1/2 particles, say electrons, and if we are only interested in the spin-orientation or polarization of this beam, then we have a system with two degrees of freedom. Thus the system is completely specified by two complex numbers a,b, i.e. the wavefunction can be written

$$\chi = a \binom{1}{0} + b\binom{0}{1} \tag{21.80}$$

where

$$|a|^2 + |b|^2 = 1. \tag{21.81}$$

We now describe this system in terms of a density matrix. This is possible since the physical system is completely specified by the <u>three</u> components of polarization

$$\vec{p} = \langle\vec{\sigma}\rangle \tag{21.82}$$

and χ also involves only <u>three</u> real parameters. A general 2×2 hermitean matrix can be written

$$\rho = u_o\, 1 + \vec{u}\cdot\vec{\sigma} \tag{21.83}$$

where $(u_o,\vec{u})$ are real parameters. Requiring that

$$\mathrm{Tr}\ \rho = 1 = 2u_o \tag{21.84}$$

leaves us with only three independent real parameters. These are related to the polarization through 21.82.

$$\vec{p} = \langle\vec{\sigma}\rangle = \mathrm{Tr}\ \rho\vec{\sigma} = \mathrm{Tr}[u_0\vec{\sigma} + \vec{\sigma}(\vec{u}\cdot\vec{\sigma})] = 2\vec{u}\ . \qquad 21.85$$

Thus,

$$\rho = 1/2[1 + \vec{p}\cdot\vec{\sigma}]\ . \qquad 21.86$$

To relate ρ directly to χ we consider the matrix

$$\rho' = \chi\chi^{\dagger} = \binom{a}{b}(a^*, b^*) = \begin{pmatrix} |a|^2 & ab^* \\ a^*b & |b|^2 \end{pmatrix} \qquad 21.87$$

$$\mathrm{Tr}\ \rho' = |a|^2 + |b|^2 = 1\ .$$

From 21.80 we get

$$\langle\vec{\sigma}\rangle = \vec{p} = (2\ \mathrm{Re}\ a^*b,\ 2\ \mathrm{Im}\ a^*b,\ |a|^2 - |b|^2). \qquad 21.88$$

But,

$$\mathrm{Tr}(\rho'\vec{\sigma}) = (2\ \mathrm{Re}\ a^*b,\ 2\ \mathrm{Im}\ a^*b,\ |a|^2 - |b|^2). \qquad 21.89$$

Thus $\rho' = \rho$.

In this form it is easy to see that

$$\rho\chi = \chi \qquad 21.90$$

and

$$\rho^2 = \rho\ , \qquad 21.91$$

so that ρ corresponds to the pure state with polarization $\vec{p}$. This is seen from

$$\vec{p}\cdot\vec{\sigma}\ \chi = \chi\ . \qquad 21.92$$

If our beam of particles passes through a magnetic field $\vec{B}$, the polarization will be changed due to the connection between magnetic moment and spin. The corresponding hamiltonian* is

$$H = -\vec{\mu}\cdot\vec{B} = -1/2\,\frac{g\,e\,\hbar}{m\,c}\,\vec{\sigma}\cdot\vec{B}\;. \qquad 21.93$$

More generally the hamiltonian would simply be a hermitean 2×2 matrix

$$H = 1/2(A_o 1 + \vec{A}\cdot\vec{\sigma})\;. \qquad 21.94$$

We then get

$$i\hbar\,\frac{d\vec{p}}{dt} = i\hbar\,\frac{d}{dt}\,\mathrm{Tr}(\rho\vec{\sigma}) = i\hbar\,\mathrm{Tr}(\frac{\partial\rho}{\partial t}\,\vec{\sigma})$$

$$= \mathrm{Tr}([H,\rho]\vec{\sigma}) = \mathrm{Tr}(H\rho\vec{\sigma} - \rho H\vec{\sigma})$$

$$= \mathrm{Tr}(\rho[\vec{\sigma},H]) \qquad 21.95$$

where we have used 21.79 and the cyclic property of the trace. Applying this result to the specific hamiltonian 21.94 we get

$$i\hbar\,\frac{d\vec{p}}{dt} = 1/2\;\mathrm{Tr}(\rho[\vec{\sigma},\vec{A}\cdot\vec{\sigma}])$$

$$= i\,\vec{A}\times\vec{p}$$

so that

$$\hbar\,\frac{d\vec{p}}{dt} = \vec{A}\times\vec{p}\;. \qquad 21.96$$

---

* Actually H is not a hamiltonian in the sense of total energy of the system since there is no kinetic energy term. Nevertheless, H is the generator of time translation for the "observable" $\vec{\sigma}$.

This equation is similar to the Euler equation for a symmetrical top in classical mechanics. Furthermore, a magnetic field can only rotate the spin. This is easily seen by considering

$$\frac{d}{dt}\vec{p}^{\,2} = 2\vec{p}\cdot\frac{d\vec{p}}{dt} = \frac{2}{\hbar}\vec{p}\cdot(\vec{A}\times\vec{p}) = 0 \ .$$

The polarization vector is thus seen to maintain its length and is only rotated.

For cases of spin greater than 1/2 as occurs for certain nuclei, the discussion must be generalized. If the spin is j, then the density matrix will be a hermitean $(2j+1)$ by $(2j+1)$ matrix with unit trace. Thus it requires more than the 3 components of the polarization vector to specify the density matrix. For spin 1, (see problem 21.6) the density matrix can be determined in terms of the 3 components of the polarization vector $\vec{p}$ and the 5 components of the quadrupole polarization tensor Q_{ij} .

21.6 Composite Systems

The main purpose of this section is to develop the density matrix formalism for composite systems and to show under what circumstances the density matrix of the composite system is determined by the density matrices of the individual systems.

We consider two systems S_1 and S_2 with respectively k and ℓ degrees of freedom and corresponding coordinates $q_1 \ldots q_k$ and $x_1 \ldots x_\ell$. The composite system

$$S_c = S_1 \otimes S_2$$

therefore has $n = k+\ell$ degrees of freedom. The inner products in the corresponding hilbert spaces are

$$(\phi,\psi)_1 = \int \phi^*(q_i)\psi(q_i)dq_1\ldots dq_k \qquad 21.97$$

$$(\phi,\psi)_2 = \int \phi^*(x_i)\psi(x_i)dx_1\ldots dx_\ell \qquad 21.98$$

and

$$(\phi,\psi)_c = \int \phi^*(q_i,x_j)\psi(q_i,x_j)dq_1\ldots dq_k\, dx_1\ldots dx_\ell \qquad 21.99$$

The corresponding observables are labelled $A^{(1)}$, $A^{(2)}$ and $A^{(c)}$.

Any observable $A^{(1)}$ in S_1 is naturally also an observable in S_c. The same is true for any $A^{(2)}$ in S_2. If $|1,m\rangle$, and $|2,n\rangle$ form bases in the hilbert spaces of S_1 and S_2 respectively, then,

$$|c;\, m,n\rangle = |1,m\rangle|2,n\rangle \qquad 21.100$$

forms a basis for the hilbert space of S_c.

The matrix correspondence between $A^{(1)}$ considered as an element of S_1 or S_c is now given by

$$\begin{aligned}\langle c;\, m,n|A^{(1)}|c;\, m',n'\rangle &= \langle 1,m|A^{(1)}|1,m'\rangle\langle 2,n|2,n'\rangle \\ &= \langle 1,m|A^{(1)}|1,m'\rangle\delta_{n,n'}\ .\end{aligned} \qquad 21.101$$

The correspondence for $A^{(2)}$ is analogous.

We now consider density matrices in S_1, S_2 and $S_1 \otimes S_2$ labelled $\rho^{(1)}$, $\rho^{(2)}$ and $\rho^{(c)}$ repectively. Now any density matrix $\rho^{(c)}$ in S_c determines a density matrix $\rho^{(i)}$ in S_i $(i = 1,2)$. To see how this correspondence is made we consider matrix elements of the density matrices referred to the bases we have given. Then using $\langle\langle\cdot\rangle\rangle$ to indicate statistical averages we have:

$$\langle\langle A^{(i)}\rangle\rangle = \mathrm{Tr}\left(\rho^{(i)}\, A^{(i)}\right) = \sum_{m,m'} \rho^{(i)}_{m,m'}\, A^{(i)}_{m',m} \qquad 21.102$$

for the averages evaluated in S_i. Evaluated in S_c we get

$$\langle\langle A^{(1)}\rangle\rangle = \mathrm{Tr}\left(\rho^{(c)} A^{(1)}\right) = \sum_{\substack{m,m'\\n,n'}} \rho^{(c)}_{m,n;\ m',n'} A^{(1)}_{m',m}\, \delta_{n',n}$$

$$= \sum_{m,m'} \left(\sum_n \rho^{(c)}_{m,n;\ m',n}\right) A^{(1)}_{m',m} \ . \qquad 21.103$$

So we have

$$\rho^{(1)}_{m,m'} = \sum_n \rho^{(c)}_{m,n;\ m',n} \qquad 21.104$$

and analogously

$$\rho^{(2)}_{n,n'} = \sum_m \rho^{(c)}_{m,n;\ m,n'} \ . \qquad 21.105$$

Thus, given a density matrix in the composite system determines uniquely a density matrix in the subsystem. We now consider the converse question. Under what circumstances do the density matrices $\rho^{(1)}$ and $\rho^{(2)}$ determine the density matrix $\rho^{(c)}$ uniquely? That there is always at least one solution is clear, for if we define

$$\tilde{\rho}_{m,n;\ m',n'} = \rho^{(1)}_{m,m'}\, \rho^{(2)}_{n,n'} \qquad 21.106$$

then clearly (problem 21.8) $\tilde{\rho}$ satisfies all the conditions of a normalized density matrix and furthermore 21.104 and 21.105 are also satisfied. This solution is, however, not unique except when either $\rho^{(1)}$ or $\rho^{(2)}$ corresponds to a pure state (problem 21.9). Since there are no correlations between S_1 and S_2 in the solution 21.106, this is not at all unexpected. For, if this solution were unique, we could never have statistical correlations between subsystems of a system. Of course, if one of the subsystems is a pure state then all the correlations are trivial and 21.106 is unique. This is easy to prove.

Let $\rho^{(1)}$ correspond to a pure state. Then using the same basis as before we can write without loss of generality

$$\rho^{(1)} = \sum_{m,m'} |1,m\rangle\delta_{1,m}\delta_{1,m'}\langle 1,m'| \\ = |1,1\rangle\langle 1,1| \, . \tag{21.107}$$

Then, we get from 21.104

$$\sum_n \rho^{(c)}_{m,n;\; m',n} = \delta_{1m}\, \delta_{1m'} \quad . \tag{21.108}$$

Thus because $\rho^{(c)}$ is a non-negative operator $\rho^{(c)} \geq 0$ we have for $m,m' \neq 1$ that

$$\rho^{(c)}_{m,n;m',n} = 0 \quad m,m' \neq 1 \, . \tag{21.109}$$

If $m = m' = 1$ we get

$$\rho^{(c)}_{1,n;\; 1,n'} = \sum_m \rho^{(c)}_{m,n;\; m,n'} = \rho^{(2)}_{n,n'} \tag{21.110}$$

where we have used 21.105. Thus all matrix elements of $\rho^{(c)}$ are uniquely determined and hence $\rho^{(c)}$ is unique. To summarize we have the following theorem

<u>Theorem</u>

A density matrix $\rho^{(c)}$ corresponding to a composite system $S_1 \otimes S_2$ is uniquely determined by

$$\rho^{(c)}_{m,n;\; m',n'} = \rho^{(1)}_{m,m'}\, \rho^{(2)}_{n,n'} \tag{21.111}$$

if and only if either $\rho^{(1)}$ or $\rho^{(2)}$ corresponds to a pure system. In this case $\rho^{(1)}$ and $\rho^{(2)}$ are called the projections of $\rho^{(c)}$ in S_1 and S_2 respectively. The "only if" part of this theorem is proved in problem 21.9. We use this theorem in the next section.

21.7 The Quantum Theory of Measurement

As a final application of the density matrix formalism we give a brief discussion of von Neumann's theory of measurement as elaborated by F.W. London and E. Bauer (ref. 21.2, 21.3). This is not strictly within the bounds of the Copenhagen interpretation. In the viewpoint adopted by N. Bohr, (the strict Copenhagen interpretation) a formal quantum theory of measurement is not required. This is a consequence of Bohr's belief that all measurements must reduce to classical concepts which are not themselves further reducible since they constitute the ultimate data of sense experience.

The approach taken by von Neumann is closer to the approach we sketched in chapter 7, and will now be described in more detail.

A state $|\Psi\rangle$ evolves in general in two distinctly different ways. If H is the total hamiltonian of the system then $|\Psi\rangle$ evolves according to the time-dependent Schrödinger equation.

$$i\hbar \frac{\partial}{\partial t} |\Psi\rangle = H|\Psi\rangle \qquad 21.112$$

or

$$|\Psi(t)\rangle = e^{-iHt/\hbar}|\Psi(0)\rangle \; . \qquad 21.113$$

This evolution is purely causal. Corresponding to this time development, the density matrix evolves according to

$$\rho_t = e^{-iHt/\hbar}\, \rho \, e^{iHt/\hbar} \; . \qquad 21.114$$

The evolution 21.113 can be written in the density matrix formalism by choosing

$$\rho = |\Psi(0)\rangle\langle\Psi(0)| \qquad 21.115$$

$$\rho_t = |\Psi(t)\rangle\langle\Psi(t)| \qquad 21.116$$

On the other hand during a measurement process, $|\Psi\rangle$ will change discontinuously and non-causally. Thus if $\{|k\rangle\}$ is a complete set of eigenkets corresponding to the eigenvalues of the observable measured, then during the measurement process, ρ goes over into

$$\tilde{\rho}_t = \sum_{k=1}^{\infty} |k\rangle|\langle\Psi|k\rangle|^2\langle k| \quad . \qquad 21.117$$

The two processes

$$\rho \rightarrow \rho_t$$

$$\rho \rightarrow \tilde{\rho}_t$$

correspond to fundamentally different situations.

A measurement can always be considered to involve a "system", an "apparatus" and an "observer". According to von Neumann the measurement is completed when the "observer" has made a certain subjective observation. To quote from von Neumann, "Indeed experience only makes statements of this type: an observer has made a certain (subjective) observation; and never like this: a physical quantity has a certain value." This viewpoint is taken in order to avoid an infinite regression. For without this viewpoint we could always attempt to analyze the measurement process further. For example, we might decide to stop when light reflected from the dial of the apparatus hits the retina of the observer's eye. On the other hand we might wish to consider the eye as part of the apparatus and stop with the electrical signal from the optic nerve reaching the brain. Indeed we could go on and consider the chemical changes occurring in the observer's brain and on and on. With the assumption made above, we can stop with any point called the observer. However, to justify this viewpoint it is necessary to show that it does not matter at which stage we stop the analysis. Regardless of how much of the inner workings of the observer are included with the "apparatus" we must always be able to obtain the same result for a measurement as long as it terminates with the subjective awareness of the observer.

The proof of this possibility consists of showing that we can always lump either the "system" and "apparatus" or else the "apparatus" and "observer" together and obtain equivalent results. Before proving these results we dispose of another possibility - whether the statistical character of measurements can depend on the state of knowledge of the observer.

The statistical character of a measurement cannot be due to the observer's lack of knowledge of his (or his apparatus') initial state, since the probability of obtaining a result corresponding to a state $|n\rangle$ is given by $|\langle\Psi|n\rangle|^2$ and is completely determined by the state $|\Psi\rangle$. Thus it does not depend on the state of the apparatus or the observer. We therefore assume that an observer making an (ideal) measurement is completely aware of his initial state. This simply means that he knows that he is seeing a dial pointing at a given place. Furthermore the apparatus (dial) is also in a pure state.

If apparatus and observer are lumped then the evolution is according to 21.117 and the probability of observing the state $|k\rangle$ is $|\langle\Psi|k\rangle|^2$. If, however we lump the system and apparatus, then evolution according to 21.117 should occur only when the observer intervenes.

Thus we consider a composite system with the original system in an unknown state given by

$$\rho^{(s)} = |\Psi\rangle\langle\Psi| . \qquad 21.118$$

The apparatus is in a known pure state given by

$$\rho^{(a)} = |a\rangle\langle a| \ . \qquad 21.119$$

Let $\{|a_n\rangle\}$ be a complete set of eigenstates of an observable A corresponding to dial readings $\{a_n\}$. The possible eigenstates of the original system corresponding to the observable S being measured are $|k\rangle$, with eigenvalues s_k. The numbering is such that the eigenvalue s_n corresponds to the dial reading a_n. Thus the composite system (apparatus + system) is in a state given by

$$\rho^{(c)} = \rho^{(s)} \otimes \rho^{(a)} = |\psi\rangle|a\rangle\langle a|\langle\psi| . \qquad 21.120$$

Now the measurement (by the apparatus on the system) corresponds to a unitary evolution 21.114

$$\rho_t^{(c)} = e^{-iHt/\hbar}\, \rho^{(c)}\, e^{iHt/\hbar} \ . \qquad 21.121$$

Thus, according to the observer, a measurement is only performed if he measures the eigenvalues of the simultaneously measurable observables S and A. These pairs of variables s_m, a_n have a probability 0 for $m \neq n$ and $|\langle\psi|n\rangle|^2$ for $m = n$. This last requirement is dictated by quantum mechanics. In this case the evolution is of the type 21.117

$$\rho_t^{(c)} \to \tilde{\rho}_t^{(c)} = \sum |a_k\rangle|k\rangle|\langle\psi|k\rangle|^2\langle k|\langle a_k| . \qquad 21.122$$

If all this holds then the measuring process, so far as it occurs in the apparatus, is explained because the split: system,(apparatus + observer) can also be viewed as the split: (system + apparatus), observer.

Thus our problem reduces to the following. Given a basis set $\{|k\rangle\}$ for our system, find a basis set $\{|a_n\rangle\}$ and state $|a\rangle$ for the apparatus, together with a hamiltonian H of the form

$$H = H_{\text{system}} + H_{\text{apparatus}} + H_{\text{interaction}} \tag{21.123}$$

and a time interval t such that the following holds. If $|\phi\rangle$ is an arbitrary state of the system and

$$|\Phi',a'\rangle = e^{iHt/\hbar}\,|\phi\rangle|a\rangle \;. \tag{21.124}$$

Then we have

$$|\Phi',a'\rangle = \sum_m \langle m|\phi\rangle|m\rangle|a_m\rangle \;. \tag{21.125}$$

Notice that instead of finding a hamiltonian H it is sufficient to find a unitary operator U, since Stone's Theorem (Chapter 6) then guarantees the existence of H.

For convenience in proving the existence of U we assume

$$|a\rangle = |a_o\rangle \;. \tag{21.126}$$

We now <u>define</u> the obviously unitary operator U by

$$U \sum_{m,n=-\infty}^{\infty} f_{mn}|m\rangle|a_n\rangle = \sum_{m,n=-\infty}^{\infty} f_{mn}|m\rangle|a_{m+n}\rangle. \tag{21.127}$$

But

$$U|\phi\rangle|a\rangle = U\sum_{m=-\infty}^{\infty} \langle m|\phi\rangle|m\rangle|a_0\rangle \tag{21.128}$$

$$= \sum_{m=-\infty}^{\infty} \langle m|\phi\rangle|m\rangle|a_m\rangle \;. \tag{21.129}$$

Thus the proof is completed. This yields a completely consistent quantum theory of measurement.

21.9 Conclusion

We have now completed a development of non-relativistic quantum mechanics. The theory is consistent and even complete. Nevertheless, from its inception attempts have been made to show that quantum mechanics is only a superficial theory based on a more fundamental hidden variables theory. In the last few years, however, experiments designed to test Bell's inequality (reference 21.4) have forced proponents of hidden variable theories to retreat further and further. It thus seems very unlikely that modifications of the quantum concept will be required.

There still remains the problem of providing a successful union of quantum mechanics and relativity theory. The groundwork for this was laid in the early thirties. Although the product of this unification, modern quantum field theory, has been highly successful we can not yet claim to know that quantum field theory is a completely consistent theory. This statement is made in spite of the great advances of the last two decades.

On the more philosophical level some difficult epistemological questions remain. In this book we have employed what is essentially the Copenhagen interpretation of quantum mechanics. There are at least two rival interpretations: the Statistical (ref 21.5) and the Many Worlds (ref. 21.6) interpretations. They are radically different in the pictures of reality that they present, yet they agree on all experimental predictions. Thus they are, in principle, experimentally indistinguishable from the Copenhagen interpretation. At this stage it is a matter of personal preference which interpretation is employed. This means that a very large element of subjectivity has entered physics at a very fundamental level.

Notes, References and Bibliography

21.1 A rather lengthy review article, treating many different aspects of the density matrix with numerous references is:
D. Ter Haar - Theory and Applications of the Density Matrix - Reports on Progress in Physics 24, 304-362 (1961).

A very enjoyable little book on statistical thermodynamics with a viewpoint different from ours is:
E. Schrödinger - Statistical Thermodynamics - Cambridge University Press (1962).

Two of the more standard texts on statistical mechanics are:
L.D. Landau and E.M. Lifshitz - Statistical Physics - Addison-Wesley Publishing Co., Inc., U.S.A. (1958), K. Huang - Statistical Mechanics - John Wiley & Sons, Inc., New York (1963).

One of the original books treating the density matrix and discussing the quantum theory of measurement is: J. von Neumann - Mathematical Foundations of Quantum Mechanics - Princeton University Press (1955).

21.2 Further theories of measurement with references to the original papers are to be found in:
M. Jammer - The Philosophy of Quantum Mechanics - John Wiley and Sons, Inc., New York (1974).

21.3 F. London and E. Bauer - La Théorie de l'Observation en Mécanique Quantique - Hermann & Cie, Paris (1939)

For a different approach to polarization see:
W.H. McMaster - Polarization and the Stokes Parameters - Amer. J. Phys. 22, 351(1954).

21.4 S.J. Freedman and J.F. Clauser - Experimental test of local hidden-variable theories - Phys. Rev. Lett. 28 938-941 (1972)

21.5 L.E. Ballentine - The Statistical Interpretation of Quantum Mechanics - Rev. Mod. Phys. 42, 358-381, (1970).

21.6 Bryce S. De Witt and Neill Graham - The Many-Worlds Inpterpretation of Quantum Mechanics - Princeton University Press (1973).

Problems Chapter 21

21.1 Compute the average energy of an assembly of identical simple harmonic oscillators using:

a) the microcanonical ensemble

b) the canonical ensemble

c) the grand canonical ensemble.

21.2 Prove the following properties of a density matrix.

a) $\rho^2 \leq \rho$ This implies $\rho \geq 0$

b) $Tr[\rho, A] = 0$.

To see that this is not trivial, consider $Tr[x,p]$.

21.3 In a gas of electrons, a fraction p are known to have their z-component of spin in the up direction. Assume the remainder are random. What is the average value of s_x, s_y and s_z? If nothing is known about the spins of the remaining 1-p electrons what are the maximum values of $\langle\langle s_x \rangle\rangle$, $\langle\langle s_y \rangle\rangle$ and $\langle\langle s_z \rangle\rangle$?

21.4 Verify equation 21.26.

21.5 Complete the steps in going from 21.94 to 21.95.

21.6 Show that for spin 1, the density matrix can be completely specified by the polarization vector $\vec{p}$, and the quadrupole polarization tensor Q_{ij} defined by

$$\vec{p} = \langle \vec{J} \rangle / j\hbar$$

$$Q_{ij} = \frac{\langle J_i J_j + J_j J_i \rangle}{j(j+1)\hbar^2} - \frac{2}{3}\delta_{ij} .$$

21.7 Show that for the case of general spin j, if we define the polarization vector by

$$\vec{p} = \langle\vec{J}\rangle / j\hbar$$

and are given a "hamiltonian"

$$H = -\gamma\ \vec{J}\cdot\vec{B}$$

where $\vec{B}$ is a magnetic field, then we have

$$\frac{d\vec{p}}{dt} = \gamma\ \vec{p}\times\vec{B}$$

$$\frac{d\vec{p}^{\,2}}{dt} = 0.$$

21.8 Show that $\tilde{\rho}$ defined by 21.101 satisfies the general properties of a density matrix

$$\tilde{\rho}^{\dagger} = \tilde{\rho}$$
$$\mathrm{Tr}\ \tilde{\rho} = 1$$
$$\tilde{\rho} \geq 1,$$

as well as the equations 21.99 and 21.100.

21.9 Show that if we have

$$\rho^{(1)} = \alpha_1\sigma^{(1)} + \beta_1\tau^{(1)}\ ,\quad \alpha_1 + \beta_1 = 1 \quad \alpha_1,\beta_1 > 0$$
$$\rho^{(2)} = \alpha_2\sigma^{(2)} + \beta_2\tau^{(2)}\ ,\quad \alpha_2 + \beta_2 = 1 \quad \alpha_2,\beta_2 > 0$$
$$\mathrm{Tr}\ \sigma^{(i)} = \mathrm{Tr}\ \tau^{(i)} = 1.$$

Then any combination

$$\tilde{\rho} = \alpha\sigma^{(1)}\ \ \sigma^{(2)} + \beta\alpha^{(1)}\ \ \tau^{(2)} + \gamma\tau^{(1)}\ \ \sigma^{(2)} + \delta\tau^{(1)}\ \ \tau^{(2)}$$

with

$$\alpha + \beta = \alpha_1 \qquad \alpha + \gamma = \alpha_2$$
$$\gamma + \delta = \beta_1 \qquad \beta + \delta = \beta_2$$
$$\alpha,\beta,\gamma,\delta > 0$$

satisfies 21.99 and 21.100 and is a possible density matrix for the composite system. This establishes the necessity of the condition that $\rho^{(1)}$ and $\rho^{(2)}$ correspond to pure states in order that 21.101 give a unique solution for a density matrix for the composite system.

21.10 Consider a "gas" of Bose particles with energy either $+E$ or $-E$. The hamitonian for this system is

$$H = E(a_2^\dagger a_2 - a_1^\dagger a_1)$$

where a_i , $a_i^\dagger$ $(i=1,2)$ are the usual annihilation and creation operators for bosons.

a) Show that the canonical partition function is given by

$$Z_N = \frac{\sinh \beta E(N+1)}{\sinh \beta E}$$

and that the grand canonical partition function is given by

$$Z_G = [1-2z \cosh \beta E + z^2]^{-1}$$

b) Compute the internal energy U, and the average number of particles $\langle\langle N\rangle\rangle$ and express U as a function of β and $\langle\langle N\rangle\rangle$ rather than as a function of β and z.

Index

Accidental degeneracy, 305
Action, 38
Action-angle variable, 52
Action integral, 50
Addition of
  angular momenta, 474-478
  spin 1/2 and orbital, 480, 481
  two spin 1/2, 479, 480
Addition Theorem for
  angular momenta, 478
  spherical harmonics, 531
Adiabatic approximation, 427-433, 439
Algebra of operators, 75
Alkali atoms, 471-473
Allowed region, 393, 396, 399
Angular momenta addition of, 474-478
Angular momentum, orbital, 261

Anharmonic Oscillator, 345, 346
Annihilation operator, 246, 258
  for boson, 587
  for fermions, 591, 592
Anti-commutator, 592
Anti-linear operator, 113
Antiperiodic boundary conditions, 180
Apparatus, 636, 637
Associated Laguerre polynomials, 295, 296, 299-304
Asymptotic state, 491, 492, 500-502
Asymptotic trajectory, 491

Basis, 148
  set, 312
  vectors, 312
Bell's inequality, 639

Bessel functions, 288, 526, 527
  spherical, 287, 527-531
Blackbody
  cavity, 2
  radiation, 2, 23
  spectrum, 4
Bogoliubov transformation, 607
Bohr, 28, 50
  magneton, 227
  radius, 305, 574
Bohr-Sommerfeld, 50, 53
  quantization, 404
Boltzmann
  constant, 4, 612
  factor, 23, 612
Born approximation, 521-524
Boson, 571, 580
Bound state, 86
Boundary condition, 96
Bra defined, 317, 318
Bragg
  formula, 19, 20
  reflection, 19
Brillouin-Wigner perturbation theory, 356, 357

Canonical
  ensemble, 612, 613
  momentum, 457
  transformations, 45, 50
Cauchy
  criterion, 149
  sequence, 149, 152
Cavity, 2
  blackbody, 24
Cayley transform, 161
Center of mass coordinates, 306
Characteristic
  equation, 369, 376
  function, 59
Chemical potential, 613, 614
Classically allowed region, 393, 396, 399
Classically, forbidden region, 122, 393, 399
Clebsch-Gordon coefficients, 475
  for 2 spin 1/2, 480
C^n, 223
c-number, 198
Commutator, 76
Complete
  set of dynamical variables, 35
  set of vectors, 148
  vector spaces, 149
Completeness relation, 317, 322,
  occupation number space, 588
Compton
  effect, 15
  wavelength, 17
Conjugate momentum, 42, 80
Connection formulas for W.K.B., 400, 401
Conservation of particles, 67
Constant of motion, 205, 206, 212, 459

Contact transformations, 45
Continuity for S', 224
Convergence
  strong, 152
  weak, 185
Copenhagen interpretation, 634, 639
Correspondence principle, 28, 52
Covalent bonding, 576
Creation operator 241,246, 258
  for bosons, 587
  for fermions, 591, 592
Cross-section, 502
Coulomb interaction integral, 576

Davisson-Germer experiment, 18
De Broglie, 14
  relation, 15
Debye
  model, 26, 27
  temperature, 27
Deficiency indices, 170
  for L_z, 249,
  for p, 171-173
Deficiency subspaces, 170
Degeneracy
  Hydrogen atom, 305
  isotropic harmonic oscillator, 291
Degenerate
  eigenvalue, 194
  subspaces, 369, 370, 372
Degree of freedom, 4
Delta function, 132, 133
Dense set of vectors, 157
Density of final states, 422-424
Density matrix, 610-615
  composite system for, 630-634
  equation of motion for, 626
  polarization and, 626-630
  properties of, 623-625
Derivative of a distribution, 226
Deuteron, 445
Differential cross-section, 502
Dirac
  notation, 316-323
  picture, 333-337
Distribution, tempered, 223,233
Domain of an operator, 153
Doublets in Alkali, 472, 473
Dual space, 223, 317

Effective
  potential, 283
  range, 556
  range approximation, 551-556
Eigenbasis, 189
  common, 212
Eigenfunction, 167
Eigenket, 318
Eigenvalue, 167
Eigenvector, generalized, 230
Eikonal approximation, 557-566
Einstein model, 24, 25
Elliptical coordinates, 574
Ensemble
  canonical, 612, 613
  grand canonical, 613-615
  micro-canonical, 610-612

Entropy, 615
Equipartition principle, 4
Essentially self-adjoint, 160
Euler-Lagrange equation, 40
Euler's constant, 578
Evolution
  operator, 204, 412
  time of observable, 208
  time of operator, 204
Exchange
  energy, 576
  integral, 577
  operator, 571, 580
Exponential integral, 578
Extremum principle, 382

Fermion, 571, 580
Fermi's golden rule, 422, 426
Field
  equation, 602
  operators, 593, 594
Fine structure constant, 299
Forbidden region, 393, 395
Fourier transform of
  pulse, 420
  S , 227-229
  S', 229
Franck-Hertz effect, 21
Fugacity, 614
Functional, 220, 382, 383
  linear, 220, 222, 317
  value of, 220
Frobenius, method of, 127

Gauge function, 452
Gauge transformations
  time-dependent, 447-449
  time-independent, 450
Generalized eigenvector, 230
Generating function, 46, 47
Generator of
  rotations, 271
  translations, 175
Geometrical optics approximation, 390
g-factor, 276
Glauber approximation, 557-566
Green's functions, 496-500, 516-518
Group, 174
  velocity, 73
Gyromagnetic ratio, 275, 470

Hall
  current, 454
  effect, 452, 454-456
Hamilton W.R., 37
Hamiltonian
  point particle, 37, 41
  system, 42
  total energy, 44
Hamilton-Jacobi, 44
  equation, 47, 391
  harmonic oscillator, for, 50
  hydrogen atom, for 53
Hamilton's characteristic function, 391

Harmonic oscillator, 125-131
displaced, 350
three dimensional, 289, 396
Heisenberg
equations, 330
equations for field operators, 601
microscope, 199
picture, 328, 330
Helium, 346-349
ground state, 387, 388
Hermitean operator, 158
Hermite functions, 243
Hidden variable Theories, 639
Hilbert space, 145-153
rigged, 232
Hydrogen
atom, 53, 296-306
molecule, 572

Idempotent operator, 624
Identical particles
two, 570-572
N, 579-582
Impact parameter, 559
Incoherent superpositon, 611
Ideterminacy principle, see Uncertainty principle
Index of refraction, 389
Integral representation for
Bessel function, 564
Laguerre polynomials, 303
associated Laguerre polynomials, 304
Interaction picture, see Dirac picture
Internal energy, 613, 615
ideal gases for, 619-621
Intertwining property of Møller operators, 494

Källén-Yang-Feldman equations, 501
Ket, defined, 316-318

L_2, 150
Lagrangian, 38, 40, 41
for electromagnetic interaction, 80
Laguerre polynomials, 303
associated, 295, 296, 299-304
Landau
gauge, 452, 455
Levels, 452
Laplace operator in spherical coordinates, 285
Larmor frequency, 63, 453
Legendre
associated functions, 257, 265
polynomials, 265
transformation, 42, 47
Levinson's Theorem, 548
Liouville's theorem, 626
Lippmann-Schwinger equations, 503-506, 516-520, 549-551
Local gauge transformation, 449, 550

Logarithmic derivative, 96
Lorentz force, 80
Lowering operator, 246
  for L_z, 258

Magnetic length, 454
Many Worlds Interpretation, 639
Matrix mechanics, 322
Measurement , 189, 635
  quantum theory of, 634-638
  repeated, 194
Mechanical momentum, 457, 458
Micro-canonical ensemble, 610-612
Minimum principle, 383, 385
Møller wave operators, 493-496
  defined, 493
Momentum
  canonical, 457
  conjugate, 42, 80
  generalized, 42
  mechanical, 457, 458
  transfer, 523

Neumann series, 521
No-particle state, 595
Norm of a vector, 147
Null space, 221
Number operator for
  bosons, 587
  fermions, 590, 591, 593

Observables, 188, 191
  compatible, 197-199
Occupation number space, 616
  bosons, 586-589
  fermions, 589-593
Operator
  annihilation, 246, 258
  anti-linear, 113
  creation, 241, 246, 258
  domain of, 75
  evolution, 204, 412
  hermitean, 158
  idempotent, 624
  linear, 75
  lowering, 246, 258
  raising, 240, 241, 258
  self-adjoint, 159, 160, 167
  step down, 246, 462
  step up, 246, 462
  symmetric, 158
  unitary, 161, 315
Optical theorem, 539-541
Orbital angular momentum, 261
Orbits, elliptical 54
Orthogonal
  complement, 221
  matrix, 267, 313
  transformation, 313
Orthogonality, 147
Orthonormal, 147
Overlap integral, 574

Parity, 91-93, 181
  and angular momentum, 292
  and spherical harmonics, 293
  operator, 268
  transformation in spherical coordinates, 292, 293
Partial wave
  analysis, 534-537
  decomposition, 531, 532
  Green's functions, 551
Particle
  density operator, 595
  exchange operator, 571, 580
Partition function, 613
  grand, 613
Pauli
  exclusion principle, 582, 590
  spin matrices, 274
Periodic boundary conditions, 173
Permanent, 585
Phase shift, 537-539
  in one dimension, 139
Phase velocity, 71
Photoelectric, 10
Photoelectron, 10
Photo-ionization of hydrogen atom, 422
Planck
  frequency relation, 604
  radiation law, 23
Poisson equation, 577
Polarization, 627-631
  vector, 641, 642
Principal function, 59
Probability
  conservation, 106
  current, 104
  density, 69, 104
Projection operator, 185, 354
Propagator, 496-500
  free particle, 215
Pure state, 624, 625

q-number, 198
Quadrupole polarization tensor, 603, 641
Quantum state, 188

Raising operator, 240
  for L_z, 258
Random phases, 610-612
Range of an operator, 156
Rayleigh-Jeans law, 2, 24
Rayleigh-Ritz, 383
Rayleigh-Schrödinger perturbation theory, 340-343
Reduced mass, 307, 308
Reflection
  coefficient, 125
  probability, 107
Refraction, 13
  index of, 14
Relative
  coordinates, 306
  momentum, 307
Resolvent operator, 185, 516
Riesz-Fischer theorem 149, 152
Riesz representation theorem, 220

Ritz, 9
RMS deviations, 193
Rodrigues formula for
  Hermite polynomials, 129, 467
  Laguerre polynomials, 303
  Legendre polynomials, 265, 532
Rotation group, 269
Rotations, 267-271
  improper, 268
  proper, 268
Rotator, 247
  one-dimensional, 249, 253
  three-dimensional, 254, 256
Rutherford
  cross-section, 525
  model, 6
Rydberg, 9
Rydberg-Ritz combination, 10, 28

Scattering, 86
  amplitude, 509, 513-515
  length, 556
Schmidt orthogonalization
  procedure, 152
Schrödinger
  picture, 327, 328
  representation 214
Schwartz space, 223
Schwarz inequality, 151, 152, 190, 202
Self-adjoint extension, 160
  of $\vec{L}$, 250-252
  of p, 171, 173
Self-adjoint operator, 159
  properties of, 167
Set, complete of
  dynamical variables, 35
  vectors, 148
Shape-independent approximation, 556
Singlet state, 573
Slater determinant, 584
S matrix 506-510
  defined, 495
Snell's law, 12
Sodium D lines, 473
Sommerfeld, 50
  polynomial method of, 127, 294, 298
S operator, 506-510
  defined, 495
Space
  dual, 223
  null, 221
  tempered distributions of 223
  test-functions of, 223
Specific heat
  Einstein model, 24, 25
  Debye model, 26, 27
Spectral resolution of operator, 185
Spectral theorem, 168
Spherical Bessel functions, 287, 527-531
  asymptotic behaviour of, 533
Spherical harmonics, 261-266
  table of, 266

Spin angular momentum, 272-275
  and magnetic moment, 275
Spin and statistics theorem, 580
Spinor, 272
Spin-orbit interaction, 471
Spontaneously broken symmetry, 179
  defined, 180
Stark effect, 361-363
State
  asymptotic, 491, 492, 500-502
  classical, 188
  pure, 624, 625
  quantum, 188
  stationary, 28, 86
Stationary principle, 382
Statistical Interpretation, 639
Step down operator, 246, 462
Step up operator, 264, 462
Stone's theorem, 174, 638
  and rotations, 271
Strong limit, 185, 492
Subspace, degenerate, 369, 370, 372
Sudden approximation, 433-435, 441
Sum rule, 357-361
Superposition principle, 70
Symmetric
  gauge, 451, 457, 469
  operator, 158

Tempered distribution, 223, 233
Test function, 223
  space $\mathcal{D}$, 233
  space $\mathcal{S}$, 223
Thermodynamic limit, 619
Thomas precession, 471
Thomson model, 6
Time-evolution of
  expectation values, 205
  probability amplitudes, 210-212
  state, 203-205
Time reversal, 111-113, 181
T matrix, 509
Topology of
  $\mathcal{S}$, 224
  space, 222
Transition
  amplitude, 415,416
  probability, 415, 419, 421
Translation operator, 176
Transmission
  coefficient, 125
  probability, 108
Trial wavefunction, 386
Triplet state, 573
Tunneling, 121, 405-408
Turning points, 395-401
Two-body interactions, 600

Uncertainty principle, 197-203
Uncertainty relation, 73
  time-energy, 206-210
Unitarity bound, 540

Unitary
  operator, 161, 315
  transformation, 313

Vacuum, 595
Value of
  function, 219
  functional, 220
Variational principle
  classical, 45
  quantum mechanical, 382
Vector
  contravariant, 77
  covariant, 77
Velocity of
  center of mass, 142
  particle, 61
Viral Theorem 246
Volterra equation, 560

Wavenumber, 3
  local, 561
Wave packet, 67, 70, 596, 597
  and scattering, 100
Wave-particle duality, 12
Weak convergence, 185
Wigner, 269
Wigner-Eckart theorem, 487
Wigner time reversal, 113
Wronskian for spherical Bessel
  functions, 542
WKB approximation, 64, 389-401

Yang-Feldman equations,
  see Källén-Yang-Feldman
  equations, 501
Young's double slit experiment, 69
Yukawa potential, 524, 567

Zeeman effect, 481